Combinatorial Geometry

Combinatorial Geometry

JÁNOS PACH
City College, New York and
Hungarian Academy of Sciences

PANKAJ K. AGARWAL
Duke University
Durham, North Carolina

A Wiley-Interscience Publication
JOHN WILEY & SONS, INC.
New York • Chichester • Brisbane • Toronto • Singapore

QA167
. P33
1995

Library of Congress Cataloging in Publication Data:

Pach, János.
    Combinatorial geometry / by János Pach and Pankaj Agarwal.
       p. cm.—(Wiley-Interscience series in discrete mathematics and
optimization)
    "A Wiley-Interscience publication."
    Includes bibliographical references (p.) and index.
    ISBN 0-471-58890-3 (acid-free)
    1. Combinatorial geometry. I. Agarwal, Pankaj K. II. Title.
III. Series.
QA167.P33 1995   94-48203
516'.13–dc20

Printed in the United States of America

10 9 8 7 6 5 4 3 2

To Paul Erdős, László Fejes Tóth, and C. Ambrose Rogers

Contents

Preface

The "crisis of the foundations", the "Entscheidungsproblem", Gödel's theorems certainly belonged to the hottest scientific subjects in the first few decades of this century. They led to spectacular new discoveries which have permeated vast areas of mathematics and fertilized many fields that had been thought to be "dead" before. Intuitive (elementary) geometry was one of the losers. This field was by and large neglected, while the more "abstract" areas of geometry, such as topology and differential geometry, flourished and had a great impact on our views of the physical universe. *Down with Euclid! Death to the triangles!*—burst out J. Dieudonné at a meeting 35 years ago, and his sentiments were shared by the majority.

However, since then we witnessed a revival of intuitive geometry. The subject received an infusion of new blood from several sources. The works of László Fejes Tóth and C. Ambrose Rogers initiated new combinatorial approaches to some classical questions studied by Newton, Gauss, Minkowski, Hilbert, and Thue. They laid the foundations of the theory of packing and covering. At the same time, Paul Erdős continued bombarding the world with new questions of combinatorial geometry that even Euclid would appreciate. Many of these problems turned out to be crucially important in coding theory, combinatorial optimization, computational geometry, robotics, computer graphics, etc. The explosive development of computer technology presented a powerful new source of inspiration for many areas of pure and applied mathematics. Combinatorial geometry is one of the fields that benefited most from this source.

Most questions in this area are about arrangements of points, lines, circles, spheres, that is, about the most fundamental objects of Euclidean geometry. Many of them have a strong intuitive appeal and can be explained to a layman. For instance, how many unit balls can be packed into a large box of a fixed volume? What is the maximum number of incidences between n points and n lines in the plane? The aim of this book is to offer a self-contained introduction to some important results on arrangements of convex bodies (Part I) and arrangements of points and lines (Part II). In spite of the elementary nature of its subject, almost half of the material presented in the book has been discovered during the past 20 years, and has not yet appeared in any textbook or monograph. Some other books and surveys on related subjects are listed below.

Geometry of Numbers: J. Cassels (1959); P. Gruber and C. Lekkerkerker (1987); Erdős, P. Gruber, and J. Hammer (1989); M. Deza, V. Grishukhin and M. Laurent (1993).
Theory of Packing and Covering: C. A. Rogers (1964); L. Fejes Tóth (1964, 1972); G. Fejes Tóth and W. Kuperberg (1993a, 1993d).
Coding Theory: I. Csiszár and J. Körner (1981); J. van Lint (1982); J. Conway and N. Sloane (1988).
Convexity: T. Bonnesen and W. Fenchel (1934); L. Danzer, B. Grünbaum, and V. Klee (1963); B. Grünbaum (1967); I. Yaglom and V. Boltyansky (1951); R. Schneider (1993).
Combinatorial Geometry: H. Hadwiger, H. Debrunner, and V. Klee (1964); V. Boltyansky and I. Gohberg (1985); P. Erdős and G. Purdy (1989); W. Moser and J. Pach (1986); V. Klee and S. Wagon (1991).
Computational Geometry: F. Preparata and M. Shamos (1985); H. Edelsbrunner (1987); K. Mehlhorn (1985); K. Mulmuley (1994); J. O'Rourke (1994), M. Sharir and P. Agarwal (1995).
Linear Programming: Chvátal (1983); *Convex Optimization:* M. Grötschel, L. Lovász, and A. Schrijver (1985).

The present book is based on the material of two courses I gave at the Courant Institute of New York University. I am very much indebted to all my students and colleagues who attended my lectures and actively participated in the discussions. It is a source of great satisfaction for me that some of the results in this book have been obtained by the participants during and after my lectures. My gratitude is due to Boris Aronov, Vasilis Capoyleas, Mikhael Gorbunov, Bud Mishra, Marco Pellegrini, Richard Pollack, Nagabhushana Prabhu, Micha Sharir, Joel Spencer, Marek Teichmann, and Chee Yap.

I am especially grateful to Pankaj K. Agarwal who took notes of my lectures, put my manuscript in TEX, carefully read and revised all the material, completed the bibliography, provided many hints to the solutions of the exercises, and prepared all the figures. He wrote the first drafts of Chapters 11 and 13. Without his enthusiastic and tireless support this book would never have come to birth.

I am also indebted to Boris Aronov, Peter Brass, György Csizmadia, Herbert Edelsbrunner, György Elekes, Gábor Fejes Tóth, Zoltán Füredi, János Komlós, Wlodzimierz Kuperberg, David Larman, Endre Makai, Jiři Matoušek, Richard Pollack, Günter Rote, Jason Rush, Micha Sharir, Torsten Thiele, Géza Tóth, György Turán, and Emo Welzl, for carefully reading and criticizing various portions of a preliminary version of the manuscript, and using it as a text for undergraduate and graduate courses. In its present form, the book should be understandable to any undergraduate student with a solid background in calculus and with some familiarity with the basic concepts of combinatorics and probability theory. It can also serve as a source of unsolved research problems for graduate students as well as professional and amateur mathematicians who want to explore this fascinating field.

Finally, I would like to express my thanks to my teachers and friends, Paul Erdős, László Fejes Tóth, and C. Ambrose Rogers, who shaped my mathematical thinking. Most of the basic results in this book are either due to them, or were directly influenced by their research.

In the Spring of 1990, in a lecture given at Courant Institute, I. M. Gelfand said: "The older I get, the more I believe that at the bottom of most deep mathematical questions there is a combinatorial problem." This book focuses on a rapidly developing field in which combinatorics has certainly played a pioneering role.

JÁNOS PACH

Budapest, Hungary
July, 1995

Part I

Arrangements of Convex Sets

1

Geometry of Numbers

The geometry of numbers is a 100-year-old discipline that emerged from number theory. Minkowski (1896) made the fruitful observation that many important results in diophantine approximation and in other central fields of number theory can be established by easy geometric arguments. The starting point of this theory is an ingenious statement formulated by Minkowski (Theorem 1.7), which can be regarded as a trivial extension of the pigeonhole principle to measurable sets. The aim of this chapter is to present some immediate consequences of this result, including the elegant proofs of the two- and four-squares theorems of Fermat.

LATTICES

A fundamental concept in the geometry of numbers is the following.

Definition 1.1. Given d linearly independent vectors (points) $u_1, \ldots, u_d$ in the d-dimensional Euclidean space $\mathbb{R}^d$, let the *lattice* Λ generated by them be defined as

$$\Lambda(u_1, \ldots, u_d) = \{m_1 u_1 + \cdots + m_d u_d \mid m_1, \ldots, m_d \in \mathbb{Z}\},$$

where $\mathbb{Z}$ is the set of integers.

The set $\{u_1, \ldots, u_d\}$ is called a *basis* of Λ. The parallelepiped P induced by the 2^d vertices of the form $m_1 u_1 + \cdots + m_d u_d$, where $m_i \in \{0, 1\}$ for every i, is said to be the *fundamental parallelepiped* (or *cell*) of Λ. Obviously,

$$\text{Vol } P = |\det(u_1, \ldots, u_d)|.$$

The same lattice can, of course, be generated in many different ways; i.e., Λ has several bases. Consequently, Λ has several fundamental parallelepipeds. However, all of them have the same volume. Indeed, let P be a fundamental parallelepiped of Λ, and let $B^d(R)$ denote the ball of radius R in $\mathbb{R}^d$ centered at the origin. If R is very large, then those translates of P which are of the form $P + u$ for some $u \in \Lambda \cap B^d(R)$ do not overlap and "almost completely" cover $B^d(R)$.

3

Definition 1.2. Let $\det \Lambda$ be defined as the volume of any fundamental parallelepiped of Λ. If $\det \Lambda = 1$, then Λ is called a *unit lattice*.

Next we illustrate with a couple of easy statements the close relationship among lattices, sphere packings, and diophantine approximation. For the sake of simplicity, we consider the planar case $d = 2$.

Theorem 1.3. *Let Λ be a unit lattice in $\mathbb{R}^2$. Then there are two lattice points whose distance is at most $\sqrt{2/\sqrt{3}}$.*

Proof. Let $u,v \in \Lambda$ be a pair of points whose distance $\delta^{\star}$ is minimum. Assume without loss of generality that $u = 0$. Then $kv \in \Lambda$, for every integer k, and by the minimality of $\delta^{\star}$, there is no other lattice point inside the union of the circles of radius $\delta^{\star}$ around the points kv. If ℓ denotes the straight line connecting 0 and v, then these circles cover a strip of half-width $(\sqrt{3}/2)\delta^{\star}$ around ℓ (see Figure 1.1). On the other hand, since Λ is a unit lattice, there are infinitely many lattice points on the line t parallel to ℓ at distance $1/\delta^{\star}$ from ℓ. Thus $(\sqrt{3}/2)\delta^{\star} \leq 1/\delta^{\star}$, and the result follows. It is also clear that the constant $\sqrt{2/\sqrt{3}}$ cannot be improved in general.   □

Let us draw a circular disc of radius r around each point of a lattice Λ. If these discs do not overlap, then they are said to form a *lattice packing*. The *density* of a lattice packing is $\pi r^2 / \det \Lambda$, i.e., the portion of the plane covered by the discs.

Corollary 1.4. *The density of a lattice packing of congruent circles is at most $\pi/\sqrt{12}$, and this bound can be attained.*

Proof. Assume without loss of generality that Λ is a unit lattice. Let $\delta^{\star}$ denote the same as in the proof of Theorem 1.3. Then the largest r, for which the circles of radius r do not overlap, is $\delta^{\star}/2$. Hence, by Theorem 1.3, the density of a lattice packing is

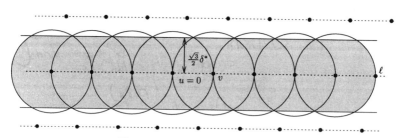

Figure 1.1. Strip covered by circles.

$$\frac{(\delta^*/2)^2\pi}{\det \Lambda} \le \frac{2}{\sqrt{3}} \cdot \frac{1}{4} \cdot \pi = \frac{\pi}{\sqrt{12}},$$

as desired. $\square$

Corollary 1.5. *Let $f(m,n) = am^2 + 2bmn + cn^2$ be a positive-definite form, $a > 0$, $ac - b^2 = 1$. Then there exist two integers, m' and n', such that at least one of them is not 0 and $f(m',n') \le 2/\sqrt{3}$.*

Proof. It is easy to see that

$$\Lambda = \left\{ \left(\sqrt{a}m + \frac{b}{\sqrt{a}}n, \frac{1}{\sqrt{a}}n \right) \in \mathbb{R}^2 \,\middle|\, m \text{ and } n \text{ are integers} \right\}$$

is a unit lattice. Thus, by Theorem 1.3, there exists a lattice point

$$\left(\sqrt{a}m' + \frac{b}{\sqrt{a}}n', \frac{1}{\sqrt{a}}n' \right)$$

other than the origin whose distance from the origin is at most $\sqrt{2/\sqrt{3}}$, i.e.,

$$\left(\sqrt{a}m' + \frac{b}{\sqrt{a}}n' \right)^2 + \left(\frac{1}{\sqrt{a}}n' \right)^2 = f(m',n') \le \frac{2}{\sqrt{3}}. \qquad \square$$

Corollary 1.6. *Let α be any irrational number. Then there exist infinitely many pairs of integers (m,n) such that*

$$\left| \alpha - \frac{m}{n} \right| \le \frac{1}{\sqrt{3}} \frac{1}{n^2}.$$

Proof. Let $\varepsilon > 0$ be fixed, and set

$$f(m,n) = \left(\frac{\alpha n - m}{\varepsilon} \right)^2 + (\varepsilon n)^2 = \frac{1}{\varepsilon^2}m^2 - \frac{2\alpha}{\varepsilon^2}mn + \left(\frac{\alpha^2}{\varepsilon^2} + \varepsilon^2 \right)n^2.$$

Then $f(m,n)$ satisfies the conditions of Corollary 1.5; hence one can pick $m',n' \in \mathbb{Z}$ (the set of integers) such that

$$f(m',n') = \left(\frac{\alpha n' - m'}{\varepsilon} \right)^2 + (\varepsilon n')^2 \le \frac{2}{\sqrt{3}}.$$

But then

$$\left| \alpha - \frac{m'}{n'} \right| = \left| \frac{\alpha n' - m'}{\varepsilon} \right| \cdot (\varepsilon n') \cdot \frac{1}{n'^2}$$

$$\leq \frac{\left(\dfrac{\alpha n' - m'}{\varepsilon} \right)^2 + (\varepsilon n')^2}{2} \cdot \frac{1}{n'^2} \leq \frac{1}{\sqrt{3}} \frac{1}{n'^2}.$$

Note that if ε is sufficiently small, then $n' \neq 0$. Similarly, by choosing smaller and smaller ε's, we obtain infinitely many pairs (m, n) satisfying the assertion.   □

One of the important observations of Minkowski, which we referred to in the beginning of this chapter, is the following.

Theorem 1.7 (Minkowski, 1896).   *Let $C \subseteq \mathbb{R}^d$ be a convex body, centrally symmetric about the origin, and let Λ be a unit lattice. If $\operatorname{Vol} C > 2^d$, then C contains at least one lattice point different from $\mathbf{0}$.*

Proof.   Consider the bodies $\frac{1}{2}C + u = \{\frac{1}{2}c + u \mid c \in C\}$, where $u \in \Lambda$. If two of them (say, $\frac{1}{2}C + u$ and $\frac{1}{2}C + v$) have a point p in common, then $p - u, p - v \in \frac{1}{2}C$. But then, $v - p \in \frac{1}{2}C$; thus $\mathbf{0} \neq v - u = (v - p) + (p - u) \in \frac{1}{2}C + \frac{1}{2}C = C$ and $v - u \in \Lambda$.

Hence, we may assume that the sets $\frac{1}{2}C + u (u \in \Lambda)$ are all disjoint, which easily implies that $\operatorname{Vol}(\frac{1}{2}C) = (1/2^d) \operatorname{Vol} C \leq 1$.   □

In fact, the same proof gives

Theorem 1.8 (Blichfeldt, 1921; van der Corput, 1936).   *Let k be a natural number, let $S \subseteq \mathbb{R}^d$ be a Jordan-measurable set with $\operatorname{Vol} S > k$, and let Λ be a unit lattice. Then there exist $s_0, s_1, \ldots, s_k \in S$ such that $s_i - s_j \in \Lambda$ for all $0 \leq i \leq j \leq k$.*

Note that using Minkowski's theorem one can easily establish a statement slightly weaker than Corollary 1.6 (see Exercise 1.2). The best possible value of the constant in Corollary 1.6 is $1/\sqrt{5}$ (cf. Hurwitz, 1891).

THE TWO- AND FOUR-SQUARES THEOREMS

In this section we use Minkowski's theorem (Theorem 1.7) to prove two classical results in number theory, which were discovered by Fermat. The first one states that every prime of the form $4m + 1$ can be expressed as the sum of two integer squares, and its first complete proof was given by Euler more than hundred years later. Euler did not succeed in proving the second theorem, which says that every positive integer is the sum of four integer squares. It was finally established by Lagrange in 1770.

To prove these results, we need some preparation. For any prime p, let $\mathbb{Z}_p = \{0, 1, \ldots, p-1\}$, which is a field under the addition and multiplication modulo p. Then $\mathbb{Z}_p^+ = \{1, 2, \ldots, p-1\}$ is a group under multiplication. [Actually, it is a cyclic group, but we shall not use this stronger property. See Exercise 1.6(i).] A number a is called a *quadratic residue* of p if

$$a \equiv z^2 \pmod{p}$$

for some $z \in \mathbb{Z}_p$.

Definition 1.9. Let p be a prime, and let $a \in \mathbb{Z}_p^+$. The *Legendre symbol* $\left(\frac{a}{p}\right)$ is defined as

$$\left(\frac{a}{p}\right) = \begin{cases} +1 & \text{if } a \text{ is a quadratic residue of } p, \\ -1 & \text{otherwise.} \end{cases}$$

Lemma 1.10. *Let p be a prime, and let $a \in \mathbb{Z}_p^+$. Then*

$$(p-1)! \equiv -\left(\frac{a}{p}\right) a^{(p-1)/2} \pmod{p}. \tag{1.1}$$

Proof. Since the equation $cz \equiv a \pmod{p}$ has a unique solution for every $c \in \mathbb{Z}_p^+ = \{1, \ldots, p-1\}$, we can assign a unique $j \in \mathbb{Z}_p^+$ to each $i \in \mathbb{Z}_p^+$ such that $ij \equiv a \pmod{p}$.

If a is not a quadratic residue of p, then $i \neq j$. Hence, there are exactly $(p-1)/2$ pairs $i, j \in \mathbb{Z}_p^+$ such that $ij \equiv a$. Thus

$$(p-1)! = \prod_{i \in \mathbb{Z}_p^+} i \equiv -\left(\frac{a}{p}\right) a^{(p-1)/2} \pmod{p},$$

because $\left(\frac{a}{p}\right) = -1$.

On the other hand, if $a \equiv c^2 \pmod{p}$ for some $c \in \mathbb{Z}_p$, then for $c' \equiv -c$, we have $cc' \equiv -a \pmod{p}$, and the elements of $\mathbb{Z}_p^+ - \{c, c'\}$ can be partitioned into $(p-3)/2$ pairs $\{i, j\}$ such that $ij \equiv a$. Hence

$$(p-1)! = cc' \cdot \prod_{\substack{i \in \mathbb{Z}_p^+ \\ i \neq c, c'}} i \equiv (-a) \cdot a^{(p-3)/2} \pmod{p}$$

$$= -\left(\frac{a}{p}\right) a^{(p-1)/2},$$

because $\left(\frac{a}{p}\right) = 1$.

If we substitute $a = 1$ in (1.1), we obtain Wilson's theorem:

$$(p - 1)! \equiv -1 \pmod{p}. \tag{1.2}$$

Corollary 1.11. -1 *is a quadratic residue of a prime p if and only if p is of the form $4m + 1$.*

Proof. By substituting $a = -1$ in (1.1), we get

$$(p - 1)! \equiv -\left(\frac{-1}{p}\right)(-1)^{(p-1)/2}.$$

Thus, in view of (1.2), -1 is a quadratic residue of p if and only if $(p - 1)/2$ is even, that is, if p is of the form $4m + 1$. $\qquad\square$

Now we are ready to prove the two-squares theorem.

Theorem 1.12 (Euler, Fermat). *Every prime p of the form $4m + 1$ can be expressed as the sum of the squares of two integers.*

Proof. If p is a prime of the form $4m + 1$ then, by Corollary 1.11, there is an integer $0 \neq z < p$ such that $z^2 \equiv -1 \pmod{p}$. It is easily seen that

$$\Lambda = \{(x, y) \in \mathbb{Z}^2 \mid y \equiv xz \pmod{p}\}$$

is a lattice in the plane with $\det \Lambda = p$ (see Exercises 1.7 and 1.8).

Let C be a disc of radius $r = \sqrt{3p/2}$ centered at the origin. Then

$$\text{Vol } C = r^2 \pi = \frac{3\pi p}{2} > 4p = 2^2 \det \Lambda.$$

So by Theorem 1.7 there exists a point $(x, y) \in \Lambda$, different from the origin, for which

$$0 \neq x^2 + y^2 \equiv x^2 + x^2 z^2 \equiv 0 \pmod{p}.$$

Since $x^2 + y^2$ is a multiple of p strictly between 0 and $2p$, $x^2 + y^2$ must be equal to p. $\qquad\square$

The same idea can be applied to obtain the four-squares theorem.

Theorem 1.13 (Fermat, Lagrange). *Every positive integer can be expressed as the sum of the squares of four integers.*

Proof. First observe that it is sufficient to show that every prime can be written as the sum of four squares. Indeed, if $n = n_1 n_2$ and

$$n_1 = x_1^2 + y_1^2 + v_1^2 + z_1^2, \qquad n_2 = x_2^2 + y_2^2 + v_2^2 + z_2^2,$$

then

$$n = (x_1^2 + y_1^2 + v_1^2 + z_1^2)(x_2^2 + y_2^2 + v_2^2 + z_2^2)$$
$$= (x_1 x_2 - y_1 y_2 - v_1 v_2 - z_1 z_2)^2 + (x_1 y_2 + y_1 x_2 + v_1 z_2 - z_1 v_2)^2$$
$$+ (x_1 v_2 - y_1 z_2 + v_1 x_2 + z_1 y_2)^2 + (x_1 z_2 + y_1 v_2 - v_1 y_2 + z_1 x_2)^2.$$

Since $2 = 1^2 + 1^2 + 0^2 + 0^2$, we have to prove the assertion only for odd primes p. Notice that a^2 (as well as $-b^2 - 1$) takes exactly $(p+1)/2$ distinct values as a (resp. b) varies over the elements of $\mathbb{Z}_p$. Thus, we can choose $a, b \in \mathbb{Z}$ such that $a^2 \equiv -b^2 - 1$, i.e.,

$$a^2 + b^2 + 1 \equiv 0 \pmod{p}.$$

Let us consider the lattice

$$\Lambda = \{(x, y, v, z) \in \mathbb{Z}^4 \mid v \equiv ax + by, z \equiv bx - ay \pmod{p}\}$$

in $\mathbb{R}^4$. It is easy to see that $\det \Lambda = p^2$. Denoting by C the four-dimensional ball of radius $r = \sqrt{1.9p}$, we obtain

$$\text{Vol } C = \frac{r^4 \pi^2}{2} = \frac{(1.9)^2 \pi^2}{2} p^2 > 2^4 \det \Lambda.$$

Hence, by Theorem 1.7, there exists a point $(x, y, v, z) \in \Lambda$ satisfying

$$0 \neq x^2 + y^2 + v^2 + z^2 \le r^2 < 2p.$$

On the other hand, modulo p we have

$$x^2 + y^2 + v^2 + z^2 \equiv x^2 + y^2 + (ax + by)^2 + (bx - ay)^2$$
$$\equiv (x^2 + y^2)(a^2 + b^2 + 1) \equiv 0.$$

Hence, $x^2 + y^2 + v^2 + z^2 = p$, completing the proof. $\qquad\square$

EXERCISES

1.1 Let P be any nondegenerate parallelepiped in $\mathbb{R}^d$, all of whose vertices belong to a lattice $\Lambda = \Lambda(u_1, \ldots, u_d)$. Prove that if P contains no point of Λ other than its vertices, then P is a fundamental parallelepiped of Λ. That is, letting $v_1, \ldots, v_d$ denote the edges of P incident with a fixed vertex x of P and oriented outward, $\Lambda = \Lambda(v_1, \ldots, v_d)$.

1.2 Use Minkowski's theorem to show that for any irrational number α there are infinitely many pairs of integers (m, n) such that

$$\left| \alpha - \frac{m}{n} \right| \le \frac{1}{n^2}.$$

1.3 Let C be a circle of unit perimeter, and let $\alpha > 0$ be an irrational number.

Show that the endpoints of infinitely many consecutive arcs of length α form an *everywhere dense* subset of the boundary of C (i.e., every arc of positive length contains at least one endpoint).

1.4 Deduce Theorem 1.7 using Theorem 1.8.

1.5 (Minkowski, 1896) Let $l_i(n_1, n_2, \ldots, n_d) = \sum_{j=1}^{d} a_{ij} n_j$ be a real linear form with d variables $(1 \leq i \leq d)$, and let $D = |\det(a_{i,j})| > 0$. Prove that for any positive reals b_i $(1 \leq i \leq d)$ satisfying $\prod_{i=1}^{d} b_i \geq D$, one can find suitable integers n_i such that not all of them are zero, and $|l_i(n_1, n_2, \ldots, n_d)| \leq b_i$ for $1 \leq i \leq d$.

1.6 Let p be a prime.

(i) Show that $\mathbb{Z}_p^+$ is a *cyclic group* under multiplication (i.e., $\mathbb{Z}_p^+ = \{a, a^2, \ldots, a^{p-1}\}$ modulo p for some $a \in \mathbb{Z}_p^+$).

(ii) Deduce from (i) that if p is of the form $4m+1$, then -1 is a quadratic residue of p.

1.7 A set $X \subseteq \mathbb{R}^d$ is called *discrete* if every ball contains only finitely many elements of X. Prove that an additive subgroup of $\mathbb{R}^d$ is a lattice if and only if it is discrete and contains d linearly independent vectors.

1.8 Let Λ be an additive subgroup of $\mathbb{Z}^d$ with index $k < \infty$. Prove that Λ is a lattice with $\det \Lambda = k$. (Recall that the *index* of a subgroup $H \subseteq G$ is the number of elements in the quotient group G/H).

1.9 Show that not every integer is the sum of three squares.

1.10 Prove the following multiplicative rule for Legendre symbols. For any prime p and $a, b \in \mathbb{Z}_p^+$,

$$\left(\frac{ab}{p}\right) = \left(\frac{a}{p}\right)\left(\frac{b}{p}\right).$$

2

Approximation of a Convex Set by Polygons

Computing (exactly or approximately) the area and the perimeter of a region is one of the oldest and toughest problems in geometry. A standard approach, which goes back at least to Archimedes, is to approximate the region with polygons. In this chapter we present some useful regularity properties of the best approximation (Dowker's theorems). We also show that from the point of view of approximation by inscribed polygons, the ellipses are the worst possible convex regions. Finally, we describe an elegant argument of Elekes to explain why it is hard to estimate the volume of high-dimensional convex bodies.

DOWKER'S THEOREMS

Let C be a *convex disc* (i.e., a convex compact set with nonempty interior) in the plane. For every $n \geq 3$, consider the smallest possible area of an n-gon circumscribed about C. The following useful theorem of Dowker (1944) states that these numbers form a *convex* sequence. More precisely, we have

Theorem 2.1 (Dowker). *Given a convex disc C in the plane, $n \geq 3$, let P_n denote an n-gon of minimum area circumscribed about C. Let $A(P_n)$ denote the area of P_n. Then*

$$A(P_n) \leq \frac{A(P_{n-1}) + A(P_{n+1})}{2} \qquad \text{for every } n \geq 4.$$

Proof. Throughout this proof we make no notational distinction between a side of a polygon and the straight line containing it. If a and b are two consecutive sides of P_{n-1} in clockwise order, we define $cap(a, b)$ as the region bounded by a, b, and the boundary of C (see Figure 2.1). By the pigeonhole principle, there are two consecutive sides a and b of P_{n-1}, and two consecutive sides s and t of P_{n+1} such that $cap(s, t) \subset cap(a, b)$.

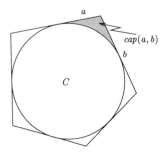

Figure 2.1. $cap(a, b)$.

Let Q denote the polygon whose $2n$ sides (in clockwise order) are

$$\underbrace{b, \ldots, a}_{\substack{\text{sides of} \\ P_{n-1}}}, \underbrace{t, \ldots, s}_{\substack{\text{sides of} \\ P_{n+1}}}.$$

That is, Q can be obtained by "switching" from a to t, and hence linking P_{n-1} and P_{n+1} to form a single polygon. Q obviously intersects itself, but its binding number at every point is at most 2. (Roughly speaking, this means that Q runs twice around C.) We call such a polygon a *double star*. It is natural to define $A(Q)$, the area of Q, with multiplicities, i.e., those regions enclosed by Q, where the binding number is two, are counted twice (see Figure 2.2). Then

$$A(Q) = A(P_{n-1}) + A(P_{n+1}) - A(T),$$

where T is the shaded region in Figure 2.2(a). Thus,

$$A(Q) \leq A(P_{n-1}) + A(P_{n+1}).$$

It is now easy to see that Q again has two pairs of consecutive sides such that $cap(s', t') \subset cap(a', b')$. Performing the same "switch" as above, we obtain two simple convex polygons P' and P'' whose total area is at most $A(Q)$. If P' and P'' are n-gons, then we obtain

$$2A(P_n) \leq A(P') + A(P'') \leq A(Q) \leq A(P_{n-1}) + A(P_{n+1}),$$

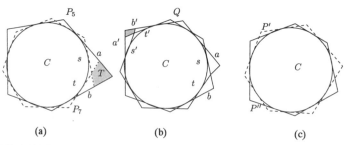

(a)   (b)   (c)

Figure 2.2. (a) Polygons P_{n-1}, P_{n+1}; (b) the double-star polygon Q; (c) polygons P', P''.

as desired. Otherwise, one of them (say, P') has a cap which is strictly contained in a cap of the other (P''). Continuing our procedure, first we get a double-star Q' with $A(Q') \leq A(P')+A(P'')$, and then two convex polygons with total area at most $A(Q')$. Since the number of those pairs of caps that are strictly contained in each other decreases at each step, our algorithm terminates in finitely many steps. That is, we obtain two circumscribing n-gons, whose total area is at most $A(P_{n-1}) + A(P_{n+1})$. $\qquad\qquad\square$

Remark 2.2. P_{n-1} and P_{n+1} together have $2n$ sides. Let us list them in a single sequence $s_1, s_2, \ldots, s_{2n}$ according to the cyclic order as they touch the boundary of C. The above argument yields the fact that the sum of the areas of the two n-gons bounded by $s_1, s_3, \ldots, s_{2n-1}$ and $s_2, s_4, \ldots, s_{2n}$ is at most $A(P_{n-1}) + A(P_{n+1})$. Moreover, our algorithm always terminates with these two polygons.

The following "dual" statement is also true.

Theorem 2.3. *Given a convex disc C in the plane, $n \geq 3$, let p_n denote an n-gon of maximum area inscribed in C. Then*

$$A(p_n) \geq \frac{A(p_{n-1}) + A(p_{n+1})}{2} \qquad \text{for every } n \geq 4.$$

This theorem can be established using almost the same argument as in Theorem 2.1, and therefore we leave its proof to the reader. Similar results can be proved for the perimeter.

Theorem 2.4 (Molnár, 1955; L. Fejes Tóth, 1959a). *Let C be a convex disc in the plane, $n \geq 3$. Let Q_n (and q_n) denote an n-gon of minimum (resp. maximum) perimeter circumscribed about C (resp. inscribed in C). Then*

$$\text{Per}(Q_n) \leq \frac{\text{Per}(Q_{n-1}) + \text{Per}(Q_{n+1})}{2} \qquad \text{for every } n \geq 4;$$

$$\text{Per}(q_n) \geq \frac{\text{Per}(q_{n-1}) + \text{Per}(q_{n+1})}{2} \qquad \text{for every } n \geq 4.$$

Using Remark 2.2 it is not hard to see that the same proof as of Theorem 2.1 yields

Theorem 2.5. *Let C be a centrally symmetric convex disc in the plane, and let $n \geq 4$ be an even integer. Then one can find convex n-gons P_n and Q_n circumscribed about C with minimum area and minimum perimeter, respectively, such that they are centrally symmetric and have the same centers as C.*

Similarly, there exist p_n and q_n, inscribed n-gons with maximum area and

maximum perimeter, such that they are centrally symmetric and have the same centers as C.

Proof. We only prove the assertion for circumscribing n-gons of minimum area (the other cases can be treated similarly). Assume without loss of generality that the center of C is the origin, and let P_n be a convex n-gon circumscribed about C with the minimum area. Then $-P_n$ is another n-gon with the same properties. Let us list the $2n$ sides of P_n and $-P_n$ in a single sequence $s_1, s_2, \ldots, s_{2n}$, according to the cyclic order in which they touch the boundary of C. As before, the sum of the areas of the n-gons bounded by $s_1, s_3, \ldots, s_{2n-1}$ and $s_2, s_4, \ldots, s_{2n}$ is at most

$$A(P_n) + A(-P_n) = 2A(P_n).$$

However, both of these n-gons are now centrally symmetric. So the area of at least one of them is at most $A(P_n)$. □

See Exercise 2.2 for a generalization of this result.

AN EXTREMAL PROPERTY OF ELLIPSES

The next theorem of Sas (1939) generalizes an observation of Blaschke (1923). Roughly speaking, it asserts that from the point of view of approximation by inscribed polygons of large area, the worst possible convex discs are the ellipses.

Theorem 2.6 (Sas). *Given a convex disc C in the plane, $n \geq 3$, let p_n denote an n-gon of maximum area inscribed in C. Then*

$$A(p_n) \geq A(C) \frac{n}{2\pi} \sin \frac{2\pi}{n},$$

with equality if and only if C is an ellipse.

Proof. We can assume without loss of generality that the diameter of C is 2, and choose a system of coordinates (x, y) such that the points $(-1, 0)$ and $(+1, 0)$ belong to C. Let us parametrize the boundary of C in the following way

$$x(\phi) = \cos \phi, \tag{2.1}$$

$$y(\phi) = g(\phi) \sin \phi, \tag{2.2}$$

where $g(\phi) > 0$ is a continuous periodic function with period 2π. Set

$$\phi_1 = \phi, \quad \phi_2 = \phi + \frac{2\pi}{n}, \quad \phi_3 = \phi + 2\frac{2\pi}{n}, \quad \ldots, \quad \phi_n = \phi + (n-1)\frac{2\pi}{n},$$

and let $p_n^\star(\phi)$ denote the inscribed n-gon whose vertex set is

$$\{(x(\phi_i), y(\phi_i)) | 1 \le i \le n\}.$$

Evidently,

$$A[p_n^\star(\phi)] = \sum_{i=1}^{n} A\left[\triangle(0,0)\left(x(\phi_i), y(\phi_i)\right)\left(x(\phi_{i+1}), y(\phi_{i+1})\right)\right]$$

$$= \sum_{i=1}^{n} \frac{1}{2}\left(g(\phi_{i+1}) \sin \phi_{i+1} \cos \phi_i - g(\phi_i) \sin \phi_i \cos \phi_{i+1}\right)$$

$$= \sum_{i=1}^{n} \frac{1}{2} g(\phi_i) \sin \phi_i (\cos \phi_{i-1} - \cos \phi_{i+1})$$

$$= \sin \frac{2\pi}{n} \sum_{i=1}^{n} g(\phi_i) \sin^2 \phi_i.$$

($\phi_{n+1} = \phi_1$ and $\triangle abc$ denotes the triangle with vertices a, b, c.) Hence,

$$\frac{1}{2\pi} \int_0^{2\pi} A[p_n^\star(\phi)] \, d\phi = \frac{n}{2\pi} \sin \frac{2\pi}{n} \int_0^{2\pi} g(\phi) \sin^2 \phi \, d\phi$$

$$= A(C) \frac{n}{2\pi} \sin \frac{2\pi}{n} \quad \text{(see Exercise 2.6)}.$$

Since, for a suitable choice of ϕ, $A[p_n^\star(\phi)]$ is at least as large as its mean value, the first part of the theorem follows. The discussion of the case of equality is left to the reader (Exercise 2.7). $\qquad\square$

Remark 2.7. It might seem natural to believe that an analogous statement is true for *circumscribing* polygons (polytopes). However, this is not the case. Easy calculations show that, for example, the smallest possible area of a triangle circumscribed about a unit square is strictly larger than the minimum area of a triangle containing a circle of area 1. Nevertheless, one can prove a somewhat weaker assertion establishing a link between the best inscribed and circumscribing n-gons approximating a convex disc C.

Theorem 2.8 (Lázár, 1947). *Let C be a convex disc in the plane, $n \ge 3$. Then there exist an n-gon P_n circumscribed about C and an n-gon p_n inscribed in C such that*

$$\frac{A(P_n) - A(p_n)}{A(P_n)} \le \sin^2 \frac{\pi}{n}.$$

This bound is tight if C is an ellipse.

Schneider (1967, 1971) proved that from the point of view of approximation by inscribed polygons of maximum *perimeter* (as well as by circumscribing polygons of minimum perimeter), the worst possible convex discs are the ellipses. More precisely, the following is true.

Theorem 2.9 (Schneider). *Let C be a convex disc in the plane, $n \geq 3$. Let Q_n (resp. q_n) denote an n-gon of minimum (resp. maximum) perimeter circumscribed about C (resp. inscribed in C). Then*

$$\text{Per}(Q_n) \leq \text{Per}(C)\frac{n}{\pi}\tan\frac{\pi}{n},$$

$$\text{Per}(q_n) \geq \text{Per}(C)\frac{n}{\pi}\sin\frac{\pi}{n}.$$

Both inequalities are tight if C is an ellipse.

APPROXIMATION OF A CONVEX BODY BY POLYTOPES

Theorem 2.6 was generalized to higher dimensions by Macbeath (1951).

Theorem 2.10 (Macbeath). *Let C be a convex body in $\mathbb{R}^d$, and let B^d denote a ball with $\text{Vol}(B)^d = \text{Vol}(C)$. Then C contains a convex polytope with n vertices, whose volume is at least as large as the maximal volume of a polytope of n vertices inscribed in B^d.*

Unfortunately, this result has no simple quantitative form similar to the inequality in Theorem 2.6, because it is extremely hard to estimate the volume of the largest polytope of n vertices, inscribed in a d-dimensional ball. If n is not too large compared to d, then the following ingenious argument of Elekes (1986) provides a nontrivial upper bound.

Theorem 2.11 (Elekes). *Let p_n^d denote a convex polytope with n vertices inscribed in a d-dimensional ball B^d. Then*

$$\text{Vol}(p_n^d) \leq \text{Vol}(B^d)\frac{n}{2^d}.$$

Proof. Let $v_1, v_2, \ldots, v_n$ be the vertices of p_n^d and c be the center of B^d. The ball of diameter cv_i will be denoted by B_i. Obviously, $\text{Vol}(B_i) = \text{Vol}(B^d)/2^d$, for $1 \leq i \leq n$. We claim that

$$p_n^d \subseteq \bigcup_{i=1}^{n} B_i,$$

which clearly implies the theorem.

Assume, in order to obtain a contradiction, that some $u \in p_n^d$ is not covered by any of the (closed) balls B_i. Then the angle $cuv_i < \pi/2$ for every i. This implies that every vertex v_i lies in the same open half-space bounded by the hyperplane passing through u and orthogonal to cu. Hence, u cannot be contained in p_n^d (the convex hull of $v_1, \ldots, v_n$, a contradiction. $\square$

Elekes's theorem has some interesting algorithmic consequences. It implies immediately that there is no polynomial time algorithm estimating the volume of a d-dimensional convex set C up to a constant factor under the following model of computation.

Let $C \subseteq \mathbb{R}^d$ be a convex body. Assume that our algorithm has access to an *oracle* that can answer in unit time whether a given point u is contained in C. For technical reasons, we also assume that the oracle can give us in advance two concentric balls B and B' such that $B \subseteq C \subseteq B'$.

Corollary 2.12. *Let $0 < \gamma < 1$ be fixed, and let $C \subseteq \mathbb{R}^d$ be a convex body. Assume that an algorithm using the above oracle can determine a number V such that*

$$\gamma V \leq \mathrm{Vol}(C) \leq V.$$

Then the time complexity of this algorithm is at least $\gamma 2^d - d - 1$.

Proof. Let $v_1, v_2, \ldots, v_{d+1}$ be the vertices of a regular simplex inscribed in B' and let B be a ball inscribed in this simplex. Assume that we have the additional information that $v_i \in C$ for every $1 \leq i \leq d + 1$. Suppose further that whenever we ask the oracle whether $u \in C$, the answer is "yes" if and only if $u \in B'$.

After having called the oracle k times, all we have is a set S of at most $k + d + 1$ points of C, and C can be as small as their convex hull. On the other hand, we cannot rule out the possibility that $C = B'$. Thus, if after k steps we can find a number V satisfying the requirements, then

$$\frac{\mathrm{Vol}(\mathrm{conv}\, S)}{\mathrm{Vol}(B')} \geq \gamma.$$

Since, by Elekes's theorem,

$$\frac{\mathrm{Vol}(\mathrm{conv}\, S)}{\mathrm{Vol}(B')} \leq \frac{k + d + 1}{2^d},$$

we can conclude that $k \geq \gamma 2^d - d - 1$. $\square$

For further results of this kind, see Bárány and Füredi (1987), Dyer and Frieze (1988), Dyer et al. (1991), Lovász and Simonovits (1990), Khachiyan (1993), and Applegate and Kannan (1991).

EXERCISES

2.1 Prove Theorem 2.4.

2.2 (G. Fejes Tóth and L. Fejes Tóth, 1973a)   Let C be a convex disc with *k-fold rotational symmetry*; i.e., assume that the rotation with angle $2\pi/k$ around O carries C into itself. Show that for any integer n divisible by k, the class of all n-gons of minimal (maximal) area circumscribed about (resp. inscribed in) C contains an element with k-fold rotational symmetry. Similar assertions hold for the perimeter.

2.3 Does Theorem 2.5 remain valid if central symmetry is replaced by axis symmetry?

2.4 (Eggleston, 1957a, 1957b; G. Fejes Tóth, 1972, 1977a)   Let $0 < w < 1$ be fixed. The *(weighted) area deviation* of D from C is defined as

$$\operatorname{dev}_A (C, D) = wA(C - D) + (1 - w)A(D - C),$$

where $C - D$ is the set of points of C not belonging to D. Let C be a strictly convex set in the plane, and let Q_n be a convex n-gon whose area deviation from C is minimum. Show that

(i) every edge of Q_n with length l is divided by the boundary of C into three segments of lengths $(w/2)l$, $(1 - w)l$, and $(w/2)l$, respectively;

(ii) if C is a circular disc, then Q_n is a regular n-gon;

(iii) if C is a circular disc, then

$$\operatorname{dev}_A (C, Q_n) \le \frac{\operatorname{dev}_A (C, Q_{n-1}) + \operatorname{dev}_A (C, Q_{n+1})}{2}$$
$$\text{for every } n \ge 4;$$

(iv)⋆ the last inequality holds for any convex disc C.

2.5 (Eggleston, 1957a, 1957b)   The *perimeter deviation* of two convex discs C and D is defined as

$$\operatorname{dev}_P (C, D) = \operatorname{Per} (C \cup D) - \operatorname{Per} (C \cap D).$$

Let q_n be a convex n-gon whose perimeter deviation from C is minimal. Show that

(i) q_n is inscribed in C;

(ii) $\operatorname{dev}_P (C, q_n) \le \dfrac{\operatorname{dev}_P (C, q_{n-1}) + \operatorname{dev}_P (C, q_{n+1})}{2}$ for every $n \ge 4$.

2.6 Prove that with the notation used in the proof of Theorem 2.6,

$$\int_0^{2\pi} g(\phi) \sin^2 \phi \, d\phi = A(C).$$

2.7 Prove that if equality holds in Theorem 2.6, then C is an ellipse.

2.8 Let C be a convex disc in the plane, $n \geq 3$. Prove, without using Theorem 2.9, that there exist an n-gon Q_n circumscribed about C and an n-gon q_n inscribed in C such that

$$\frac{\text{Per}(Q_n) - \text{Per}(q_n)}{\text{Per}(Q_n)} \leq 2 \sin^2 \frac{\pi}{2n}.$$

This is tight if C is an ellipse.

2.9 (Elekes, 1986) Let $\frac{1}{2} < \gamma < 1$ be fixed, and let $C \subseteq \mathbb{R}^d$ be a convex body. Assume that an algorithm using the same oracle as in Corollary 2.12, can determine a number W such that

$$\gamma W \leq \text{width}(C) \leq W.$$

Then the running time of this algorithm is at least $\frac{1}{2}(2\gamma)^d - d - 1$. (The *width* of C is the minimum distance between two parallel hyperplanes supporting C.)

2.10 (Bárány and Füredi, 1987)

(i) Let S be a simplex inscribed in the unit ball $B^d \subseteq \mathbb{R}^d$, and let k be an integer, $0 \leq k < d$. Prove that for any point $x \in S$, there exists a k-dimensional face F_x^k and a point $y_x \in F_x^k$ such that $x - y_x$ is orthogonal to F_x^k and $|x - y_x| \leq \sqrt{(d-k)/(dk)}$.

(ii) Let p_n^d denote a convex polytope with $n \leq d^\alpha$ vertices (for some fixed $\alpha > 0$) inscribed in the unit ball $B^d \subseteq \mathbb{R}^d$. Show that

$$\text{Vol}(p_n^d) \leq \text{Vol}(B^d) \left(\frac{10\alpha \ln d}{d} \right)^{d/2},$$

provided that d is sufficiently large.

3

Packing and Covering with Congruent Convex Discs

As we have seen (Corollary 1.4), it is easy to show that the density of any lattice packing C of congruent circles (i.e., the portion of the plane covered by the elements of C) is at most $\pi/\sqrt{12}$. In his lecture at the Scandinavian Natural Science Congress in 1892, A. Thue extended this result to any packing of congruent circles. In other words, he proved that the densest packing of equal circles is necessarily lattice-like. Thue's theorem, as well as its dual counterpart for covering (established by Kershner in 1939), remained largely unnoticed until the 1950s, when L. Fejes Tóth discovered many far-reaching generalizations of these results. His seminal book (L. Fejes Tóth, 1953) laid the foundations of a rich discipline: the theory of packing and covering. The purpose of this chapter is to introduce the basic concepts of this theory and to present some generalizations of the above results to packing and covering with arbitrary convex discs.

PACKING OF CONVEX DISCS

We need some basic definitions and notations which will be used often in the next five chapters.

Definition 3.1. Let $C = \{C_1, C_2, \dots\}$ be a collection of convex discs in the plane and let D be a domain. C is called a *covering* of D if $\bigcup_i C_i \supseteq D$. On the other hand, if $\bigcup_i C_i \subseteq D$ and no two of them have an interior point in common, then C is said to form a *packing* in D.

If D is a bounded domain, then the *density* of the collection C with respect to D is defined as

$$d(C, D) = \frac{\sum_i A(C_i)}{A(D)},$$

where the sum is taken over all i for which $C_i \cap D \neq \emptyset$. If D is the whole plane, then we define the *upper* and *lower densities*, denoted by $\overline{d}$ and $\underline{d}$, respectively,

21

as follows. Let $D(r)$ denote the circular disc of radius r centered at a fixed point O (the origin). Let

$$\overline{d}(C, \mathbb{R}^2) = \lim_{r \to \infty} \sup d(C, D(r)) \quad \text{and}$$

$$\underline{d}(C, \mathbb{R}^2) = \lim_{r \to \infty} \inf d(C, D(r)).$$

If these two numbers coincide, then their common value is called the *density* of the collection C in the plane, and is denoted by $d(C, \mathbb{R}^2)$.

Observe that $\overline{d}(C, \mathbb{R}^2)$, $\underline{d}(C, \mathbb{R}^2)$, and hence $d(C, \mathbb{R}^2)$, are independent of the choice of origin O (see Exercise 3.1).

At the end of the last century, Thue (1892, 1910) proved that the (upper) density of the packing C of congruent circles in the plane is at most $\pi/\sqrt{12} \approx 0.907$, i.e.,

$$\overline{d}(C, \mathbb{R}^2) \leq \frac{\pi}{\sqrt{12}}.$$

The dual counterpart of this theorem was established much later by Kershner (1939). He showed that the (lower) density of a covering C of the plane with congruent circles is at least $2\pi/\sqrt{27} \approx 1.209$, i.e.,

$$\underline{d}(C, \mathbb{R}^2) \geq \frac{2\pi}{\sqrt{27}}.$$

Both of these bounds can be attained for suitable lattice arrangements (see Figure 3.1).

The following theorem of L. Fejes Tóth (1950) leads to a far-reaching generalization of Thue's result. The proof presented here is based on a theorem of Dowker (1944) discussed in Chapter 2.

Theorem 3.2. (L. Fejes Tóth).   *Let H be a convex hexagon, let C be a convex disc, and let P_6 denote a hexagon of minimum area circumscribed about C. If*

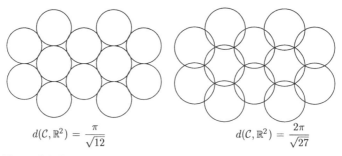

$$d(C, \mathbb{R}^2) = \frac{\pi}{\sqrt{12}} \qquad\qquad\qquad d(C, \mathbb{R}^2) = \frac{2\pi}{\sqrt{27}}$$

Figure 3.1. Densest packing and thinnest covering with congruent circles.

n congruent (nonoverlapping) copies of C are packed in H, then

$$n \le \frac{A(H)}{A(P_6)}.$$

We need the following technical lemma, which is a simple consequence of *Euler's formula* for polytopes. More precisely, we will use the fact that every planar graph with n vertices has at most $3n - 6$ edges.

Lemma 3.3. *Let $C_1, \ldots, C_n$ be a system of nonoverlapping convex discs arranged in a convex hexagon H. Then we can find a system of nonoverlapping convex polygons $R_1, \ldots, R_n \subseteq H$ such that $R_i \supseteq C_i$ for every i, and*

$$\sum_{i=1}^{n} s_i \le 6n,$$

where s_i denotes the number of sides of R_i.

Proof. Let us start "growing" the sets C_i within H so that they remain convex and their interiors remain disjoint. It is easy to see that in a position when none of them can be increased, all the expanded sets $R_i \supseteq C_i$ are convex polygons. Note that they do not necessarily fill H. Assume, for the sake of simplicity, that any two sets R_i and R_j are either disjoint or share a common boundary segment. (The general case can be treated similarly.) Construct a planar graph with $n + 1$ vertices which correspond to $R_1, \ldots, R_n$ and $\overline{H}$ (the complement of H), two vertices being joined by an edge if and only if the corresponding faces touch each other. If e denotes the number of edges in this graph, then

$$3(n + 1) - 6 \ge e \ge \frac{1}{2}\left(\sum_{i=1}^{n} s_i + q - 6 \right),$$

where q denotes the number of points on the boundary of H that belong to more than one R_i. From this the lemma follows; see Cassels (1959), Heppes (1964), and Edelsbrunner et al. (1990) for more detailed arguments. $\qquad\square$

Proof of Theorem 3.2. Assume that $C_1, \ldots, C_n$ are congruent copies of C which form a packing in H, and for $1 \le i \le n$, let $R_i \supseteq C_i$ denote the polygons constructed above. Furthermore, let P_s be an s-gon of minimum area circumscribed about C. Then

$$A(H) \ge \sum_{i=1}^{n} A(R_i) \ge \sum_{i=1}^{n} A(P_{s_i}).$$

On the other hand, by Theorem 2.1, there exists a monotone decreasing convex function $a(x)$ such that $a(s) = A(P_s)$ for any integer $s \geq 3$. Hence, by Jensen's inequality and Lemma 3.3,

$$\sum_{i=1}^{n} A(P_{s_i}) = \sum_{i=1}^{n} a(s_i) \geq n \cdot a\left(\sum_{i=1}^{n} s_i \bigg/ n\right) \geq n \cdot a(6). \qquad \square$$

In other words, Theorem 3.2 states that the density of any packing C of congruent copies of C in a convex hexagon H,

$$d(C, H) \leq \frac{A(C)}{A(P_6)}.$$

This immediately yields the same upper bound for the packing density in the *plane* (see Exercise 3.3).

Corollary 3.4.   *Given a packing C with congruent copies of a convex disc C in the plane,*

$$\overline{d}(C, \mathbb{R}^2) \leq \frac{A(C)}{A(P_6)},$$

where P_6 denotes a hexagon of minimum area circumscribed about C.

In particular, if C is a circle, then P_6 is a regular hexagon circumscribed about C; hence

$$\overline{d}(C, \mathbb{R}^2) \leq \frac{\pi}{\sqrt{12}},$$

which is Thue's theorem mentioned above.

Before we formulate another important consequence of Theorem 3.2, we have to introduce some notation. Recall that for any lattice Λ in the plane and for any convex disc C, the collection of translates $C = \{C + u \mid u \in \Lambda\}$ is called a *lattice arrangement*. If, in addition, C is a packing (covering), then it is called a *lattice packing* (resp. *lattice covering*).

Definition 3.5.   Given any convex disc $C \subseteq \mathbb{R}^2$, let

$$\delta(C) = \sup_{C \text{ packing}} \overline{d}(C, \mathbb{R}^2),$$

$$\delta_L(C) = \sup_{\substack{C \text{ lattice} \\ \text{packing}}} \overline{d}(C, \mathbb{R}^2) = \sup_{\substack{C \text{ lattice} \\ \text{packing}}} d(C, \mathbb{R}^2),$$

where the supremum is taken over all packings (resp. lattice packings) in the plane with congruent copies of C. Evidently, $\delta_L(C) \leq \delta(C)$.

It is not hard to prove (Exercise 3.2) that

$$\delta(C) = \max_{C \text{ packing}} d(C, \mathbb{R}^2),$$

$$\delta_L(C) = \max_{\substack{C \text{ lattice} \\ \text{packing}}} d(C, \mathbb{R}^2),$$

where the maximum is taken over all packings (resp. lattice packings) with congruent copies of C, whose densities exist.

Corollary 3.6. *Let C be a centrally symmetric convex disc in the plane. Then $\delta(C) = \delta_L(C)$.*

Proof. By Corollary 3.4, $\delta(C) \leq A(C)/A(P_6)$. In view of Theorem 2.5, P_6 can be chosen to be centrally symmetric. Fig. 3.2 shows that one can easily tile the whole plane with translates of a centrally symmetric convex hexagon (P_6) in a lattice-like fashion. If we inscribe a translate of C into each cell of this tiling, then we obtain a lattice packing in the plane, whose density is obviously equal to $A(C)/A(P_6)$. Thus $\delta_L(C) \geq A(C)/A(P_6) \geq \delta(C)$. $\qquad\square$

A somewhat closer examination of the proof of Theorem 3.2 shows that the statement can be generalized to packings of *similar* copies of a convex disc, provided that the sizes of these copies do not differ too much. The following result is due to K. Böröczky (see L. Fejes Tóth, 1972, p. 194; and Blind, 1969). The simple proof presented below was found by G. Fejes Tóth (1972).

Theorem 3.7. *Let H be a convex hexagon, let C be a convex disc, and let P_s denote a convex s-gon of minimum area circumscribed about C ($s \geq 3$). If $C = \{C_1, C_2, \ldots, C_n\}$ is a packing of similar copies of C in H, and*

$$\frac{A(C_i)}{A(C_j)} \leq \frac{A(P_5) - A(P_6)}{A(P_6) - A(P_7)} \qquad \text{for all } i \text{ and } j,$$

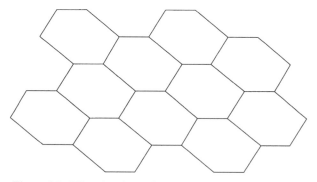

Figure 3.2. Tiling the plane with centrally symmetric hexagons.

then

$$d(C, H) \leq \frac{A(C)}{A(P_6)}.$$

Proof. Let $R_i \supseteq C_i$ denote the same polygon of s_i sides, as in Theorem 3.2. Then

$$A(H) \geq \sum_{i=1}^{n} A(R_i) \geq \sum_{i=1}^{n} A(P_{s_i}) \frac{A(C_i)}{A(C)}. \tag{3.1}$$

Suppose that $s_i > 6$ for some i. Since $\sum_{i=1}^{n} s_i \leq 6n$, there exists a j for which $s_j < 6$. By Theorem 2.1,

$$A(P_{s_j}) - A(P_{s_j+1}) \geq A(P_5) - A(P_6)$$
$$\geq A(P_6) - A(P_7)$$
$$\geq A(P_{s_i-1}) - A(P_{s_i}).$$

Thus, by our assumptions,

$$\frac{A(C_i)}{A(C_j)} \leq \frac{A(P_5) - A(P_6)}{A(P_6) - A(P_7)} \leq \frac{A(P_{s_j}) - A(P_{s_j+1})}{A(P_{s_i-1}) - A(P_{s_i})},$$

which implies that

$$A(P_{s_i}) \frac{A(C_i)}{A(C)} + A(P_{s_j}) \frac{A(C_j)}{A(C)} \geq A(P_{s_i-1}) \frac{A(C_i)}{A(C)} + A(P_{s_j+1}) \frac{A(C_j)}{A(C)}. \tag{3.2}$$

So we can decrease the right-hand side of (3.1) by replacing the ith and jth terms with the right-hand side of (3.2). Repeating this process until we get rid of all P_{s_i}'s with $s_i > 6$, we obtain that

$$A(H) \geq \sum_{i=1}^{n} A(P_6) \frac{A(C_i)}{A(C)}.$$

Equivalently,

$$d(C, H) = \frac{\sum_{i=1}^{n} A(C_i)}{A(H)} \leq \frac{A(C)}{A(P_6)}. \qquad \square$$

COVERING BY CONVEX DISCS

Next we will try to deduce some analogous results for *coverings*.

Two convex discs C and C' are said to *cross* each other if $C - C'$ and $C' - C$ are disconnected.

Theorem 3.8 (L. Fejes Tóth). *Let H be a convex hexagon completely covered by n congruent copies of a convex disc C, which do not cross each other. Then*

$$n \geq \frac{A(H)}{A(p_6)},$$

where p_6 denotes a hexagon of maximum area inscribed in C.

Instead of Lemma 3.3, we can now prove

Lemma 3.9. *Let $C_1, \ldots, C_n$ be a system of noncrossing convex discs which cover a convex hexagon H. Then one can find nonoverlapping convex polygons $R_i \subseteq C_i$, for $1 \leq i \leq n$, which altogether cover H, and*

$$\sum_{i=1}^{n} s_i \leq 6n,$$

where s_i denotes the number of sides of R_i.

Proof. Consider any pair of discs C_i, C_j whose interiors intersect (but do not cross each other). Let a and b denote the intersection points of their boundaries. We can easily make the interiors of C_i and C_j disjoint by shrinking them until they share the common boundary segment ab. In this way we obtain two new convex discs C_i' and C_j' such that $C_i' \cup C_j' = C_i \cup C_j$. The trouble is that in our new system of sets some of the discs may cross each other (see Figure 3.3). We can avoid this difficulty by observing that whenever C_j' and C_k cross each other, then $C_i \cap C_k \subset C_i \cap C_j$. Hence, if we start our shrinking procedure with a pair (C_i, C_j) for which $C_i \cap C_j$ is minimal, then we cannot get into trouble.

Thus, our algorithm produces a system of nonoverlapping convex polygons $R_1 \cdots R_n$ in at most $\binom{n}{2}$ steps, with the property that $\bigcup_{i=1}^{n} R_i = H$. The last part of the lemma follows from Euler's formula in exactly the same way as in the proof of Lemma 3.3 (see Bambah and Rogers, 1952, for details). $\square$

Proof of Theorem 3.8. Let $C_1, \ldots, C_n$ be a system of noncrossing congruent copies of C, and let R_i, s_i denote the same as in Lemma 3.9.

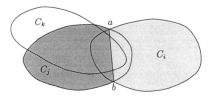

Figure 3.3. Shrinking C_i and C_j.

Furthermore, let p_s denote an s-gon of maximum area inscribed in C. Now we have

$$A(H) = \sum_{i=1}^{n} A(R_i) \leq \sum_{i=1}^{n} A(p_{s_i}).$$

In view of Theorem 2.3, there exists a monotone increasing function $a(x)$ such that $a(s) = A(p_s)$ for every integer $s \geq 3$. Hence, by Jensen's inequality and Lemma 3.9,

$$\sum_{i=1}^{n} A(p_{s_i}) = \sum_{i=1}^{n} a(s_i) \leq n \cdot a\left(\frac{\sum_{i=1}^{n} s_i}{n} \right)$$
$$\leq n \cdot a(6) = n \cdot A(p_6). \qquad \square$$

In other words, Theorem 3.8 states that the density of any covering C of a convex hexagon H with noncrossing congruent copies of C,

$$d(C, H) \geq \frac{A(C)}{A(p_6)}.$$

As in the case of packings, the last bound continues to hold for coverings with noncrossing *similar* copies of C, whose sizes are not too different (Exercise 3.5).

Corollary 3.10.   *Given a covering C of the whole plane with noncrossing congruent copies (e.g. translates) of C,*

$$\underline{d}(C, \mathbb{R}^2) \geq \frac{A(C)}{A(p_6)},$$

where p_6 denotes a hexagon of maximum area inscribed in C.

In particular, if C is a circle (or an ellipse), then

$$\underline{d}(C, \mathbb{R}^2) \geq \frac{2\pi/6}{\sin(2\pi/6)} = \frac{2\pi}{\sqrt{27}},$$

which is Kershner's result mentioned earlier.

Definition 3.11.   Given any convex disc $C \subseteq \mathbb{R}^2$, let

$$\vartheta(C) = \inf_{C \text{ covering}} \underline{d}(C, \mathbb{R}^2),$$

$$\vartheta_L(C) = \inf_{\substack{C \text{ lattice} \\ \text{covering}}} \underline{d}(C, \mathbb{R}^2) = \inf_{\substack{C \text{ lattice} \\ \text{covering}}} d(C, \mathbb{R}^2),$$

where the infimum is taken over all coverings (resp. lattice coverings) of the plane with congruent copies of C. Similarly, let

$$\vartheta^\star(C) = \inf_{\substack{C \text{ noncrossing} \\ \text{covering}}} \underline{d}(C, \mathbb{R}^2).$$

Obviously,

$$\vartheta(C) \le \vartheta^\star(C) \le \vartheta_L(C).$$

As in the case of packing densities, it is not hard to show that

$$\vartheta(C) = \min_{C \text{ covering}} d(C, \mathbb{R}^2),$$

$$\vartheta_L(C) = \min_{\substack{C \text{ lattice} \\ \text{covering}}} d(C, \mathbb{R}^2),$$

$$\vartheta^\star(C) = \min_{\substack{C \text{ noncrossing} \\ \text{covering}}} d(C, \mathbb{R}^2),$$

where the minimum is taken over those coverings only, whose densities exist.

The following fact can be deduced from Corollary 3.10 and Theorem 2.5.

Corollary 3.12. *Let C be a centrally symmetric convex disc in the plane. Then $\vartheta^\star(C) = \vartheta_L(C)$.*

One of the major unsolved problems in this area is whether or not $\vartheta(C) = \vartheta_L(C)$ for every centrally symmetric convex C. (It is felt that the condition requiring the copies of C to be *noncrossing* in Theorem 3.8 is merely a technical necessity.)

Another interesting consequence of above results is that any centrally symmetric convex disc permits a lattice covering at least as economical as the thinnest covering with circles. More precisely, we have

Corollary 3.13. *For any centrally symmetric convex disc C, we have $\vartheta_L(C) \le 2\pi/\sqrt{27}$ with equality if and only if C is an ellipse.*

Proof. By Corollaries 3.10 and 3.12,

$$\vartheta_L(C) = \frac{A(C)}{A(p_6)}.$$

In view of Theorem 2.6, this implies that

$$\vartheta_L(C) \le \frac{2\pi}{6} \bigg/ \sin \frac{2\pi}{6} = \frac{2\pi}{\sqrt{27}}. \qquad \square$$

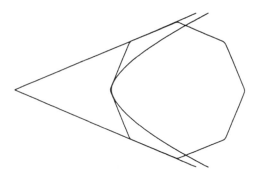

Figure 3.4. Smoothed octagon.

It is conjectured that $\vartheta(C) \leq 2\pi/\sqrt{27} \approx 1.2092$ is valid for any convex disc C. The best result of this type is due to W. Kuperberg (1989), who showed that

$$\vartheta(C) \leq \frac{8(2\sqrt{3}-3)}{3} < 1.238.$$

Note that the dual of Corollary 3.13 is not true. It is not hard to find a centrally symmetric convex disc that does not permit a lattice packing of density $\geq \pi/\sqrt{12} \approx 0.9069$ (the density of the most economical packing with equal circles). This fact is in perfect accordance with Remark 2.7, that is, with the observation that a statement similar to Theorem 2.6 cannot hold for circumscribing n-gons (see Exercise 3.6).

A famous conjecture of Reinhardt (1934) states that every centrally symmetric convex disc C permits a lattice packing of density

$$\delta_L(C) = \delta(C) \geq \frac{8 - 4\sqrt{2} - \ln 2}{2\sqrt{2}-1} = 0.9024\ldots$$

with equality only for the so-called "smoothed octagon" (see Figure 3.4). Mahler (1946a) and L. Fejes Tóth established the weaker bound $\delta_L(C) \geq \sqrt{3}/2 = 0.8660\ldots$, which was later improved by Ennola (1961) and Tammela (1970) to about 0.8925. Let us remark that, in fact, it is not known whether there exists any (not necessarily symmetric) convex disc C whose packing density $\delta(C) < 0.9024$. (See Blind, 1983, for some likely candidates.)

BETWEEN PACKING AND COVERING

We end this chapter by establishing a link between packing and covering problems. Suppose that we have a large regular hexagon H and a collection of unit circular discs of total area $\rho A(H)$, where $\pi/\sqrt{12} < \rho < 2\pi/\sqrt{27}$. It follows from the results of the preceding two sections that these discs cannot be arranged so as to form a packing in H or a covering of H. The question arises: What is the maximum fraction of the area of H that can be covered by our discs?

The following theorem of L. Fejes Tóth (1972, p. 80) provides an asymptotically tight answer to this question. The proof presented here is due to G. Fejes Tóth (1972); it generalizes to arrangements of noncrossing congruent copies of any convex disc C.

Theorem 3.14 (L. Fejes Tóth). *Let H be a regular hexagon, and let $C = \{C_1, C_2, \ldots, C_n\}$ be a collection of congruent circles. Furthermore, let D denote a circle of area $\sum_{i=1}^{n} A(C_i)$, concentric with H. Then*

$$A\left(\bigcup_{i=1}^{n} C_i \cap H \right) \leq A(D \cap H).$$

Given two convex sets $C, D \subseteq \mathbb{R}^2$ and a constant $0 < w < 1$, the *weighted area deviation* of D from C is defined by

$$\mathrm{dev}_A (C, D) = wA(C - D) + (1 - w)A(D - C),$$

where $C - D$ denotes the set of points of C not belonging to D. (Notice that for $w \neq \frac{1}{2}$, $\mathrm{dev}_A(C, D)$ is not necessarily symmetric in C and D!)

If C is a strictly convex disc, and Q_n is an n-gon whose deviation from C is minimum, then any edge of Q_n of length l is divided by the boundary of C into three segments of lengths $(w/2)l$, $(1 - w)l$ and $(w/2)l$, respectively. In particular, if C is a circle, then it follows that Q_n is a regular n-gon concentric with C.

Furthermore, if $a(n) = \mathrm{dev}_A(C, Q_n)$ denotes the minimum deviation of an n-gon from C, then the following Dowker-type result holds:

$$a(n) \leq \frac{a(n - 1) + a(n + 1)}{2} \qquad \text{for all } n \geq 4$$

(see Exercise 2.4).

We also need the following common generalization of Lemmas 3.3 and 3.9.

Lemma 3.15. *Let $C_1, \ldots, C_n$ be a system of noncrossing convex discs, and let H be a fixed hexagon. Assume that each C_i has an interior point that belongs to H but not to any other C_j ($j \neq i$). Then one can find nonoverlapping convex polygons $R_i \subseteq H$ ($1 \leq i \leq n$) such that*

$$\bigcup_{i=1}^{n} C_i \cap H = \bigcup_{i=1}^{n} (C_i \cap R_i).$$

and $\sum_{i=1}^{n} s_i \leq 6n$, where s_i denotes the number of sides of R_i.

Proof of Theorem 3.14. Let D_1 and D_2 denote the incircle and circumcircle of H, respectively. Clearly, we can assume that

$$A(D_1) < A(D) = \sum_{i=1}^{n} A(C_i) < A(D_2), \tag{3.3}$$

for otherwise there is nothing to prove. We may also assume without loss of generality that every C_i has an interior point in common with H that does not belong to any other C_j ($j \neq i$). Thus, we can apply Lemma 3.15 to find some suitable polygons R_i meeting the requirements. By the definitions, we have

$$A\left(\bigcup_{i=1}^{n} C_i \cap H \right) = A\left(\bigcup_{i=1}^{n} (C_i \cap R_i) \right)$$

$$= \sum_{i=1}^{n} \left(A(C_i) - A(C_i - R_i) \right).$$

Similarly,

$$A\left(\bigcup_{i=1}^{n} C_i \cap H \right) = \sum_{i=1}^{n} \left(A(R_i) - A(R_i - C_i) \right).$$

Hence,

$$A\left(\bigcup_{i=1}^{n} C_i \cap H \right) = wA\left(\bigcup_{i=1}^{n} C_i \cap H \right) + (1 - w)A\left(\bigcup_{i=1}^{n} C_i \cap H \right)$$

$$= w\sum_{i=1}^{n} A(C_i) + (1 - w)\sum_{i=1}^{n} A(R_i) - \sum_{i=1}^{n} \text{dev}_A\,(C_i, R_i)$$

$$\leq wA(D) + (1 - w)A(H) - \sum_{i=1}^{n} a(s_i),$$

where $a(s_i)$ denotes the minimum area deviation of an s_i-gon from C_i. Again using Jensen's inequality, we obtain

$$A\left(\bigcup_{i=1}^{n} C_i \cap H \right) \leq wA(D) + (1 - w)A(H) - na(6).$$

Let Q_6 denote a regular hexagon with minimum deviation from $C_1 = C$. Observe that by varying w in $(0, 1)$, Q_6 will take all possible positions between the inscribed and the circumscribing hexagons of C. In view of (3.3), for some choice of w,

$$a(6) = \text{dev}_A\,(C, Q_6) = \text{dev}_A\,(D, H)\frac{A(C)}{A(D)}$$

$$= \frac{1}{n}\text{dev}_A(D, H).$$

However, in this case, we get

$$A\left(\bigcup_{i=1}^{n} C_i \cap H\right) \le wA(D) + (1 - w)A(H) - \text{dev}_A\,(D, H)$$

$$= A(D \cap H).$$

This inequality is obviously tight for $n = 1$. It is also tight asymptotically for large values of n. To see this, scale down D and H to obtain a disc D' and a hexagon H' with $A(D') = A(D)/n = A(C_i)$ and $A(H') = A(H)/n$. Arrange in a lattice-like fashion n copies of H' to form an approximate tiling of H. Putting a concentric copy of D' around each copy of H', we obtain a collection of circles $\{C_1, \ldots, C_n\}$ for which the bound in the theorem is almost tight.   □

In Chapter 5 we shall see that in the case of circle arrangements there is a very easy construction to find some polygons R_i satisfying the conditions in Lemma 3.15. The generalization of Theorem 3.14 to arbitrary convex discs is left to the reader (Exercise 3.9).

Most of the results discussed in this chapter suggest that under sufficiently general circumstances, the most economical packings and coverings with congruent copies of a convex set are "automatically" *lattice-like*. This explains why lattices, the geometry of numbers, and regular and symmetric configurations play a special role in this field. For recent surveys, see G. Fejes Tóth (1983), L. Fejes Tóth (1984a), G. Fejes Tóth and W. Kuperberg (1993a, 1993d), and Moser and Pach (1986).

EXERCISES

3.1 Prove that if C is any arrangement of convex discs in the plane, then $\bar{d}(C, \mathbb{R}^2)$ and $\underline{d}(C, \mathbb{R}^2)$ are independent of the choice of the origin (cf. Definition 3.1).

3.2 Prove that one can always find a packing C of the plane with congruent copies of a convex disc C whose density $d(C, \mathbb{R}^2)$ exists and is equal to $\delta(C)$. Similarly, show that there is a lattice packing C with $d(C, \mathbb{R}^2) = \delta_L(C)$.

3.3 Given a convex polygonal disc P containing the origin in its interior, let $P(r) = rP = \{rp \mid p \in P\}$. For any packing C of congruent copies of a given convex disc C, let

$$\overline{d}_P(C, \mathbb{R}^2) = \lim_{r \to \infty} \sup d(C, P(r)).$$

Show that

$$\sup_C \overline{d}_P(C, \mathbb{R}^2) = \sup_C d(C, \mathbb{R}^2)(= \delta(C)),$$

and use this fact to deduce Corollary 3.4 from Theorem 3.2 (cf. Definitions 3.1 and 3.5).

3.4 Construct a packing C of unit discs and two convex discs P, P' such that

$$\lim_{r \to \infty} d(C, P(r)) \neq \lim_{r \to \infty} d(C, P'(r)).$$

3.5 (Böröczky)   Let H be a convex hexagon, let C be a convex disc, and let p_s denote a convex s-gon of maximum area inscribed in C ($s \geq 3$). Show that if $C = \{C_1, C_2, \ldots, C_n\}$ is a noncrossing covering of H with similar copies of C such that

$$\frac{A(C_i)}{A(C_j)} \leq \frac{A(p_6) - A(p_5)}{A(p_7) - A(p_6)} \qquad \text{for all } i \text{ and } j,$$

then

$$d(C, H) \geq \frac{A(C)}{A(p_6)}.$$

3.6 (Mahler, 1946a)   Find a centrally symmetric convex disc C with $\delta_L(C) < \pi/\sqrt{12}$.

3.7 A convex disc C is called a *tile* if the plane can be filled with nonoverlapping congruent copies of C without gaps. Prove that the following three conditions are equivalent.
  (i) C is a tile
  (ii) $\delta(C) = 1$
  (iii) $\vartheta(C) = 1$.

3.8 (G. Fejes Tóth and L. Fejes Tóth, 1973b)
  (i) Let Q be a convex quadrilateral which is cut into convex pieces (cells) by a finite number of lines. For any collection $Q_1, Q_2, \ldots, Q_n$ of these cells, decompose Q into nonoverlapping convex polygons $R_1, R_2, \ldots, R_n$ such that $Q_i \subseteq R_i$ for every n, and $\sum_{i=1}^{n} s_i \leq 4n$, where s_i denotes the number of sides of R_i.

  (ii) A packing C of convex discs is called *totally separable* if any two members of C can be separated by a straight line avoiding all of the other members. Show that if C is a totally separable packing of n congruent copies of a convex disc C arranged in a convex quadrilateral Q, then

$$n \le \frac{A(Q)}{A(P_4)},$$

where P_4 is a convex quadrilateral of minimum area circumscribed about C.

(iii) Show that for any centrally symmetric convex disc C, the maximum density of a totally separable packing of congruent copies of C in the plane is $A(C)/A(P_4)$.

3.9 (G. Fejes Tóth, 1972)   Given a convex disc $C \subseteq \mathbb{R}^2$ with $A(C) = 1$, let $f(x)$ denote the maximum area of the part of C that can be covered by a hexagon of area x. Furthermore, let F be the smallest concave function with $F(x) \ge f(x)$ for every $x \ge 0$. Show that for any collection $\{C_1, \ldots, C_n\}$ of noncrossing congruent copies of C, and for any convex hexagon H,

$$A\left(\bigcup_{i=1}^{n} C_i \cap H \right) \le nF\left(\frac{A(H)}{n} \right).$$

4

Lattice Packing and Lattice Covering

Given a convex disc C and a lattice Λ, the collection $C = \{C + \lambda \mid \lambda \in \Lambda\}$ of all translates of C by vectors in Λ is called a *lattice arrangement*. In the special case when the members of C do not have interior points in common (resp. the members of C cover the whole plane), C is said to be a *lattice packing* (resp. *lattice covering*). The main conclusion of Chapter 3 was that under fairly general conditions, the density of a packing of congruent copies of C in the plane cannot exceed the density of the densest lattice packing of C. Similarly, no covering with (noncrossing) copies of C can be thinner than the thinnest lattice covering (cf. Corollaries 3.6 and 3.12). In this chapter we discuss how economical the best lattice packings and coverings are.

FÁRY'S THEOREM

The following well-known result was proved by Fáry (1950). (For an elegant alternative proof of the first part of the theorem, see Courant, 1965.)

Theorem 4.1 (Fáry). *Given a convex disc C, let $\delta_L(C)$ and $\vartheta_L(C)$ denote the density of the densest lattice packing and the density of the thinnest lattice covering of C in the plane, respectively. Then*

(i) $\delta_L(C) \geq \frac{2}{3}$,
(ii) $\vartheta_L(C) \leq \frac{3}{2}$,

with equality if and only if C is a triangle.

The lattice arrangements for which equality holds in Theorem 4.1 are shown in Figure 4.1. Observe that the fact that Figure 4.1 shows a lattice arrangement of an *equilateral* triangle T is an inessential restriction, for a suitable affine transformation of the plane can take T into any other triangle T' without violating the lattice structure or changing the densities. (A nonsingular linear

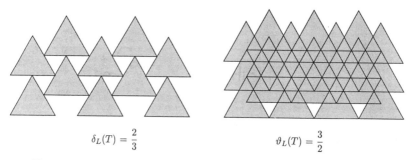

$$\delta_L(T) = \frac{2}{3} \qquad\qquad\qquad \vartheta_L(T) = \frac{3}{2}$$

Figure 4.1. The most economical lattice packing and covering with a triangle T.

mapping $f : \mathbb{R}^2 \to \mathbb{R}^2$ is called an *affine transformation* if it preserves the orientation of a triangle.)

Before we turn to the proof of Theorem 4.1, we have to introduce a new notion which plays a key role in the sequel.

Definition 4.2. Given a convex disc C in the plane, the *difference region* of C is defined as

$$D(C) = C + (-C) = \{c - c' \mid c, c' \in C\}.$$

Evidently, $D(C)$ is convex and centrally symmetric about the origin O (Exercise 4.2).

A convex hexagon which is the image of a regular hexagon under an affine transformation is said to be *affinely regular*. It is obvious that a convex hexagon with vertices $p_1, \ldots, p_6$ is affinely regular if and only if (a) it is centrally symmetric, and (b) $\overrightarrow{p_2 p_1} + \overrightarrow{p_2 p_3} = \overrightarrow{p_3 p_4}$ (where $\overrightarrow{p_i p_j} = p_j - p_i$ denotes the vector from p_i to p_j).

Lemma 4.3. *Any centrally symmetric convex disc D contains an inscribed affinely regular hexagon H with the same center. Moreover, we are free to choose the direction of one side of H.*

Proof. Given any $p \in \mathrm{Bd}\,D$, let $p' \in \mathrm{Bd}\,D$ denote the point antipodal to p. Fix any $p_1 \in \mathrm{Bd}\,D$. One can easily find a chord $p_2 p_3$ of D, which is parallel to $\overrightarrow{p_1 p_1'}$; moreover, $\overrightarrow{p_2 p_3} = \frac{1}{2} \overrightarrow{p_1 p_1'}$. The points $p_1, p_2, p_3, p_1', p_2', p_3'$ obviously span an affinely regular hexagon inscribed in D. $\qquad\square$

Proof of Theorem 4.1. First we prove (i). Assume, for simplicity, that the boundary of C is smooth. Let $p_1, \ldots, p_6$ be the vertices of an affinely regular

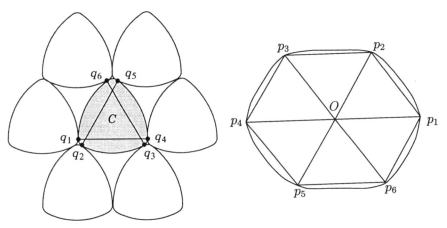

Figure 4.2. C and $D(C)$.

hexagon inscribed in the difference region $D(C)$ (see Figure 4.2). It follows immediately from the definition of $D(C)$ that one can find some points $q_1, \ldots, q_6$ on the boundary of C such that

$$\overrightarrow{q_1 q_4} = \overrightarrow{Op_1} = p_1,$$

$$\overrightarrow{q_2 q_5} = \overrightarrow{Op_2} = p_2,$$

$$\overrightarrow{q_3 q_6} = \overrightarrow{Op_3} = p_3,$$

where $p_2 = p_1 + p_3$. (Note that the lines tangent to C at q_1 and q_4 are parallel to each other and to the tangents of $D(C)$ at p_1 and p_4. This yields, in particular, that $q_1 q_4$ is the longest chord of C parallel to $\overrightarrow{Op_1} = p_1$. It also implies that the points $q_1, \ldots, q_6$ follow each other in the order shown in Figure 4.2.) It is now clear that

$$C = \{C + mp_1 + np_2 \mid m, n \in \mathbb{Z}\}$$

is a lattice packing.

By the second part of Lemma 4.3, we are free to choose the direction of p_1. We claim that for a proper choice of p_1, the lines $q_2 q_6$ and $q_3 q_5$ become parallel. Assume that originally they are not parallel to each other; say, they meet on the left-hand side of $q_1 q_4$. If we rotate p_1 (and hence $\overrightarrow{q_1 q_4}$) by π, then q_i and q_{i+3} will change places ($i = 1, 2, 3$); thus $q_2 q_6$ and $q_3 q_5$ will meet on the right-hand side of $q_1 q_4$. This proves our claim, by continuity. Furthermore, applying (if necessary) an affine transformation, we can assume without loss of generality that $q_1 q_4$ is perpendicular to $q_2 q_6$ (and $q_3 q_5$). Let $q_i' = q_i + \overrightarrow{q_1 q_4}$, $1 \le i \le 6$ (see Figure 4.3), and let $d(C, \mathbb{R}^2)$ denote the density of C.

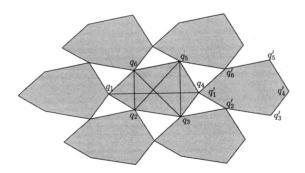

Figure 4.3. Six translates of C.

Then

$$
\begin{aligned}
d(C, \mathbb{R}^2) &= \frac{A(C)}{A(q_2 q_2' q_5' q_5)} \\
&= \frac{A(C)}{A(q_1 q_2 q_3 q_2' q_4 q_6' q_5 q_6)} \\
&\geq \frac{A(C)}{A(C) + A(q_3 q_2' q_4) + A(q_4 q_6' q_5)}
\end{aligned}
$$

(see Exercise 4.1). On the other hand,

$$
\begin{aligned}
A(q_3 q_2' q_4) &\leq \tfrac{1}{2} A(q_1 q_2 q_3 q_4), \\
A(q_4 q_6' q_5) &\leq \tfrac{1}{2} A(q_1 q_4 q_5 q_6);
\end{aligned}
$$

hence $A(q_3 q_2' q_4) + A(q_4 q_6' q_5) \leq \tfrac{1}{2} A(C)$, and $d(C, \mathbb{R}^2) \geq \tfrac{2}{3}$.

To prove (ii), we need the following variant of Lemma 4.3, whose proof is left as an exercise (Exercise 4.3): Any convex disc C has an inscribed affine regular hexagon H, whose vertices are denoted by $p_1, \ldots, p_6$.

Let us consider a tiling of the plane with translates of H (see Figure 4.4), and circumscribe a translate of C about each copy of H. Let us denote this lattice covering by C, and let $R_1, \ldots, R_6$ be the (darkly shaded) components of $C - H$.

It is easy to see that

$$
A(R_1) + A(R_2) \leq \frac{A(H)}{6}.
$$

Indeed, if R_1' denotes the reflection of R_1 about p_2, then $R_1' \cap R_2 = \varnothing$ and both regions are contained in a little triangle constituting one sixth of the area of H. Similarly,

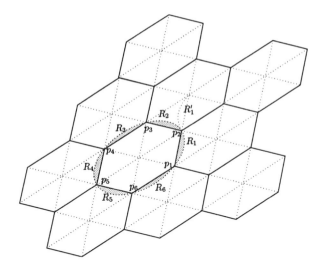

Figure 4.4. Tiling the plane with H.

$$A(R_i) + A(R_{i+1}) \le \frac{A(H)}{6} \qquad \text{for every } 1 \le i \le 6;$$

hence $\sum_{i=1}^{6} A(R_i) \le A(H)/2$. This implies that

$$d(C, \mathbb{R}^2) = \frac{A(C)}{A(H)} = \frac{A(H) + \sum_{i=1}^{6} A(R_i)}{A(H)} \le \frac{3}{2}.$$

It is clear that (i) and (ii) hold with equality only when C is a triangle.   □

DOUBLE-LATTICE PACKING

In the second section of Chapter 3 we pointed out that it is an open problem to determine which convex disc C has the smallest possible packing density $\delta(C)$. The answer is obviously not a triangle T, because [although $\delta_L(T) = 2/3 = \min_C \delta_L(C)$] the plane can be tiled with congruent copies of T, i.e., $\delta(T) = 1$ (see Figure 4.5). A major obstacle in solving the above problem is that, in general, it is very hard to construct dense packings that do not have a lattice structure.

    The tiling of the plane with triangles may give us a clue how to proceed, for it has a *double-lattice* structure.

Definition 4.4.   A packing $\mathcal{C}$ of congruent copies of C is said to be a *double-lattice packing* if $\mathcal{C}$ is the union of two lattice packings $\mathcal{C}_1$ and $\mathcal{C}_2$

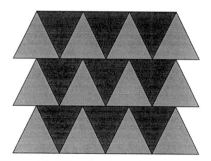

Figure 4.5. Tiling the plane with triangles.

such that C_2 can be obtained from C_1 by rotating it by π about some point of the plane.

The lattices Λ_1 and Λ_2 underlying C_1 and C_2, respectively, are obviously the same, and if P denotes a fundamental parallelogram of Λ_1, then

$$d(C, \mathbb{R}^2) = d(C_1, \mathbb{R}^2) + d(C_2, \mathbb{R}^2) = 2\, \frac{A(C)}{A(P)}.$$

The following result is due to G. Kuperberg and W. Kuperberg (son and father) (1990). It generalizes a theorem of Mahler (1946a), who established the same result for centrally symmetric convex discs.

Theorem 4.5 (G. Kuperberg and W. Kuperberg). *For every convex disc C, there is a double-lattice packing C of congruent copies of C with $d(C, \mathbb{R}^2) \geq \sqrt{3}/2$. Hence, $\delta(C) \geq \sqrt{3}/2$.*

For the proof we need some preparations.

Definition 4.6. Given a convex disc C, a parallelogram Q inscribed in C is called *extensive* if the length of each of its sides is at least half of the length of the longest chord parallel to it.

It is clear that each extensive parallelogram $Q = q_1 q_2 q_3 q_4$ inscribed in C generates a double-lattice packing with congruent copies of C in the following way.

Assume without loss of generality that $q_1 = O$, and let

$$\Lambda = \{2m\, \overrightarrow{q_1 q_2} + 2n\, \overrightarrow{q_1 q_4} \mid m, n \in \mathbb{Z}\} = \{2mq_2 + 2nq_4 \mid m, n \in \mathbb{Z}\}.$$

Then $C = C_1 \cup C_2$, where

$$C_1 = \{C + \lambda \mid \lambda \in \Lambda\}, \quad C_2 = \{-C + \lambda \mid \lambda \in \Lambda\},$$

obviously forms a double-lattice packing (Figure 4.6).

In the sequel we will only be concerned with double-lattice packings generated by extensive parallelograms.

Definition 4.7. Let C be a convex disc, and let v be a fixed direction. The *length* (resp. *width*) of C in direction v is defined as the length of the longest chord parallel to v (resp. the distance between the two supporting lines parallel to v) and is denoted by $\ell_v(C)$ (resp. $w_v(C)$).

Let c_1 and c_2 be two equal chords of length $\ell_v(C)/2$ parallel to v. The parallelogram determined by them is called the *half-length parallelogram* in the direction v.

Let c_1' and c_2' be two equal chords parallel to v such that the distance between their lines is $w_v(C)/2$. The parallelogram determined by them is called the *half-width parallelogram* in the direction v.

It is easy to see (Exercise 4.4) that the half-length and half-width parallelograms are always extensive.

Lemma 4.8. *Given any convex disc C and direction v, the area of either the half-length or the half-width parallelogram in direction v is at most $A(C)/\sqrt{3}$.*

Proof. Let p_1 and p_2 denote the endpoints of a (longest) chord parallel to v whose length is $\ell_v(C)$, and let s_1 and s_2 be two parallel lines supporting C at p_1 and p_2, respectively. Let t_1 and t_2 be two lines parallel to v supporting C at some points r_1 and r_2, respectively. Further, let $\ell_1, \ell_2, \ell_3, \ell_4$ and w_1, w_2, w_3, w_4 denote

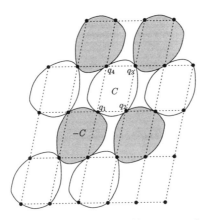

Figure 4.6. Double-lattice packing generated by an extensive parallelogram.

the vertices of the half-length and the half-width parallelograms in direction v, respectively. We can assume without loss of generality (by possibly applying an affine transformation and scaling) that the lines s_1, s_2, t_1, t_2 enclose a unit square (see Figure 4.7).

Denoting the length of the segment w_1w_2 by y and the altitude of the half-length parallelogram (corresponding to the sides of length $1/2$) by x yields

$$A(\ell_1\ell_2\ell_3\ell_4) = \frac{x}{2}, \quad A(w_1w_2w_3w_4) = \frac{y}{2}.$$

On the other hand,

$$\begin{aligned}
A(C) &\geq (A(\ell_1\ell_2r_1) + A(\ell_3\ell_4r_2)) + (A(w_1w_2\ell_2\ell_1) + A(w_3w_4\ell_4\ell_3)) \\
&\quad + A(p_1w_1w_2p_2w_3w_4) \\
&= \frac{1-x}{4} + \frac{(y+\frac{1}{2})(x-\frac{1}{2})}{2} + \frac{y+1}{4} \\
&= \frac{3}{8} + \frac{xy}{2} \\
&= \left(\frac{3}{8\sqrt{xy}} + \frac{\sqrt{xy}}{2}\right)\cdot\sqrt{xy} \geq \frac{\sqrt{3}}{2}\sqrt{xy} \\
&= \sqrt{3}\sqrt{\frac{x}{2}\cdot\frac{y}{2}}.
\end{aligned}$$

Hence, either $x/2$ or $y/2$ is at most $A(C)/\sqrt{3}$, as desired.   □

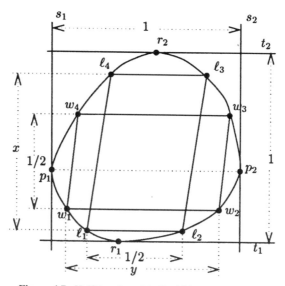

Figure 4.7. Half-length and half-width parallelograms.

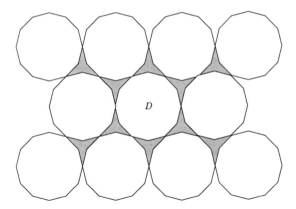

Figure 4.8. Maximum density double-lattice packing with regular dodecagons under the constraint that one basis vector is parallel to the diagonal.

Proof of Theorem 4.5. In view of Lemma 4.8, there is an extensive parallelogram Q inscribed in C (in any direction) with $A(Q) \leq A(C)/\sqrt{3}$. The density of the double-lattice packing generated by Q is clearly equal to $A(C)/2A(Q) \geq \sqrt{3}/2$. $\qquad\qquad\square$

Any convex disc C permits a double-lattice packing of maximum density, which is generated by an extensive parallelogram (Exercise 4.6). On the other hand, a regular dodecagon D does not have any extensive parallelogram of area less than $A(D)/\sqrt{3}$, which has a side parallel to a main diagonal of D. Thus, Lemma 4.8 cannot be improved, and Theorem 4.5 is also the best possible in the following sense. There exists no double-lattice packing of D, whose density is larger than $\sqrt{3}/2$, and one of the basis vectors of the underlying lattice is parallel to a main diagonal of D. (See Figure 4.8.)

The above argument can easily be turned into a linear time algorithm for computing the densest double-lattice packing of an n-sided convex polygonal disc C (see Mount, 1991, and Exercise 4.6).

EXERCISES

4.1 Let $C = \{C + \lambda \mid \lambda \in \Lambda\}$ be a lattice packing of C, and let P be the fundamental parallelogram of Λ. Prove that

$$d(C, \mathbb{R}^2) = A(C)/A(P).$$

4.2 Prove that for any convex disc C, the difference region $D(C) = C + (-C)$ is convex and centrally symmetric about the origin. Furthermore, if p is a boundary point of $D(C)$, then there exist $q, q' \in \text{Bd } C$ such that $p = q - q'$.

4.3 Show that any convex disc permits an inscribed affinely regular hexagon.

4.4 Prove that the half-length and half-width parallelograms are extensive.

4.5 (Rogers, 1951) Prove that the (upper) density of a packing of translates of a convex disc C in the plane is at most $\delta_L(C)$, the density of the densest lattice packing of C.

4.6 (Kuperberg and Kuperberg, 1990) Show that for any disc C, there is a densest double-lattice packing of C which is generated by an extensive parallelogram (cf. Definition 4.6).

5

The Method of
Cell Decomposition

Thue's theorem, discussed in Chapter 3, states that the (upper) density of a packing of congruent circles in the plane cannot exceed $\pi/\sqrt{12}$, the density of the densest lattice packing. The proof enabled us to generalize this result for packings of congruent copies of any convex disc (Theorem 3.2). However, in the special case when all members of the packing are equal circles, our proof can be much simplified. The simplification is due to the application of a concept which turns out to be extremely useful in the investigation of many other problems in this field.

DIRICHLET–VORONOI CELLS

Definition 5.1. Given n points $O_1, O_2, \ldots, O_n$ in the interior of a convex polygonal domain D, let D_i denote the set of those points of D, which are at least as close to O_i as to any other O_j, i.e.,

$$D_i = \left\{ x \in D \mid \min_{1 \le j \le n} |x - O_j| = |x - O_i| \right\}, \quad 1 \le i \le n.$$

D_i is called the *Dirichlet cell* or *Voronoi region* of O_i (see Figure 5.1).

It is easy to see that $O_i \in D_i$ and each cell D_i is a convex polygon (Exercise 5.2). Clearly, no two distinct D_i's have an interior point in common, and altogether they cover the whole domain D. In particular, if $D = H$ is a convex hexagon, $C = \{C_1, \ldots, C_n\}$ is a packing of unit circles in D, and O_i denotes the center of C_i ($1 \le i \le n$), then the Dirichlet cell D_i obviously contains C_i.

Let s_i denote the number of sides of D_i. The same analysis as in the proof of Lemma 3.3 now gives

$$\sum_{i=1}^{n} s_i \le 6n.$$

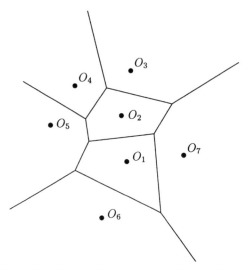

Figure 5.1. Dirichlet–Voronoi regions of a set of points.

In view of the fact that the area of any s-gon containing a unit circle is at least $s \tan (\pi/s)$ (the area of a circumscribed regular s-gon), we obtain

$$A(H) = \sum_{i=1}^{n} A(D_i) \geq \sum_{i=1}^{n} s_i \tan \frac{\pi}{s_i}$$

$$\geq n \cdot 6 \tan \frac{\pi}{6},$$

where the last inequality follows from the convexity of the function $s \tan (\pi/s)$, for $s > 0$ (which can be checked by simple differentiation). Hence, the density of the packing C in H satisfies

$$d(C,H) = \frac{\sum_{i=1}^{n} A(C_i)}{A(H)} \leq \frac{n\pi}{n \cdot 6 \tan (\pi/6)} = \frac{\pi}{\sqrt{12}}.$$

The Dirichlet–Voronoi cell decomposition has a remarkable property, which allows us to give a completely different proof of Thue's theorem. Namely, the following is true.

Lemma 5.2. *With the notation used before,*

$$\frac{A(C_i)}{A(D_i)} \leq \frac{\pi}{\sqrt{12}} \qquad \text{for all } 1 \leq i \leq n.$$

Notice that this immediately implies that

$$d(C,H) = \frac{\sum_{i=1}^{n} A(C_i)}{\sum_{i=1}^{n} A(D_i)} \le \frac{\pi}{\sqrt{12}}.$$

(cf. Exercise 5.1).

Proof (Sketch). Let H_i denote a regular hexagon circumscribed about C_i, and let C_i' denote the circle circumscribed about H_i. Assume that the Dirichlet cell D_i has s_i sides, and let the lines containing them be denoted by $\ell_1, \dots, \ell_{s_i}$. Further, let $x_1, \dots, x_{s_i}$ be the orthogonal projections of O_i onto $\ell_1, \dots, \ell_{s_i}$, respectively. Since the reflection of O_i about x_j is the center of some other circle belonging to C, two such centers are at least distance 2 apart. Therefore, we obtain

$$|x_j - x_k| \ge 1 \qquad \text{for all } 1 \le j \ne k \le s_i.$$

Using the fact that the side length of a regular heptagon inscribed in C_i' is about 1.002, it is not difficult to show that C_i' contains at most seven distinct x_j's.

If C_i' contains exactly seven distinct x_j's, then by a similar argument one can show that all of them must be "extremely close" to the boundary of C_i', so that the corresponding lines ℓ_j chop off "very small caps" (of area at most 0.1) from C_i'. More precisely,

$$A(D_i) \ge A(D_i \cap C_i') \ge A(C_i') - 7 \cdot \frac{1}{10}$$

$$= \frac{4}{3}\pi - \frac{7}{10} > \sqrt{12} \qquad \text{(Exercise 5.3)}.$$

If C_i' contains at most 6 distinct x_j's, then each corresponding line ℓ_j chops off from C_i' a cap of area at most $(A(C_i') - A(H_i))/6$; thus

$$A(D_i) \ge A(D_i \cap C_i') \ge A(C_i') - 6 \cdot \frac{A(C_i') - A(H_i)}{6}$$

$$= A(H_i) = \sqrt{12}. \qquad \square$$

Recall that the proof of Theorem 3.14 was based on a nontrivial statement (Lemma 3.15), which allowed us to assign suitable polygons R_i to each member C_i of a collection of discs. We needed this lemma only in the special case when the C_i's were equal circles centered at O_i ($1 \le i \le n$). However, in this case we can choose R_i to be $D_i \cap H$, the intersection of the Dirichlet cell of O_i with the hexagon H.

Theorem 3.14 can be applied to solve the following extremal problem on Dirichlet cell decomposition.

Theorem 5.3 (L. Fejes Tóth). *Let $O_1, \ldots, O_n$ be n points in the plane, let H be a regular hexagon, and let f be a monotone increasing function. Then*

$$\int_H f\left(\min_{1 \leq i \leq n} |x - O_i| \right) dx \geq n \int_{H'} f(|x|) \, dx,$$

where H' is a regular hexagon with $A(H') = A(H)/n$, centered at $\mathbf{0}$.

Proof. It is sufficient to prove the assertion for step functions of the form

$$f(t) = \begin{cases} -1 & \text{if } 0 \leq t \leq r, \\ 0 & \text{if } t > r, \end{cases}$$

for some $r > 0$. However, in this case the statement reduces to (-1 times) the inequality in Theorem 3.14. Indeed, let C_i and C denote the circles of radius r centered at O_i and $\mathbf{0}$, respectively. Then

$$\int_H f\left(\min_{1 \leq i \leq n} |x - O_i| \right) dx = -A\left(\bigcup_{i=1}^{n} C_i \cap H \right),$$

$$n \int_{H'} f(|x|) \, dx = -nA(C \cap H') = -A(D \cap H),$$

and the claim follows. □

An interesting interpretation of this result is the following. Assume that in a uniformly populated region we want to distribute a fixed large number of producers, each of which supplies those people who live nearer to it than to any other producer. In other words, each producer supplies those who live in its Dirichlet cell. The transportation cost is supposed to be an increasing function of the distance. Theorem 5.3 implies that the arrangement of producers that minimizes the average cost of transportation is roughly an equilateral triangular lattice.

SHADOW CELLS

Next we will introduce a completely different kind of cell decomposition (L. Fejes Tóth, 1983) which has similar applications.

Definition 5.4. Let the convex discs $C_1, C_2, \ldots$ form a packing in the plane, and let v be a nonzero vector. For every i, let S_i be defined as the set of those points $x \in \mathbb{R}^2$, which are either in C_i or for which the first intersection point of the ray parallel to v and starting at x with the set $C_1 \cup C_2 \cup \cdots$ belongs to C_i. S_i is called the *shadow cell* of C_i.

We show the power of the shadow cell decomposition method by giving an elegant alternative proof of the following theorem of Rogers (1951) (see also Exercise 4.5).

Theorem 5.5 (Rogers).   *The (upper) density of a packing of translates of a convex disc C cannot exceed $\delta_L(C)$, the density of the densest lattice packing of C.*

Observe (as in the solution of Exercise 4.5) that it is sufficient to prove this theorem for centrally symmetric convex discs, because of the following easy observation.

Lemma 5.6.   *Given a convex disc C, let*

$$C^{\star} = \frac{1}{2}(C + (-C)) = \left\{ \frac{c - c'}{2} \;\middle|\; c, c' \in C \right\}$$

(cf. Definition 4.2). Let L denote a set of vectors in the plane. Then

$$L + C = \{ \lambda + C \mid \lambda \in L \}$$

is a packing (of translates of C) if and only if $L + C^{\star} = \{ \lambda + C^{\star} \mid \lambda \in L \}$ is a packing (of translates of $C^{\star}$).

Thus, $L + C$ is a packing (of translates of C) with maximum density if and only if $L + C^{\star}$ is a packing (of translates of $C^{\star}$) with maximum density. This, in turn, implies that Theorem 5.5 is true for C if and only if it is true for $C^{\star}$.

For the same reason, it is clearly sufficient to verify Theorem 5.5 for a class of discs C with the property that the corresponding $C^{\star}$ exhaust the family of all centrally symmetric convex discs. Next we exhibit such a class.

Definition 5.7.   A convex disc C contained in a convex hexagon $p_1 q_1 p_2 q_2 p_3 q_3$ with pairwise parallel opposite sides is called *trigonal* if p_1, p_2, and p_3 are on the boundary of C. p_1, p_2, and p_3 are called the vertices of C (see Figure 5.2).

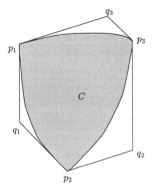

Figure 5.2. A trigonal disc with vertices p_1, p_2, p_3.

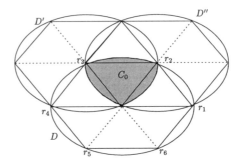

Figure 5.3. Trigonal C_0 for which $C_0 + (-C_0) = D$.

Lemma 5.8. *For any centrally symmetric convex disc D, there exists a trigonal disc C with $C^\star = D$, where $C^\star = \frac{1}{2}(C + (-C))$.*

Proof. Let O be the center of D, and let $r_1, \ldots, r_6$ be the vertices of an affinely regular hexagon inscribed in D, which is centrally symmetric about O (see Lemma 4.3). Let D' and D'' denote the discs obtained from D by translating it with $\overrightarrow{r_1 r_2}$ and $\overrightarrow{r_4 r_3}$, respectively (see Figure 5.3). Observe now that the shaded region in the figure, $C_0 = D \cap D' \cap D''$ is a trigonal disc such that $C_0 + (-C_0) = D$. Hence, $C = 2C_0$ meets the requirements. $\qquad\square$

Proof of Theorem 5.5 (L. Fejes Tóth).   It is sufficient to prove the assertion for the case when C is a trigonal disc with vertices p_1, p_2, p_3. Let C' and C'' denote the translates of C by $\overrightarrow{p_1 p_2}$ and $\overrightarrow{p_1 p_3}$, respectively, and let T denote the region enclosed by C, C', and C'' (see Figure 5.4). Suppose without loss of generality that the opposite sides of the hexagon containing C and incident to p_2 and p_3, respectively, are vertical.

Let $C_1 (= C), C_2, C_3, \ldots$ form a packing. We consider two cases.

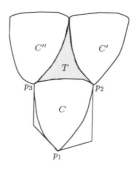

Figure 5.4. Trigonal C and its translates.

Case A: No C_i contains any interior point of T.

Let v be a vector pointing vertically downward. Then the shadow cell S_1 associated with $C = C_1$ in the packing contains $T \cup C_1$. Thus,

$$A(S_1) \geq A(T) + A(C),$$

and C, C', C'' generate a lattice packing, in which the area of every shadow cell (as well as the area of the fundamental parallelogram) is equal to $A(T) + A(C)$.

Case B: There is a C_i which overlaps T.

Then there are two disjoint translates $\hat{C}$ and $\overset{*}{C}$ of C, touching C and C_i simultaneously, such that the translation carrying C to $\hat{C}$ will take $\overset{*}{C}$ to C_i (see Figure 5.5).

Let $\hat{T}$ denote the region enclosed by C, $\hat{C}$, C_i, and $\overset{*}{C}$. It is now clear that no member of the packing overlaps $\hat{T}$; hence

$$A(S_1) \geq A(\hat{T}) + A(C),$$

and these four discs again generate a lattice packing, in which the area of every shadow cell (and the area of the fundamental parallelogram) is equal to $A(\hat{T}) + A(C)$.

From this the theorem follows immediately (cf. Exercise 5.1).   □

It would be interesting to extend Theorem 5.5 to classes of *nonconvex* domains. A disc C for which the upper density of any packing of translates of C is at most the density of the densest lattice packing is called a *Rogers domain*. Bezdek and Kertész (1987) (see also Heppes, 1990; and Schmitt, 1991) constructed a connected disc C which is not Rogers domain. On the other hand, Kertész (1987) proved that the union of two intersecting translates of a convex disc is always a Rogers domain. Except for these results almost nothing is known about this problem (see also Exercise 5.4).

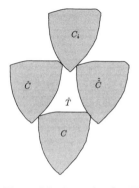

Figure 5.5. C_i overlapping T.

EXERCISES

5.1 Let C be a packing of convex discs in $\mathbb{R}^2$, and assume that the plane is subdivided into Jordan measurable sets (cells) $S_1, S_2, \ldots$ so that every S_i contains at most one disc $C_i \in C$ in its interior, $A(C_i)/A(S_i) \le \delta$, and the diameter of S_i is at most Δ (where Δ and δ are constants not depending on i). Prove that the upper density of C in $\mathbb{R}^2$ is at most δ.

5.2 Prove that the Dirichlet–Voronoi cells of a finite point system in the plane are convex polygons (some of them will be unbounded).

5.3 Provide the details of the proof of Lemma 5.2.

5.4 (L. Fejes Tóth, 1985, 1986)   Let S_1 and S_2 be two vertical straight lines supporting a convex disc D at the points p_1 and p_2, respectively, and let p_3 be any point on the boundary of D, lying on the lower arc determined by p_1, p_2.

Let D' and D'' be the translates of D by $\overrightarrow{p_3 p_1}$ and $\overrightarrow{p_3 p_2}$, respectively. The region enclosed by D', D'' and the lower arc of D (determined by p_1 and p_2) is denoted by R.

A disc C bounded by the lower arc of D and any simple Jordan arc connecting p_1 and p_2 within the region R is called *semiconvex* (see Figure 5.6).

Prove that every semiconvex disc C is a Rogers domain.

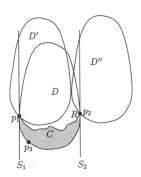

Figure 5.6. A semiconvex disc C.

6

Methods of Blichfeldt and Rogers

All notions and questions about packing and covering densities can readily be generalized to higher dimensions (see Definitions 3.1, 3.5, and 3.11). However, it is a surprisingly difficult problem to find a reasonable upper bound for the density $\delta(C)$ of the densest packing of congruent copies of a compact convex set $C \subseteq \mathbb{R}^d$, even in the special case when C is the unit ball B^d.

In a review written in 1831, Gauss (1836) proved the density of the densest *lattice packing* with balls in $\mathbb{R}^3$, $\delta_L(B^3) = \pi/\sqrt{18} \approx 0.7404$, but it is still not known (although undoubtedly true) whether $\delta(B^3) = \delta_L(B^3)$. (For a recent attempt to establish this equality, see Hsiang, 1993. Cf. Figure 6.1.)

In higher-dimensional spaces the situation is even more complicated. This can be explained by the fact, supported by many "experiments" that it is very unlikely that $\delta(B^d) = \delta_L(B^d)$ would hold for all d. Actually, there is no reason to believe that there always exists a ball packing of maximum density which is *periodic* (i.e., invariant under a suitable translation). The aim of this chapter is to present some upper bounds on $\delta(B^d)$.

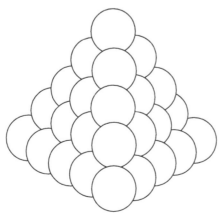

Figure 6.1. The densest lattice packing of balls in $\mathbb{R}^3$.

BLICHFELDT'S METHOD OF ENLARGEMENT

The first nontrivial upper bound for $\delta(B^d)$ was given by Blichfeldt (1929). His idea was to replace each ball of a packing with a larger concentric ball having a nonuniform mass distribution so that, at any point of the space, the total density of mass is at most 1. Then we have the following.

Theorem 6.1 (Blichfeldt). *Let $D(r) \geq 0$ be a density function which is continuous in the interval $[0, r_0]$ and vanishes for $r > r_0$. Suppose that for any packing $\{B^d + c_i \mid i = 1, 2, \ldots\}$ of unit balls and for any $x \in \mathbb{R}^d$,*

$$\sum_{i=1}^{\infty} D(|x - c_i|) \leq 1.$$

Then the density of the densest packing of unit balls satisfies

$$\delta(B^d) \leq \frac{1}{d \int_0^{r_0} r^{d-1} D(r) \, dr}.$$

Proof. Let $B^d(R) = \{x \in \mathbb{R}^d \mid |x| \leq R\}$ denote the ball of radius R around the origin [i.e., $B^d(R) = RB^d(1) = RB^d$]. Let $\{B^d + c_i \mid i \in I\}$ be a packing of unit balls in $B^d(R)$. (Observe that I must be a finite set.) Then every enlarged ball $B^d(r_0) + c_i$ is entirely contained in $B^d(R + r_0 - 1)$ and the condition in the theorem implies that

$$\begin{aligned}
\mathrm{Vol}\,(B^d(R + r_0 - 1)) &= \int_{B^d(R + r_0 - 1)} 1 \, dx \\
&\geq \sum_{i \in I} \int_{B^d(R + r_0 - 1)} D(|x - c_i|) \, dx \\
&= |I| \int_{\mathbb{R}^d} D(|x|) \, dx \\
&= |I| \int_0^{r_0} r^{d-1} s_{d-1} D(r) \, dr,
\end{aligned}$$

where s_{d-1} denotes the surface area of the unit ball B^d (see Figure 6.2). Evidently, $\mathrm{Vol}\,B^d = s_{d-1}/d$. Therefore, the density of the packing $\{B^d + c_i \mid i \in I\}$ in $B^d(R)$ is at most

$$\begin{aligned}
\frac{|I|\,\mathrm{Vol}\,B^d}{\mathrm{Vol}\,(B^d(R))} &= \frac{\mathrm{Vol}\,(B^d(R + r_0 - 1))}{\mathrm{Vol}\,(B^d(R))} \cdot \frac{|I|\,\mathrm{Vol}\,(B^d)}{\mathrm{Vol}\,(B^d(R + r_0 - 1))} \\
&\leq \left(1 + \frac{r_0 - 1}{R}\right)^d \frac{|I| s_{d-1}/d}{|I| \int_0^{r_0} r^{d-1} s_{d-1} D(r) \, dr}
\end{aligned}$$

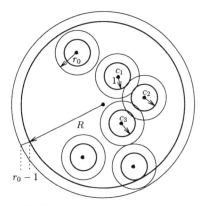

Figure 6.2. Illustration for the proof of Theorem 6.1.

$$= \left(1 + \frac{r_0 - 1}{R}\right)^d \frac{1}{d \int_0^{r_0} r^{d-1} D(r)\, dr}.$$

Taking the limit as $R \to \infty$, we obtain the desired upper bound on $\delta(B^d)$. $\square$

Lemma 6.2 (Blichfeldt's inequality). *For any points* $x, c_1, c_2, \ldots, c_n \in \mathbb{R}^d$,

$$\sum_{i=1}^{n} \sum_{j=1}^{n} |c_i - c_j|^2 \le 2n \sum_{i=1}^{n} |x - c_i|^2.$$

Proof.   We can assume without loss of generality that $x = 0$; otherwise, translate every point by $-x$.

The assertion is obviously true for $d = 1$, because then

$$\sum_{i=1}^{n} \sum_{j=1}^{n} |c_i - c_j|^2 = \sum_{i=1}^{n} \sum_{j=1}^{n} (c_i^2 + c_j^2 - 2c_i c_j)$$

$$= 2n \sum_{i=1}^{n} c_i^2 - 2\left(\sum_{i=1}^{n} c_i\right)\left(\sum_{j=1}^{n} c_j\right)$$

$$= 2n \sum_{i=1}^{n} c_i^2 - 2\left(\sum_{i=1}^{n} c_i\right)^2$$

$$\le 2n \sum_{i=1}^{n} |c_i|^2.$$

However, this yields the inequality for any d, because if $c_i = (c_{i1}, c_{i2}, \ldots, c_{id})$, then

$$\sum_{i=1}^{n}\sum_{j=1}^{n}|c_i - c_j|^2 = \sum_{k=1}^{d}\sum_{i=1}^{n}\sum_{j=1}^{n}(c_{ik} - c_{jk})^2$$

$$\le \sum_{k=1}^{d} 2n \sum_{i=1}^{n} c_{ik}^2$$

$$= 2n \sum_{i=1}^{n} |c_i|^2. \qquad \square$$

Corollary 6.3. *Let $\delta(B^d)$ denote the density of the densest packing of unit balls in $\mathbb{R}^d$. Then*

$$\delta(B^d) \le \frac{d+2}{2} 2^{-d/2}.$$

Proof.   Set

$$D(r) = \begin{cases} 1 - \frac{r^2}{2} & \text{if } r \le \sqrt{2}, \\ 0 & \text{if } r > \sqrt{2}. \end{cases}$$

It is easy to verify that this density function satisfies the condition of Theorem 6.1. Indeed, let $\{B^d + c_i \mid i = 1, 2, \dots\}$ be any packing of unit balls, and let $x \in \mathbb{R}^d$. Without loss of generality assume that c_i's are numbered in increasing order of their distances from x; that is, $|x - c_i| \le \sqrt{2}$ if and only if $i \le n$ (for some integer n). Then, by Lemma 6.2,

$$\sum_{i=1}^{\infty} D(|x - c_i|) = \sum_{i=1}^{n} D(|x - c_i|)$$

$$= n - \frac{1}{2}\sum_{i=1}^{n} |x - c_i|^2$$

$$\le n - \frac{1}{4n}\sum_{i=1}^{n}\sum_{j=1}^{n} |c_i - c_j|^2$$

$$\le n - \frac{1}{4n}n(n-1)4 = 1.$$

On the other hand,

$$d\int_0^{\sqrt{2}} r^{d-1} D(r)\, dr = \left[r^d - \frac{d}{2(d+2)} r^{d+2} \right]_0^{\sqrt{2}} = \frac{2}{d+2} 2^{d/2},$$

and the result follows by Theorem 6.1. $\qquad \square$

The following generalization of Theorem 6.1, due to G. Fejes Tóth and W. Kuperberg (1993b), allows us to extend Blichfeldt's enlargement method to establish nontrivial upper bounds for the density of a packing of congruent copies of a large class of convex bodies $C \subseteq \mathbb{R}^d$.

Definition 6.4. Given a compact convex set $C \subseteq \mathbb{R}^d$, let $r(C)$ denote the radius of the largest ball contained in C. For any $\rho \le r(C)$, let $C_{-\rho}$ denote the *inner parallel body* of C with radius ρ; i.e., the set of all points $c \in C$ with the property that the ball of radius ρ around c is entirely contained in C.

$$C_{-\rho} = \{c \in C \,|\, c + \rho B^d \subseteq C\}.$$

Furthermore, for any point $x \in \mathbb{R}^d$, let $\eta(C, x)$ denote the (unique) point of C nearest to x (see Exercise 6.2).

Theorem 6.5 (G. Fejes Tóth and W. Kuperberg).   *Let $D(r) \ge 0$ be a density function, which is continuous in the interval $[0, r_0]$ and vanishes for $r > r_0$. Suppose that for any packing $\{B^d + c_i \,|\, i = 1, 2, \dots\}$ of unit balls, and for any $x \in \mathbb{R}^d$,*

$$\sum_{i=1}^{\infty} D(|x - c_i|) \le 1.$$

Furthermore, let $C \subseteq \mathbb{R}^d$ be a compact convex set, let $\delta(C)$ be the density of the densest packing of congruent copies of C in $\mathbb{R}^d$, and let $f : \mathbb{R}^d \to \mathbb{R}$ be defined by

$$f(x) = D\left(\left|\frac{x - \eta(C_{-\rho}, x)}{\rho}\right|\right),$$

where $\rho \le r(C)$ is fixed. Then

$$\delta(C) \le \frac{\mathrm{Vol}\, C}{\int_{\mathbb{R}^d} f(x)\, dx}.$$

The proof of this theorem can be obtained by an easy generalization of the argument proving Theorem 6.1, and is left as an exercise to the reader (see Exercise 6.3).

ROGERS' SIMPLEX BOUND

Next we describe an ingenious improvement of Blichfeldt's upper bound on the density of the sphere packing in $\mathbb{R}^d$, due to Rogers (1958, 1964). Rogers' method is based on Blichfeldt's inequality (cf. Lemma 6.2) and a closer analysis of the Dirichlet–Voronoi cell decomposition discussed in Chapter 3. An interesting feature of the method is that it reproduces Thue's theorem (Corollary 3.4, when C is a circle) in the plane, and it also provides an extremely good bound in $\mathbb{R}^3$.

Let $c_1, c_2, \ldots$ be an infinite set of points in $\mathbb{R}^d$ so that the unit balls $\{B^d + c_i \mid i = 1, 2, \ldots\}$ form a packing. For technical reasons, we shall assume throughout this chapter that this packing is reasonably dense in the sense that any ball of radius R contains at least one c_i, where $R > 0$ is a constant. It is easy to see that under these circumstances the *Dirichlet cells*

$$D(c_i) = \left\{x \in \mathbb{R}^d \mid \min_j |x - c_j| = |x - c_i|\right\}$$

are (bounded) convex polytopes of diameter at most $2R$, which fill the whole space without gaps and (full-dimensional) overlapping.

Theorem 6.6. *Let $\{B^d + c_i \mid i = 1, 2, \ldots\}$ be a packing of unit balls in $\mathbb{R}^d$, and let $D(c_i)$ denote the Dirichlet cell of c_i. Then the distance between c_i and any flat induced by a $(d - k)$-dimensional face of $D(c_i)$ is at least*

$$\sqrt{2k/(k + 1)}, \quad 1 \le k \le d.$$

Proof. Let x be a point on a flat induced by a $(d - k)$-dimensional face of $D(c_i)$. Then x is equidistant from at least $k + 1$ centers c_j including c_i. Assume without loss of generality that these centers are $c_1, c_2, \ldots, c_{k+1}$ and $c_i = c_1$. By Blichfeldt's inequality (Lemma 6.2),

$$2(k + 1)^2 |x - c_1|^2 = 2(k + 1) \sum_{j=1}^{k+1} |x - c_j|^2$$

$$\ge \sum_{j=1}^{k+1} \sum_{h=1}^{k+1} |c_j - c_h|^2$$

$$\ge 4k(k + 1).$$

Thus,

$$|x - c_i| = |x - c_1| \ge \sqrt{\frac{2k}{k + 1}},$$

which completes the proof.   □

Next we study a single Dirichlet cell, say, $D(c_1)$, and we dissect it into a finite number of simplices, in the following way. Given any sets (or points) $A, B, C, \ldots$ in $\mathbb{R}^d$, let conv $\{A, B, C, \ldots\}$ ($=$ conv $(A \cup B \cup C \cup \cdots)$) denote the convex hull of $A \cup B \cup C \cup \cdots$.

Let $v_0 = c_1$, and let us dissect $D(c_1)$ into the pyramids conv $\{v_0, F_{d-1}\}$, where F_{d-1} is a $(d-1)$-dimensional face (a facet) of $D(c_1)$.

We proceed by induction on k. Assume that for some $1 \le k < d$, we have already subdivided $D(c_1)$ into polytopes of the form

$$\text{conv } \{v_0, v_1, \ldots, v_{k-1}, F_{d-k}\},$$

where F_{d-k} is a $(d-k)$-dimensional face of $D(c_1)$, and every point v_j $(1 \le j \le k-1)$ lies in a $(d-j)$-dimensional face of $D(c_1)$ containing

$$\text{conv } \{v_j, v_{j+1}, \ldots, v_{k-1}, F_{d-k}\},$$

and v_j is the closest point of this face to c_1.

Let us choose v_k to be the point of F_{d-k} closest to c_1; v_k is not necessarily a vertex of $D(c_1)$. If $k < d$, then we decompose conv $\{v_0, v_1, \ldots, v_{k-1}, F_{d-k}\}$ into smaller parts conv $\{v_0, v_1, \ldots, v_k, F_{d-k-1}\}$, where F_{d-k-1} runs over all $(d-k-1)$-dimensional faces lying in F_{d-k} that do not contain v_k.

Finally, for $k = d$, we obtain a dissection of $D(c_1)$ into simplices:

$$\text{conv } \{v_0, v_1, \ldots, v_{d-1}, F_0\} = \text{conv } \{v_0, v_1, \ldots, v_d\}$$

satisfying the following properties.

Lemma 6.7. *The Dirichlet cell $D(c_1)$ can be decomposed into simplices of the form*

$$\text{conv } \{v_0, v_1, \ldots, v_d\},$$

where $v_0 = c_1$, and

- **(i)** *v_k lies in a $(d-k)$-dimensional face of $D(c_1)$ containing $v_k, v_{k+1}, \ldots, v_d$, and is the nearest point of this face to c_1 $(1 \le k \le d)$.*

- **(ii)** *The scalar product*

$$\langle v_k - v_0, v_j - v_0 \rangle \ge \frac{2k}{k+1}$$

for every $1 \le k \le j \le d$.

Proof. Part (i) follows from the construction. By Theorem 6.6, (ii) is true for $k = j$.

Assume that $k < j$. Then, by (i), v_k is the unique point of a face of $D(c_1)$ containing v_j, closest to $c_1 = v_0$. Hence,

$$|v_k + \lambda (v_j - v_k) - v_0|^2 \ge |v_k - v_0|^2$$

for every $0 \le \lambda \le 1$. This yields

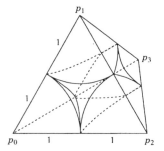

Figure 6.3. σ_d is the portion of the simplex covered by balls.

$$\langle v_k - v_0, v_j - v_k \rangle \geq -\frac{\lambda}{2}|v_j - v_k|^2$$

for any small positive λ. Thus, we can conclude that

$$\langle v_k - v_0, v_j - v_k \rangle \geq 0,$$

and

$$\langle v_k - v_0, v_j - v_0 \rangle = \langle v_k - v_0, v_k - v_0 \rangle + \langle v_k - v_0, v_j - v_k \rangle$$

$$\geq \frac{2k}{k+1} \qquad\qquad\qquad \square$$

Now we are in the position to prove Rogers' upper bound on $\delta(B^d)$, the density of the densest packing of unit balls in $\mathbb{R}^d$.

Theorem 6.8 (Rogers). *Let $S^d = \mathrm{conv}\,\{p_0, p_1, \ldots, p_d\}$ be a regular simplex in $\mathbb{R}^d$, whose side length is 2. Draw a unit ball around each vertex of S^d. Let σ_d denote the ratio of the volume of the portion of S^d, covered by balls, to the volume of the whole simplex (see Figure 6.3), and let $\delta(B^d)$ be the density of the densest packing of unit balls in $\mathbb{R}^d$. Then*

$$\delta(B^d) \leq \sigma_d.$$

Proof. Let $\{B^d + c_i | i = 1, 2, \ldots\}$ be a packing of unit balls in $\mathbb{R}^d$ with density $\delta(B^d)$, and assume that it is *maximal* in the sense that no unit ball can be added without intersecting the others. Then the Dirichlet cells $D(c_i)$ form a decomposition of the space, and the diameter of each $D(c_i)$ is at most 4.

Further, we subdivide each cell $D(c_i)$ into simplices, as in Lemma 6.7. It is sufficient to show that the density of the packing in each simplex is at most σ_d. (See Exercise 5.1 for the planar case.)

Fix such a simplex, say, $T = \mathrm{conv}\,\{v_0, v_1, \ldots, v_d\} \subseteq D(c_1)$. We are going to show that

$$\frac{\mathrm{Vol}\,((B^d + c_1) \cap T)}{\mathrm{Vol}\,T} \leq \sigma_d.$$

(Observe that $\mathrm{Vol}\,((B^d + c_i) \cap T) = 0$ if $i \neq 1$.)

Let $S = S^d$ be the regular d-dimensional simplex, whose side length is 2, embedded in $\mathbb{R}^{d+1}$ with vertices

$$(\sqrt{2}, 0, \ldots, 0), (0, \sqrt{2}, \ldots, 0), \ldots, (0, 0, \ldots, \sqrt{2}).$$

Divide S into $(d+1)!$ congruent d-dimensional simplices, as follows (see Figure 6.4). Choose u_0 to be any vertex of S. If $u_0, u_1, \ldots, u_{k-1}$ have already been selected for some $k \leq d$, then let u_k be the centroid of one of the $d - k + 1$ k-dimensional faces of S containing $u_0, u_1, \ldots, u_{k-1}$. In particular, the vertices of one of the simplices S' will be

$$u_0 = (\sqrt{2}, 0, \ldots, 0)$$

$$u_1 = \left(\frac{\sqrt{2}}{2}, \frac{\sqrt{2}}{2}, 0, \ldots, 0\right)$$

$$u_2 = \left(\frac{\sqrt{2}}{3}, \frac{\sqrt{2}}{3}, \frac{\sqrt{2}}{3}, \ldots, 0\right)$$

$$\vdots$$

$$u_d = \left(\frac{\sqrt{2}}{d+1}, \frac{\sqrt{2}}{d+1}, \frac{\sqrt{2}}{d+1}, \ldots, \frac{\sqrt{2}}{d+1}\right).$$

Now it can be easily checked that

$$\langle u_k - u_0, u_j - u_0 \rangle = \frac{2k}{k+1} \qquad (1 \leq k \leq j \leq d).$$

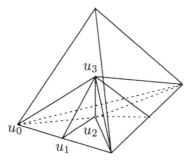

Figure 6.4. The dissection of S into $(d + 1)!$ smaller simplices.

Thus, by Lemma 6.7(ii),

$$\langle v_k - v_0, v_j - v_0 \rangle \geq \langle u_k - u_0, u_j - u_0 \rangle$$

for every $1 \leq k \leq j \leq d$.

Consider now an affine transformation $L : \mathbb{R}^d \to \mathbb{R}^{d+1}$ mapping T into S' so that $L(v_k) = u_k$ $(0 \leq k \leq d)$. That is,

$$L : v_0 + \sum_{k=1}^{d} \lambda_k(v_k - v_0) \to u_0 + \sum_{k=1}^{d} \lambda_k(u_k - u_0).$$

Given any point $v = v_0 + \sum_{k=1}^{d} \lambda_k(v_k - v_0)$ in the intersection of T and the unit ball $B^d + c_1 = B^d + v_0$, we have $\lambda_k \geq 0$ and $|v - v_0|^2 \leq 1$. Setting $u = L(v)$, we obtain

$$|u - u_0|^2 = \sum_{k=1}^{d}\sum_{j=1}^{d} \lambda_k \lambda_j \langle u_k - u_0, u_j - u_0 \rangle$$

$$\leq \sum_{k=1}^{d}\sum_{j=1}^{d} \lambda_k \lambda_j \langle v_k - v_0, v_j - v_0 \rangle$$

$$= |v - v_0|^2 \leq 1.$$

Using the fact that L preserves the ratio of two volumes, we conclude that

$$\frac{\text{Vol}((B^d + c_1) \cap T)}{\text{Vol } T} \leq \frac{\text{Vol}((B^{d+1} + u_0) \cap S')}{\text{Vol } S'}.$$

However, it follows from the congruence of the $(d+1)!$ simplices S' into which S was subdivided that

$$\frac{\text{Vol}((B^{d+1} + u_0) \cap S')}{\text{Vol } S'} = \frac{\text{Vol}\left(\bigcup_{k=0}^{d} ((B^{d+1} + p_k) \cap S)\right)}{\text{Vol } S} = \sigma_d,$$

where p_k, $0 \leq k \leq d$, denote the vertices of S. This completes the proof. $\quad\square$

It can be proved by some tedious (but routine) calculations that

$$\sigma_d = \left(\frac{1}{e} + o(1)\right) d \cdot 2^{-d/2} \qquad \text{as} d \to \infty,$$

showing that for large values of d, Rogers' bound is better than the bound of Blichfeldt (Corollary 6.3) only by a factor of $2/e$ (see Exercise 6.5).

Theorem 6.8 was conjectured by L. Fejes Tóth (1959b), and it was rediscovered by Baranovskiĭ, (1964). It was believed for a long time that for large values of d the estimates of Blichfeldt and Rogers are asymptotically not far from being optimal. However, it turned out that the situation is exactly the opposite. For low dimensions d, Rogers' bounds are quite strong and so far have not been improved by any other methods. (The only exception is the case $d = 3$, where the upper bound $\delta(B^3) \leq 0.7796\ldots$ was improved by Lindsey, 1986; Muder, 1988, 1993; and perhaps Hsiang, 1993.) For high-dimensional spaces these bounds were subsequently improved by Sidelnikov (1973, 1974) and Levenštein (1975). The currently best known asymptotical result,

$$\delta(B^d) \leq 2^{-0.599\,d + o(d)} \qquad \text{as } d \to \infty,$$

was established by Kabatjanskiĭ and Levenštein (1978).

For the minimal density of a covering of $\mathbb{R}^d$ with unit balls, a bound analogous to the one in Theorem 6.8 was established by Coxeter et al. (1959).

Theorem 6.9 (Coxeter et al., 1959). *Let $S^d = \operatorname{conv}\{p_0, p_1, \ldots, p_d\}$ be a regular simplex inscribed in the unit ball $B^d \subseteq \mathbb{R}^d$. Draw a unit ball around each vertex of S^d. Let τ_d be defined as the ratio of the sum of the volumes of the portions of these balls lying in S^d to the volume of S^d, and let $\vartheta(B^d)$ denote the density of the thinnest covering of $\mathbb{R}^d$ with unit balls. Then*

$$\vartheta(B^d) \geq \tau_d.$$

The proof of this statement is similar to the proof of Theorem 6.8. The main difference is that instead of the Dirichlet cell decomposition, we have to consider another subdivision of $\mathbb{R}^d$, called the *Delaunay triangulation*.

Definition 6.10. Let $C = \{c_1, c_2, \ldots\}$ be a set of points in $\mathbb{R}^d$ with no points of accumulation, and assume that not all of them lie in a hyperplane. For any maximal subset $C' \subseteq C$, for which there is a closed ball B containing no elements of C in its interior and $B \cap C = C'$, let $\tilde{D}(C') = \operatorname{conv} C'$. The polytope $\tilde{D}(C')$ is called the *Delaunay cell* associated with C'. If we subdivide each Delaunay cell into simplices (without introducing new vertices), then we obtain a partition of $\operatorname{conv} C$ into simplices, which is called the *Delaunay triangulation* of C (cf. Delaunay, 1934) (see Figure 6.5 for an example).

Now Theorem 6.9 can be easily deduced from the following lemma.

Lemma 6.11. *Let S be a d-dimensional simplex contained in a unit ball B^d. Then the ratio of the sum of the solid angles of S to the volume of S attains its minimum when S is a regular simplex inscribed in B^d.*

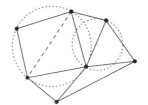

Figure 6.5. Delaunay triangulation.

It is not too hard to check that

$$\tau_d = \left(\frac{2d}{d+1} \right)^{d/2} \sigma_d = d(e^{-3/2} + o(1))$$

as $d \to \infty$ (see Exercise 6.8).

SECTIONS OF A BALL PACKING

Intersecting the densest lattice packing of balls in $\mathbb{R}^3$ by a suitably chosen plane, we obtain a packing of congruent circles with density $\sigma_2 = \pi/\sqrt{12}$. In general, any packing of congruent circles in the plane can be obtained from some packing of balls of the same radii, by intersecting them with a plane. However, there are many packings of unequal circles that cannot be constructed in this way. This follows, for example, from a theorem of G. Fejes Tóth (1980) (conjectured by L. Fejes Tóth), according to which the plane section of any packing of equal balls is a circle packing of density at most $\pi/\sqrt{12}$.

In fact, this result is a special case of the following generalization of Rogers' theorem discussed in the preceding section. As before, let S^k be a regular simplex in $\mathbb{R}^k$ with side length 2. Let σ_k denote the ratio of the volume of the part of the simplex covered by the $k + 1$ unit balls, centered at its vertices, to the volume of the whole simplex.

Theorem 6.12 (G. Fejes Tóth). *Let $\mathcal{B}$ be a packing of unit balls in $\mathbb{R}^d$, and let $\mathcal{B}'$ denote the k-dimensional ball packing obtained from $\mathcal{B}$ by intersecting its members with a k-dimensional subspace $\mathbb{R}^k \subseteq \mathbb{R}^d$ $(1 \leq k \leq d)$. Let $\overline{d}(\mathcal{B}', \mathbb{R}^k)$ denote the upper density of $\mathcal{B}'$ in $\mathbb{R}^k$. Then*

$$\overline{d}(\mathcal{B}', \mathbb{R}^k) \leq \sigma_k.$$

This result can be established by following the main steps of Rogers' proof. The crucial observation is that in the proof of Theorem 6.8 we use only the property of Dirichlet cells formulated in Theorem 6.6. Any other cell decomposition satisfying this condition would do. More precisely, we have shown the following.

Lemma 6.13. *Let $B' \subseteq \mathbb{R}^k$ be a ball of radius r centered at c'. Let $P \subseteq \mathbb{R}^k$ be a convex polytope such that the distance between c' and any point of an i-dimensional face of P is at least*

$$r\sqrt{\frac{2(k-i)}{k-i+1}} \qquad (0 \le i \le k-1).$$

Then

$$\frac{\mathrm{Vol}_k(B')}{\mathrm{Vol}_k(P)} \le \sigma_k.$$

Note that for $i = k - 1$, the condition in the lemma implies that $B' \subseteq P$.

Proof of Theorem 6.12. Without loss of generality assume that $\mathcal{B}$ is a maximal packing of unit balls (i.e., any unit ball in $\mathbb{R}^d$ intersects at least one member of $\mathcal{B}$). This implies that the diameters of the Dirichlet cells associated with the centers of the balls cannot exceed 4.

Let c denote the center of a ball $B \in \mathcal{B}$ that intersects $\mathbb{R}^k$, and let $D(c)$ denote the corresponding Dirichlet cell in $\mathbb{R}^d$.

We claim that the conditions of Lemma 6.13 are satisfied with $B' = B \cap \mathbb{R}^k$, $P = D(c) \cap \mathbb{R}^k$. Indeed, let x be any point lying on an i-dimensional face of P. Then x also belongs to a $(d - k + i)$-dimensional face of $D(c)$. Therefore, by Theorem 6.6,

$$|x - c| \ge \sqrt{\frac{2(k-i)}{k-i+1}}.$$

Let y denote the intersection point of the segment cx with the boundary of B, and let y' be the orthogonal projection of y onto the line $c'x$ (see Figure 6.6). Then we have

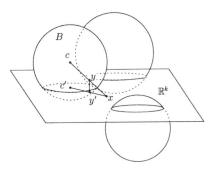

Figure 6.6. Illustration for the proof of Theorem 6.12.

$$|x - c'| = \frac{|y' - c'|}{|y - c|}|x - c|$$

$$\geq r|x - c| \geq r\sqrt{\frac{2(k - i)}{k - i + 1}},$$

as claimed.

Thus, we can apply Lemma 6.13 to conclude that

$$\frac{\text{Vol}_k (B')}{\text{Vol}_k (P)} \leq \sigma_k.$$

Using the fact that the intersection of $\mathbb{R}^k$ and the d-dimensional Dirichlet cells associated with $\mathcal{B}$ defines a decomposition of $\mathbb{R}^k$ into cells of bounded size, and each of these cells meets at most one member of $\mathcal{B}'$, the theorem follows.
□

It is not difficult to see that the analogous statement for covering is false (Exercise 6.9). To establish a positive result, we have to restrict our attention to lattice coverings (see Bezdek, 1984).

EXERCISES

6.1   (i) Let $c_1, c_2, \ldots, c_n \in \mathbb{R}^d$ be points in a ball of radius $\sqrt{2}$ such that $|c_i - c_j| > 2$. Show that $n \leq d + 1$.

(ii) Give an upper bound on $\delta(B^d)$ using the density function

$$D(r) = \begin{cases} \dfrac{1}{d + 1} & \text{if } r \leq \sqrt{2} - \varepsilon, \\ 0 & \text{if } r > \sqrt{2} - \varepsilon, \end{cases}$$

where ε is a fixed small positive number.

6.2   Prove that for any convex body $C \subseteq \mathbb{R}^d$ and for any positive ρ not exceeding the in-radius of C, the inner parallel body $C_{-\rho}$ of C with radius ρ is convex (cf. Definition 6.4).

6.3   Prove Theorem 6.5.

6.4   Let C be the Cartesian product of the $(d - 1)$-dimensional unit ball B^{d-1} and a segment of length l. Give an upper bound on $\delta(C)$, the maximal density of a packing of congruent copies of C in $\mathbb{R}^d$.

6.5⋆   (Daniels; see Rogers, 1964)   Let σ_d denote the ratio of the volume of that part of the regular d-dimensional simplex of side 2, which is covered by the system of $d + 1$ unit balls centered at its vertices, to the volume of the whole simplex. Show that

$$\sigma_d = \left(\frac{1}{e} + o(1)\right) d \cdot 2^{-d/2},$$

as $d \to \infty$.

6.6 Let C be a set of points in $\mathbb{R}^d$ with no points of accumulation. Given any maximal subset $C' \subseteq C$ (by inclusion) such that the closed Dirichlet cells corresponding to the elements of C' have at least one point in common, let $\hat{D}(C') = \text{conv } C'$. Show that $\hat{D}(C') = \tilde{D}(C')$, the Delaunay cell associated with C' (see Definitions 5.1 and 6.10).

6.7★ (Coxeter et al., 1959)   Prove Lemma 6.11.

6.8 Show that

$$\tau_d = \left(\frac{2d}{d+1}\right)^{d/2} \sigma_d = d(e^{-3/2} + o(1)).$$

(see Theorem 6.9 and Exercise 6.5).

6.9 For every $\varepsilon > 0$, construct a covering of $\mathbb{R}^3$ with unit balls, whose intersection with a fixed plane has density at most $1 + \varepsilon$.

7

Efficient Random Arrangements

In Chapter 6 we established upper bounds for the maximal (minimal) density of a packing (covering) with unit balls in d-dimensional Euclidean space. Now we discuss the opposite problem: We want to show the existence of unit ball packings with high density (coverings with low density). Our methods will be nonconstructive. We will show that some lattice arrangements generated by randomly selected basis vectors will be quite efficient, with positive probability. The idea of the proof goes back to Minkowski and was developed further by Hlawka and Rogers. We also present a slightly different probabilistic argument, due to Rush, which is based on the existence of certain codes.

MINKOWSKI–HLAWKA THEOREM

We start with an easy observation. As before, let $\delta(C)$ (and $\vartheta(C)$) denote the maximal (resp. minimal) density of a packing (resp. covering) of $\mathbb{R}^d$ with congruent copies of a convex body C.

Proposition 7.1. *For the unit ball $B^d \subseteq \mathbb{R}^d$,*

$$\delta(B^d) \geq \frac{\vartheta(B^d)}{2^d} \geq \frac{1}{2^d}.$$

Proof. We only have to prove the first inequality. Consider a *saturated* packing of unit balls $\mathcal{B} = \{B^d + c_i \mid i = 1, 2, \ldots\}$ in $\mathbb{R}^d$, that is, a packing with the property that no unit ball can be added to it without intersecting at least one element of $\mathcal{B}$. Then, by enlarging every ball of $\mathcal{B}$ by a factor of 2 about its center, we obtain a covering $\mathcal{B}' = \{2B^d + c_i \mid i = 1, 2, \ldots\}$ of $\mathbb{R}^d$. Notice that the density of $\mathcal{B}'$ in $\mathbb{R}^d$ satisfies

$$\vartheta(B^d) \leq d(\mathcal{B}', \mathbb{R}^d) = 2^d d(\mathcal{B}, \mathbb{R}^d) \leq 2^d \delta(B^d). \qquad \square$$

This argument can obviously be generalized to packings and coverings with translates of any centrally symmetric convex body $C \subseteq \mathbb{R}^d$ (see Exercise 7.1).

It is astonishing that for large values of d, the above trivial result gives the best known lower bound for the maximum density of a packing of unit balls in $\mathbb{R}^d$. However, the above argument has a serious drawback. When we saturate a packing $\mathcal{B}$ by adding extra balls, we lose control over its structure. In particular, if we start with a lattice packing, we cannot guarantee that this property will be preserved during the saturation process. As a matter of fact, in higher-dimensional Euclidean spaces, no construction of saturated lattice packings of balls is known.

Nevertheless, in this chapter, using some techniques developed by Minkowski and Hlawka, we shall be able to show that $\delta_L(B^d) > 1/2^d$; that is, there exist lattice packings of balls with densities matching (actually improving) the bound given in Proposition 7.1. In order to formulate our results in a slightly more general setting we need some definitions.

Definition 7.2. Let C be a *star body* (or *star-shaped body*); i.e., a compact subset of $\mathbb{R}^d$ containing the origin $\mathbf{0}$ in its interior such that $x \in C$ implies that $\lambda x \in C$ for every $0 \leq \lambda \leq 1$.
  A lattice

$$\Lambda = \Lambda(u_1, \ldots, u_d) = \{m_1 u_1 + \cdots + m_d u_d \mid m_1, \ldots, m_d \in \mathbb{Z}\}$$

is called *admissible* for C if it contains no interior point of C except $\mathbf{0}$.
  The *critical determinant* $\Delta(C)$ of C is defined by

$$\Delta(C) = \inf \{\det \Lambda \mid \Lambda \text{ is admissible for } C\}.$$

(cf. Definitions 1.1 and 1.2).

  An easy compactness argument shows that this infimum is attained by some admissible lattice; i.e.,

$$\Delta(C) = \min \{\det \Lambda \mid \Lambda \text{ is admissible for } C\}.$$

  This is an immediate consequence of the following simple lemma (known as the *selection theorem of Mahler*), whose proof is left as an exercise (Exercise 7.2).

Lemma 7.3 (Mahler, 1946b). *Let $\Lambda_1, \Lambda_2, \ldots$ be an infinite sequence of lattices in $\mathbb{R}^d$ with the property that there are constants $\alpha, \beta > 0$ such that for every i,*

  (i) Λ_i *is admissible for* αB^d,
  (ii) $\det \Lambda_i \leq \beta$.

Then we can select a convergent subsequence $\Lambda_{i_1}, \Lambda_{i_2}, \ldots \rightarrow \Lambda$. That is, there are suitable bases $(u_{i_j,1}, \ldots, u_{i_j,d})$ for every Λ_{i_j}, and a lattice Λ with basis $(u_1, \ldots, u_d)$ such that

$$\lim_{j \to \infty} u_{i_j,k} = u_k \quad (1 \leq k \leq d).$$

A planar version of this statement has already been used in Chapter 3, in Definitions 3.5 and 3.11 (see Exercise 3.2).

Minkowski's fundamental theorem (Theorem 1.7) can now be reformulated as follows.

Theorem 7.4. *For any centrally symmetric convex body* $C \subseteq \mathbb{R}^d$,

$$\frac{\Delta(C)}{\mathrm{Vol}\, C} \geq \frac{1}{2^d}.$$

Surprisingly, the reverse question proved to be much harder: Given a convex body (or a star body) C, how large can $\Delta(C)/\mathrm{Vol}\, C$ be? In other words, given a convex body (or star-shaped body) C, we want to find a lattice which is admissible for C and whose determinant is as small as possible. For $C = B^d$, Minkowski (1905) has already established the inequality

$$\frac{\Delta(C)}{\mathrm{Vol}\, C} \leq \frac{1}{2\zeta(d)} = \frac{1}{2\left(1 + \dfrac{1}{2^d} + \dfrac{1}{3^d} + \cdots\right)},$$

where ζ denotes *Riemann's zeta function*.

Here we present a generalization of Minkowski's result, due to Hlawka (1944). We need some preparation.

Lemma 7.5 (Davenport and Rogers, 1947). *Let* $f: \mathbb{R}^d \to \mathbb{R}$ *be a continuous function vanishing outside a bounded region, and, for any real number* γ, *set*

$$V(\gamma) = \int_{-\infty}^{+\infty} \cdots \int_{-\infty}^{+\infty} f(x_1, \ldots, x_{d-1}, \gamma)\, dx_1 \cdots dx_{d-1}.$$

Furthermore, let Λ' *be the integer lattice in the hyperplane* $x_d = 0$, *and let* $\delta > 0$ *be fixed. Given any vector* $y \in \mathbb{R}^d$ *of the form* $y = (y_1, \ldots, y_{d-1}, \delta)$, *let* Λ_y *denote the lattice in* $\mathbb{R}^d$ *generated by* Λ' *and* y. *Then*

$$\int_0^1 \cdots \int_0^1 \left(\sum_{\substack{x \in \Lambda_y \\ x_d \neq 0}} f(x) \right) dy_1 \cdots dy_{d-1} = \sum_{i \in \mathbb{Z} - \{0\}} V(i\delta).$$

Proof. Let $\Lambda' = \{(m_1, \ldots, m_{d-1}, 0) \mid m_1, \ldots, m_{d-1} \in \mathbb{Z}\}$. Then $\det \Lambda' = 1$, and

$$\sum_{\substack{x \in \Lambda_y \\ x_d \neq 0}} f(x) = \sum_{i \neq 0} \sum_{m_1, \ldots, m_{d-1} \in \mathbb{Z}} f(m_1 + iy_1, \ldots, m_{d-1} + iy_{d-1}, i\delta).$$

For a fixed $i \neq 0$, introducing new variables

$$z_1 = iy_1, \ldots, z_{d-1} = iy_{d-1}$$

we obtain

$$\int_0^1 \cdots \int_0^1 \sum_{m_1, \ldots, m_{d-1} \in \mathbb{Z}} f(m_1 + iy_1, \ldots, m_{d-1} + iy_{d-1}, i\delta) \, dy_1 \ldots dy_{d-1}$$

$$= \frac{1}{i^{d-1}} \int_0^i \cdots \int_0^i \sum_{m_1, \ldots, m_{d-1} \in \mathbb{Z}} f(m_1 + z_1, \ldots, m_{d-1} + z_{d-1}, i\delta) \, dz_1 \cdots dz_{d-1}$$

$$= \int_0^1 \cdots \int_0^1 \sum_{m_1, \ldots, m_{d-1} \in \mathbb{Z}} f(m_1 + z_1, \ldots, m_{d-1} + z_{d-1}, i\delta) \, dz_1 \cdots dz_{d-1}$$

$$= \int_{-\infty}^\infty \cdots \int_{-\infty}^\infty f(z_1, \ldots, z_{d-1}, i\delta) \, dz_1 \cdots dz_{d-1}.$$

$$= V(i\delta)$$

Hence,

$$\int_0^1 \cdots \int_0^1 \left(\sum_{\substack{x \in \Lambda_y \\ x_d \neq 0}} f(x) \right) dy_1 \cdots dy_{d-1} = \sum_{i \neq 0} V(i\delta). \qquad \square$$

Theorem 7.6 (Hlawka, 1944).   *Let* $g \colon \mathbb{R}^d \to \mathbb{R}$ *be a bounded Riemann integrable function vanishing outside a bounded region, and let* $\varepsilon > 0$. *Then there exists a unit lattice* Λ *in* $\mathbb{R}^d$ *(i.e.,* $\det \Lambda = 1$*) such that*

$$\sum_{0 \neq x \in \Lambda} g(x) < \int_{\mathbb{R}^d} g(x) \, dx + \varepsilon.$$

Proof.   Approximate g by a continuous function $f \geq g$ vanishing outside a bounded region so that

$$\int_{\mathbb{R}^d} f(x) \, dx < \int_{\mathbb{R}^d} g(x) \, dx + \frac{\varepsilon}{2}.$$

Let $\delta > 0$ be chosen so small that

(a) $f(x) = 0$ if $|x| \geq \dfrac{1}{\delta^{1/(d-1)}}$,

(b) $\quad \delta \sum_{i \neq 0} V(i\delta) = \delta \sum_{i \neq 0} \int_{-\infty}^{+\infty} \cdots \int_{-\infty}^{+\infty} f(x_1, \ldots, x_{d-1}, i\delta) \, dx_1 \cdots dx_{d-1}$

$$< \int_{\mathbb{R}^d} f(x) \, dx + \frac{\varepsilon}{2}.$$

Furthermore, let Λ' be a $(d-1)$-dimensional lattice defined by

$$\Lambda' = \left\{ \left(\frac{m_1}{\delta^{1/(d-1)}}, \ldots, \frac{m_{d-1}}{\delta^{1/(d-1)}}, 0 \right) \middle| m_1, \ldots, m_{d-1} \in \mathbb{Z} \right\}.$$

Now Lemma 7.5 states (after proper scaling) that the *mean value* of $(\sum_{x \in \Lambda_y; x_d \neq 0} f(x)) \det \Lambda'$ over the set of lattices Λ_y is equal to $\sum_{i \neq 0} V(i\delta)$. Thus, there exists a $y = (y_1, \ldots, y_{d-1}, \delta)$, $0 \leq y_i \leq 1/\delta^{1/(d-1)}$, such that

$$\sum_{\substack{x \in \Lambda_y \\ x_d \neq 0}} f(x) \leq \frac{\sum_{i \neq 0} V(i\delta)}{\det \Lambda'} = \delta \sum_{i \neq 0} V(i\delta).$$

By properties (a) and (b), we have

$$\sum_{\substack{x \in \Lambda_y \\ x \neq 0}} f(x) = \sum_{\substack{x \in \Lambda_y \\ x_d \neq 0}} f(x) \leq \delta \sum_{i \neq 0} V(i\delta)$$

$$< \int_{\mathbb{R}^d} f(x) \, dx + \frac{\varepsilon}{2}.$$

This yields

$$\sum_{\substack{x \in \Lambda_y \\ x \neq 0}} g(x) \leq \sum_{\substack{x \in \Lambda_y \\ x \neq 0}} f(x) < \int_{\mathbb{R}^d} f(x) \, dx + \frac{\varepsilon}{2}$$

$$< \int_{\mathbb{R}^d} g(x) \, dx + \varepsilon,$$

as required. $\qquad\qquad\qquad\qquad\qquad\qquad\qquad\qquad\qquad\qquad \Box$

Note that the heart of the above proof is essentially a random argument. We could have said, equivalently, that the probability that the desired inequality holds for $\Lambda = \Lambda_y$, where y is a *randomly selected* vector whose last coordinate is δ, is not zero.

Now we can easily obtain nontrivial upper bounds for the critical determinants of d-dimensional bodies.

Theorem 7.7 (Minkowski–Hlawka theorem). *Let $C \subseteq \mathbb{R}^d$ be a star body. Then*

(i) $\dfrac{\Delta(C)}{\operatorname{Vol} C} \leq 1.$

(ii) *Moreover, if C is centrally symmetric, then* $\dfrac{\Delta(C)}{\operatorname{Vol} C} \leq \dfrac{1}{2\zeta(d)}$, *where*

$\zeta(d) = 1 + \dfrac{1}{2^d} + \dfrac{1}{3^d} + \cdots$ *is Riemann's zeta function.*

Proof. To establish (i), it is sufficient to show that $\operatorname{Vol} C < 1$ implies that $\Delta(C) \leq 1$.

Let g be the *indicator function* of C, i.e.,

$$g(x) = \begin{cases} 1 & \text{if } x \in C, \\ 0 & \text{if } x \notin C, \end{cases}$$

and let $\varepsilon > 0$ be chosen sufficiently small so that

$$\int_{\mathbb{R}^d} g(x)\, dx + \varepsilon = \operatorname{Vol} C + \varepsilon < 1.$$

Then, by Theorem 7.6, there is a unit lattice Λ with $\sum_{0 \neq x \in \Lambda} g(x) < 1$, which means that C does not contain any lattice point other than $\mathbf{0}$. Hence, $\Delta(C) \leq 1$, which proves part (i) of the theorem.

To prove (ii), it suffices to show that $\operatorname{Vol} C < 2\zeta(d)$ implies $\Delta(C) \leq 1$. As before, let $g(x)$ be the indicator function of C, and put

$$f(x) = \sum_{i=1}^{\infty} \mu(i) g(ix),$$

where μ denotes the *Möbius function*, i.e.,

$$\mu(i) = \begin{cases} 1 & \text{if } i = 1, \\ 0 & \text{if } p^2 | i \text{ for some prime } p, \\ (-1)^k & \text{if } i = p_1 p_2 \cdots p_k \text{ where the } p_j\text{'s are distinct primes.} \end{cases}$$

(See Exercise 7.3 for some basic properties of the Möbius function.) We call a point $x \neq \mathbf{0}$ of a lattice Λ *primitive* if the open segment connecting $\mathbf{0}$ with x does not contain any element of Λ.

$$\sum_{\substack{0 \neq x \in \Lambda \\ x \text{ primitive}}} f(x) = \sum_{\substack{0 \neq x \in \Lambda \\ x \text{ primitive}}} \sum_{j=1}^{\infty} f(jx)$$

$$= \sum_{\substack{0 \neq x \in \Lambda \\ x \text{ primitive}}} \sum_{j=1}^{\infty} \sum_{i=1}^{\infty} \mu(i) g(ijx)$$

$$= \sum_{\substack{0 \neq x \in \Lambda \\ x \text{ primitive}}} \sum_{k=1}^{\infty} g(kx) \sum_{i \mid k} \mu(i)$$

$$= \sum_{\substack{0 \neq x \in \Lambda \\ x \text{ primitive}}} g(x),$$

and by the centrally symmetric property of C, this is a nonnegative *even* integer for every lattice Λ.

On the other hand, we have

$$\int_{\mathbb{R}^d} f(x) \, dx = \sum_{i=1}^{\infty} \mu(i) \int_{\mathbb{R}^d} g(ix) \, dx$$

$$= \sum_{i=1}^{\infty} \mu(i) \frac{\text{Vol } C}{i^d}$$

$$= \frac{\text{Vol } C}{\zeta(d)} < 2.$$

Thus, it follows from Theorem 7.6—applied to the function f—that there is a unit lattice Λ with $\sum_{0 \neq x \in \Lambda} f(x) < 2$. This, in turn, implies that

$$\sum_{0 \neq x \in \Lambda} f(x) = \sum_{\substack{0 \neq x \in \Lambda \\ x \text{ primitive}}} g(x) = 0.$$

Since C is star-shaped about $\mathbf{0}$, we can conclude that $C \cap (\Lambda - \{\mathbf{0}\}) = \varnothing$ and $\Delta(C) \leq 1$. This completes the proof of part (ii). $\qquad\square$

Note that part (i) of the above theorem remains valid for any *Jordan-measurable* set C with nonzero volume. (C is *Jordan-measurable* if for any $\varepsilon > 0$, one can find not necessarily convex polytopes P_1, P_2 so that $P_1 \subseteq C \subseteq P_2$ and $\text{Vol } P_2 - \text{Vol } P_1 < \varepsilon$.) Also, it is not difficult to generalize the Minkowski–Hlawka theorem to unbounded star-shaped bodies (see Exercise 7.4).

DENSE LATTICE PACKINGS IN SPACE

Our original goal in this chapter was to derive lower bounds for the maximal density of a lattice packing with congruent copies of a body C. To obtain such bounds, we need to generalize the notion of difference region (cf. Definition 4.2) to any star body in $\mathbb{R}^d$.

Definition 7.8. Given any star body $C \subseteq \mathbb{R}^d$ about $\mathbf{0}$, let us define its *difference region* by

$$D(C) = C + (-C) = \{c - c' \,|\, c, c' \in C\}.$$

We now observe the following.

Lemma 7.9. *Let $C \subseteq \mathbb{R}^d$ be a star body about $\mathbf{0}$, and let Λ be a lattice. Then $\Lambda + C$ is a packing if and only if Λ is admissible for $D(C)$.*

Proof.   Observe that if Λ is a lattice and $x \in \Lambda$, then $C \cap (x + C)$ has an interior point if and only if $D(C)$ contains x in its interior. This immediately implies the claim.   □

Corollary 7.10. *Let $C \subseteq \mathbb{R}^d$ be a star body about $\mathbf{0}$, let $D(C)$ be its difference region, and let $\delta_L(C)$ denote the maximal density of a lattice packing of C. Then*

(i) $\delta_L(C) \geq 2\zeta(d)\mathrm{Vol}\,C/\mathrm{Vol}\,(D(C))$.

(ii) *Moreover, if C is centrally symmetric, then $\delta_L(C) \geq \zeta(d)/2^{d-1}$.*

Proof.   By Lemma 7.9, $\delta_L(C) = \mathrm{Vol}\,C/\Delta(D(C))$. The result now follows from Theorem 7.7(ii) and from the fact that for centrally symmetric C, $\mathrm{Vol}\,(D(C)) = 2^d\,\mathrm{Vol}\,C$.   □

Next we shall make the lower bound in Corollary 7.10(i) more explicit in the case when C is a convex body. The bound

$$\delta_L(C) \geq 2\zeta(d) \Big/ \binom{2d}{d}$$

is an immediate consequence of Corollary 7.10(i) and the following result of Rogers and Shephard (1957, 1958).

Theorem 7.11 (Rogers and Shephard).   *Let $C \subseteq \mathbb{R}^d$ be a closed convex body with difference region $D(C)$. Then $\mathrm{Vol}\,(D(C)) \leq \binom{2d}{d}\,\mathrm{Vol}\,C$, with equality for any simplex.*

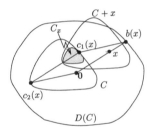

Figure 7.1. C, $C + x$, $D(C)$, and C_x.

Proof. Let $\chi(x)$ denote the indicator function of C, i.e.,

$$\chi(x) = \begin{cases} 1 & \text{if } x \in C, \\ 0 & \text{if } x \notin C. \end{cases}$$

Given any $0 \neq x = D(C)$, let $b(x)$ denote the (unique) intersection point of the boundary of $D(C)$ with the half-line $0x$. Then, $x = \rho(x)b(x)$ for some $0 < \rho(x) \leq 1$, and $b(x)$ can be written in the form

$$b(x) = \frac{x}{\rho(x)} = c_1(x) - c_2(x) \qquad \text{for some } c_1(x), c_2(x) \in C.$$

Let

$$C_x = (1 - \rho(x))C + \rho(x)c_1(x) = (1 - \rho(x))C + \rho(x)c_2(x) + x.$$

Obviously, C_x is a translate of $(1 - \rho(x))C$, and by the convexity of C,

$$C_x \subseteq C \quad \text{and} \quad C_x \subseteq C + x.$$

Thus, for any $x \in D(C)$,

$$\int_{\mathbb{R}^d} \chi(y)\chi(y - x)\,dy = \text{Vol}\,(C \cap (C + x))$$

$$\geq \text{Vol}\,C_x$$

$$= (1 - \rho(x))^d \,\text{Vol}\,C$$

$$= \text{Vol}\,C \int_{\rho(x)}^1 d(1 - r)^{d-1}\,dr.$$

On the other hand, if $x \notin D(C)$, then clearly

$$\int_{\mathbb{R}^d} \chi(y)\chi(y - x)\,dy = \text{Vol}\,(C \cap (C + x)) = 0.$$

Hence, we obtain

$$
\int_{\mathbb{R}^d}\left(\int_{\mathbb{R}^d}\chi(y)\chi(y-x)\,dy\right)dx
$$

$$
=\int_{D(C)}\left(\int_{\mathbb{R}^d}\chi(y)\chi(y-x)\,dy\right)dx
$$

$$
\geq \operatorname{Vol}C\int_{D(C)}\left(\int_{\rho(x)}^{1}d(1-r)^{d-1}\,dr\right)dx
$$

$$
=\operatorname{Vol}C\int_{0}^{1}\left(\int_{\substack{x\in D(C)\\ \rho(x)\leq r}}d(1-r)^{d-1}\,dx\right)dr
$$

$$
=\operatorname{Vol}C\int_{0}^{1}d(1-r)^{d-1}\operatorname{Vol}(rD(C))\,dr
$$

$$
=(\operatorname{Vol}C)(\operatorname{Vol}(D(C)))\int_{0}^{1}d(1-r)^{d-1}r^{d}\,dr
$$

$$
=\frac{(\operatorname{Vol}C)(\operatorname{Vol}(D(C)))}{\dbinom{2d}{d}}
$$

(see Exercise 7.5).

On the other hand, evidently,

$$
\int_{\mathbb{R}^d}\left(\int_{\mathbb{R}^d}\chi(y)\chi(y-x)\,dy\right)dx=\int_{\mathbb{R}^d}\chi(y)\left(\int_{\mathbb{R}^d}\chi(y-x)\,dx\right)dy
$$

$$
=\int_{\mathbb{R}^d}\chi(y)(\operatorname{Vol}C)\,dy
$$

$$
=(\operatorname{Vol}C)^2.
$$

Comparing the last two inequalities, we get the desired inequality

$$
\operatorname{Vol}(D(C))\leq\binom{2d}{d}\operatorname{Vol}C.
$$

It is easy to verify that if C is a simplex, then all of the above inequalities become equalities. □

LATTICE PACKING AND CODES

Rush (1989, 1992) discovered a slightly different method for the construction of efficient lattice packings. In high dimensions his approach yields almost the same lower bound for the density of the densest lattice packing with balls, as

Corollary 7.10(ii). Moreover, for many convex bodies C it provides the densest known lattice packings with C.

Let C be a compact convex body in $\mathbb{R}^d$, centrally symmetric about the origin. For any $x \in \mathbb{R}^d$, let

$$\|x\|_C = \min_{\lambda \geq 0} \{\lambda \,|\, x \in \lambda C\}.$$

It is easy to check that

(i) $\|x\|_C > 0$ for all $x \in \mathbb{R}^d$ except that $\|\mathbf{0}\|_C = 0$;

(ii) $\|\mu x\|_C = |\mu| \cdot \|x\|_C$ for all $x \in \mathbb{R}^d$ and $\mu \in \mathbb{R}$;

(iii) $\|x + y\|_C \leq \|x\|_C + \|y\|_C$ for all $x, y \in \mathbb{R}^d$.

In other words, $\|\cdot\|_C$ is a *norm*. Defining the distance of two points $x, y \in \mathbb{R}^d$ as $\|x - y\|_C$, we get the so-called *Minkowski metric* on $\mathbb{R}^d$. With respect to this metric, C is the unit ball around the origin, i.e.,

$$C = \{x \in \mathbb{R}^d | \|x\|_C \leq 1\}.$$

In the literature, $\mathbb{R}^d$ equipped with the Minkowski metric is usually called the *Minkowski space* with *gauge body C*.

The proof of the following useful observation is left to the reader (see Exercise 7.7).

Lemma 7.12. *Let* $C \subseteq \mathbb{R}^d$ *be a compact convex body containing* $\mathbf{0}$ *in its interior, and assume that C is symmetric through each of the coordinate hyperplanes. Then*

(i) *C is centrally symmetric about* $\mathbf{0}$;

(ii) *$\|x\|_C$ is a nondecreasing function of the absolute value of each coordinate of x. That is, if*

$$x = (x_1, \ldots, x_d), \ x' = (x'_1, \ldots, x'_d) \quad and \quad |x'_1| \geq |x_1|, \ldots, |x'_d| \geq |x_d|,$$

then $\|x'\|_C \geq \|x\|_C$.

Let p be an odd prime. By a somewhat unusual notation, throughout this section $\mathbb{Z}_p$ will denote the set of all integers q satisfying

$$-\frac{p-1}{2} \leq q \leq \frac{p-1}{2}.$$

Addition and multiplication in $\mathbb{Z}_p$ are taken mod p. Accordingly, $\mathbb{Z}_p^d$ consists of all integer points of $\mathbb{R}^d$, all of whose coordinates belong to $\mathbb{Z}_p$.

Rush (1989) and, in a special case, Rush and Sloane (1987) introduced the following natural norm on $\mathbb{Z}_p^d$, which can be regarded as a discrete analogue of the Minkowski metric.

Definition 7.13.   Let $C \subseteq \mathbb{R}^d$ be a compact convex body which is centrally symmetric about $\mathbf{0}$, and let p be an odd prime. For any $x \in \mathbb{Z}_p^d$, let

$$\|x\|_{C,p} = \min_{\lambda \geq 0} \{\lambda \mid x \in \lambda C + p\mathbb{Z}^d\}.$$

It can be checked that $\|\cdot\|_{C,p}$ satisfies the triangle inequality (see Exercise 7.9). Clearly, $\|x\|_{C,p} \leq \|x\|_C$. Moreover, if C is symmetric about each of the coordinate hyperplanes, then

$$\|x\|_{C,p} = \|x\|_C \qquad \text{for all } x \in \mathbb{Z}_p^d.$$

Indeed, it follows from Lemma 7.12 that in this case

$$x \in \lambda C + p\mathbb{Z}^d \implies x \in \lambda C.$$

Defining the distance between two points $x, y \in \mathbb{Z}_p^d$ as $\|x - y\|_{C,p}$, we get the *Rush metric* on $\mathbb{Z}_p^d$, which plays a crucial role in the following argument.

Let $B_{C,p}^d(r)$ denote the ball of radius r around the origin in this metric, i.e.,

$$B_{C,p}^d(r) = \{x \in \mathbb{Z}_p^d \mid \|x\|_{C,p} \leq r\}.$$

Lemma 7.14.   *Let $C \subseteq \mathbb{R}^d$ be a compact convex body which is symmetric through each of the coordinate hyperplanes, and assume that the standard basis vectors*

$$(1, 0, \ldots, 0), (0, 1, \ldots, 0), \ldots, (0, 0, \ldots, 1)$$

are on its surface. Then

$$|B_{C,p}^d(r)| \leq \left(r + \frac{d}{2}\right)^d \operatorname{Vol} C \qquad \text{for any } r \geq 0.$$

Proof.   Let K denote the unit cube defined by

$$K = \left\{x = (x_1, x_2, \ldots, x_d) \in \mathbb{R}^d \mid |x_i| \leq \tfrac{1}{2} \text{ for all } i\right\}.$$

By Lemma 7.12 we have that, for any $x \in K$,

$$
\begin{aligned}
\|x\|_C &\leq \left\|\left(\frac{1}{2}, \frac{1}{2}, \ldots, \frac{1}{2}\right)\right\|_C \\
&\leq \left\|\left(\frac{1}{2}, 0, \ldots, 0\right)\right\|_C + \left\|\left(0, \frac{1}{2}, \ldots, 0\right)\right\|_C + \cdots + \left\|\left(0, 0, \ldots, \frac{1}{2}\right)\right\|_C \\
&= \frac{d}{2}.
\end{aligned}
$$

Thus, $K \subseteq (d/2)C$, and

$$
\begin{aligned}
|B_{C,p}^d(r)| &= |\{x \in \mathbb{Z}_p^d \,|\, \|x\|_C \le r\}| \\
&\le |\{x \in \mathbb{Z}^d \,|\, \|x\|_C \le r\}| \\
&= \sum_{x \in rC \cap \mathbb{Z}^d} \mathrm{Vol}\,(K + x) \\
&\le \mathrm{Vol}\,(K + rC) \\
&\le \mathrm{Vol}\!\left(\frac{d}{2}C + rC\right) \\
&= \left(r + \frac{d}{2}\right)^d \mathrm{Vol}\, C. \qquad\qquad \square
\end{aligned}
$$

Obviously, $\mathbb{Z}_p^d$ can be viewed as a d-dimensional linear space over the field $\mathbb{Z}_p$. Any k linearly independent elements $u_1, u_2, \ldots, u_k \in \mathbb{Z}_p^d$ induce a k-dimensional subspace

$$
L = \{q_1 u_1 + \cdots + q_k u_k \,|\, q_1, \ldots, q_k \in \mathbb{Z}_p\} \subseteq \mathbb{Z}_p^d.
$$

Such a subspace is usually called a k-dimensional *linear code*, and its elements are said to be *codewords*. Given any two code words $x = (x_1, x_2, \ldots, x_d)$, $y = (y_1, y_2, \ldots, y_d) \in L$, their *Hamming distance* is defined as the number of indices i for which $x_i \ne y_i$. The larger the Hamming distance between two codewords, the easier it is to distinguish one from other. Therefore, in information theory (more precisely, in the theory of *error-correcting codes*) the most important characteristic of a code is the minimum Hamming distance between any two of its words.

Leech and Sloane (1971) have constructed many dense low-dimensional sphere packings by using various codes with large minimum Hamming distance. However, for our purposes the Hamming distance has to be replaced by the Rush metric on $\mathbb{Z}_p^d$.

Theorem 7.15. *Let $C \subseteq \mathbb{R}^d$ be a compact convex body which is centrally symmetric about the origin, let p be an odd prime, and $r \ge 0$. Then, for any*

$$
k < d + 1 - \log_p\!\left(\frac{p-1}{2} |B_{C,p}^d(r)|\right),
$$

there exists a k-dimensional linear code $L \subseteq \mathbb{Z}_p^d$ such that

$$
\min_{\substack{x,y \in L \\ x \ne y}} \|x - y\|_{C,p} = \min_{0 \ne x \in L} \|x\|_{C,p} \ge r.
$$

Proof. For $k = 0$ there is nothing to prove. Thus, we can assume that $k \ge 1$, and that we have already proved the existence of a $(k-1)$-dimensional linear code

$L' \subseteq \mathbb{Z}_p^d$ with the desired property. Let $B'(r)$ denote the set of all points in $\mathbb{Z}_p^d$ whose Rush distance from at least one element of L' is not greater than r.

We claim that the probability that a randomly chosen point $u \in \mathbb{Z}_p^d$ will not belong to the set

$$\mathbb{Z}_p B'(r) = \{qv \,|\, q \in \mathbb{Z}_p, \, v \in B'(r)\}$$

is strictly positive. To see this, it is enough to note that

$$|\mathbb{Z}_p B'(r)| = \left|\left\{qv \,\Big|\, 1 \le q \le \frac{p-1}{2}, \, v \in B'(r)\right\}\right|$$

$$\le \frac{p-1}{2}|B'(r)|$$

$$\le \frac{p-1}{2}|L'| \cdot |B_{C,p}^d(r)|$$

$$= \frac{p-1}{2} p^{k-1}|B_{C,p}^d(r)|$$

$$< p^d,$$

by the assumption of the theorem.

Suppose that $u \notin \mathbb{Z}_p B'(r)$, and let $L \subseteq \mathbb{Z}_p^d$ be the k-dimensional code generated by L' and u. That is,

$$L = \{qu + x' \,|\, q \in \mathbb{Z}_p, \, x' \in L'\}.$$

Assume, for contradiction, that L does not meet the requirements of the theorem, i.e., $\|x\|_{C,p} < r$ for some $0 \ne x = qu + x'$. Obviously, $q \ne 0$; otherwise,

$$\|x\|_{C,p} = \|x'\|_{C,p} \ge r,$$

by the induction hypothesis.

We have $qu = x - x' \in B'(r)$, because $-x' \in L'$ and $\|qu - (-x')\|_{C,p} = \|x\|_{C,p} < r$. However, this implies that

$$u \in q^{-1}B'(r) \subseteq \mathbb{Z}_p B'(r),$$

contradicting the choice of u. $\qquad\square$

Now we are ready to construct a dense lattice packing of a compact convex body $C \subseteq \mathbb{R}^d$ which is symmetric through each of the coordinate hyperplanes. Since $\delta_L(C)$ is invariant under affine transformations of $\mathbb{R}^d$, without loss of generality we can assume that the standard basis vectors $(1, 0, \ldots, 0)$, $(0, 1, \ldots, 0), \ldots, (0, 0, \ldots, 1)$ lie on the surface of C.

Let $0 \le r \le p$,

$$k = \left\lceil d - \log_p\left(\frac{p-1}{2}|B_{C,p}^d(r)|\right)\right\rceil,$$

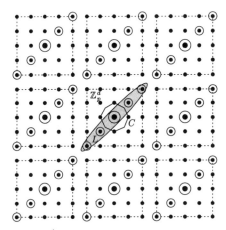

Figure 7.2. $\Lambda = L + p\mathbb{Z}^d$ ($p = 5, d = 2$); the encircled points belong to Λ.

and let $L \subseteq \mathbb{Z}_p^d$ be a k-dimensional linear code with $\min_{0 \neq x \in L} \|x\|_{C,p} \geq r$, whose existence is guaranteed by the previous theorem. Then

$$\Lambda = L + p\mathbb{Z}^d = \{x + pz \mid x \in L, z \in \mathbb{Z}^d\}$$

is a lattice (see Exercise 7.10). Furthermore, Λ is *admissible* for rC (cf. Definition 7.2). Otherwise, we could find $x \in L, z \in \mathbb{Z}^d$ such that not both of them are equal to $\mathbf{0}$, and x is in the interior of $rC + pz$. If $x \neq \mathbf{0}$, then this would imply that

$$\|x\|_{C,p} = \min_{\lambda \geq 0} \{\lambda \mid x \in \lambda C + p\mathbb{Z}^d\} < r,$$

a contradiction. If $x = \mathbf{0} \neq z$, then it is enough to observe that C is enclosed in a cube centered at $\mathbf{0}$, whose sides are parallel to the coordinate axes and have length 2. Hence, $x = \mathbf{0}$ cannot be in the interior of $rC + pz$, provided that $0 \leq r \leq p$.

Using the fact that C is centrally symmetric, we obtain that $\Lambda + (r/2)C$ is a packing (cf. Lemma 7.9). Since the volume of a fundamental parallelepiped of Λ is p^{d-k}, the density of this packing is

$$\frac{\text{Vol}\left(\frac{r}{2}C\right)}{p^{d-k}} = \frac{r^d}{2^d p^{d-k}} \text{Vol } C$$

$$\geq \frac{r^d}{2^d \left(\dfrac{p-1}{2}\right) |B_{C,p}^d(r)|} \text{Vol } C.$$

Therefore, by Lemma 7.14 (which gives away quite a lot),

$$\delta_L(C) \geq \frac{1}{2^{d-1}(p-1)\left(1 + \dfrac{d}{2r}\right)^d}.$$

For $d \geq 2$, choosing an odd prime $d^2/2 \leq p \leq d^2$ and setting $r = d^2/2$, we obtain the following.

Corollary 7.16. *Let $C \subseteq \mathbb{R}^d$ be a compact convex body, symmetric through each coordinate hyperplane. Then the maximal density of a lattice packing of C satisfies*

$$\delta_L(C) \geq \frac{1}{ed^2} \frac{1}{2^{d-1}}.$$

The lower bounds in Corollaries 7.10 and 7.16 for the density of the densest lattice packing of C are only slightly weaker than the best known estimates. Improving some earlier bounds of Rogers (1947) and Davenport and Rogers (1947), Schmidt (1963) showed that for centrally symmetric convex bodies $C \subseteq \mathbb{R}^d$,

$$\delta_L(C) > \varepsilon \frac{d}{2^d}, \tag{7.1}$$

for any $\varepsilon < \log_{10} 2$, provided that d is sufficiently large (see Exercise 7.6). In the special case when $C = B^d$, Ball (1992) established that

$$\delta_L(B^d) \geq 2\zeta(d)\frac{d-1}{2^d},$$

by using a variational argument. Unfortunately, even this bound is very far from the upper bounds discussed in Chapter 6.

THIN COVERINGS IN SPACE

The gap between the existing lower and upper bounds for the minimal density of a covering of $\mathbb{R}^d$ with congruent balls is much smaller than in the corresponding problem for packings. As we have seen in Chapter 6 (Theorem 6.9 and Exercise 6.8),

$$\vartheta(B^d) \geq \tau_d \approx \frac{d}{e^{3/2}}.$$

On the other hand, Rogers (1957) proved the following result.

Theorem 7.17 (Rogers). *Given any closed convex body $C \subseteq \mathbb{R}^d$ ($d \geq 3$), let $\vartheta_T(C)$ denote the minimal density of a covering of $\mathbb{R}^d$ with translates of C. Then*

$$\vartheta(C) \leq \vartheta_T(C) \leq d \ln d + d \ln \ln d + 5d.$$

Proof. Without loss of generality assume that the centroid of C is the origin and Vol $C = 1$. Let s be a large positive constant, and consider the lattice

$$\Lambda = \{(m_1 s, m_2 s, \ldots, m_{d-1} s) \mid m_1, m_2, \ldots, m_{d-1} \in \mathbb{Z}\}.$$

If s is sufficiently large, then $C + \Lambda = \{C + x \mid x \in \Lambda\}$ is a packing.

Select n random points $y_1, y_2, \ldots, y_n$ from the cube $K = [0, s]^d$ independently with uniform distribution. For a fixed i ($1 \le i \le n$), the expected portion of $\mathbb{R}^d$ left uncovered by the family of translates of C,

$$C + \Lambda + \{y_1, \ldots, y_i\} = \{C + x + y_j \mid x \in \Lambda, j \le i\},$$

is obviously equal to

$$\left(1 - \frac{\text{Vol } C}{\text{Vol } K}\right)^i = \left(1 - \frac{1}{s^d}\right)^i.$$

In particular, there is a specific choice of the points $y_1, y_2, \ldots, y_n$ such that the portion of the space not covered by $C_1 = C + \Lambda + \{y_1, \ldots, y_n\}$ is at most

$$\left(1 - \frac{1}{s^d}\right)^n < e^{-n/s^d}.$$

Let $0 < \varepsilon < 1/d$ be a small constant to be specified later. Let $\{z_1, z_2, \ldots, z_m\}$ be a maximal system of points in K with the property that the sets $-\varepsilon C + x + z_j$ ($x \in \Lambda, 1 \le j \le m$) have no interior points common with each other or with any member of C_1. Put $C_2 = C + \Lambda + \{z_1, \ldots, z_m\}$.

It is not necessarily true that $C_1 \cup C_2$ forms a covering of $\mathbb{R}^d$. However, we shall show that if we slightly enlarge every member of $C_1 \cup C_2$ about its centroid, then we get a covering.

More precisely, for any point $p \in \mathbb{R}^d$, at least one of the following two possibilities holds:

(a) there exists $x \in \Lambda$, $1 \le i \le n$ such that

$$p \in (1 + \varepsilon)C + x + y_i;$$

(b) there exists $x \in \Lambda$, $1 \le j \le m$ such that

$$p \in (1 + \varepsilon)C + x + z_j.$$

To prove this, fix any $p \in \mathbb{R}^d$, and consider the family of sets $-\varepsilon C + x + p$ ($x \in \Lambda$). They obviously form a packing.

By the maximal choice of m, at least one of the following two possibilities must arise.

Case 1: $-\varepsilon C + p$ has a point in common with some member of C_1. Then there are $c_1, c_2 \in C$, $1 \le i \le n$ and $x \in \Lambda$ such that

$$-\varepsilon c_1 + p = c_2 + x + y_i.$$

By the convexity of C,

$$p = (1 + \varepsilon)\left(\frac{1}{1+\varepsilon}c_2 + \frac{\varepsilon}{1+\varepsilon}c_1\right) + x + y_i$$
$$\in (1+\varepsilon)C + x + y_i,$$

and (a) holds.

Case 2: $-\varepsilon C + p$ has a point in common with some set of the form $-\varepsilon C + x + z_j$ $(x \in \Lambda, 1 \le j \le m)$. Then there are $c_1, c_2 \in C$ such that

$$-\varepsilon c_1 + p = -\varepsilon c_2 + x + z_j.$$

By a well-known result in convexity theory, $-(1/d)C \subseteq C$, for any convex body $C \subseteq \mathbb{R}^d$ whose centroid is at $\mathbf{0}$ (see, e.g., Bonnesen and Fenchel, 1934). Using the fact that $\varepsilon < 1/d$, we get that $c_3 = -\varepsilon c_2 \in C$. Thus,

$$p = -\varepsilon c_2 + \varepsilon c_1 + x + z_j$$
$$= c_3 + \varepsilon c_1 + x + z_j$$
$$= (1+\varepsilon)\left(\frac{1}{1+\varepsilon}c_3 + \frac{\varepsilon}{1+\varepsilon}c_1\right) + x + z_j$$
$$\in (1+\varepsilon)C + x + z_j,$$

and (b) holds.

Hence, the family of sets

$$C = (1+\varepsilon)C_1 \cup (1+\varepsilon)C_2$$
$$= \{(1+\varepsilon)C + x + y_i \mid x \in \Lambda,\ 1 \le i \le n\} \cup$$
$$\{(1+\varepsilon)C + x + z_j \mid x \in \Lambda,\ 1 \le j \le m\}$$

forms a covering of $\mathbb{R}^d$.

It remains to estimate the density $d(C, \mathbb{R}^d)$ of this covering. By the definition of C_1,

$$d(C_1, \mathbb{R}^d) = \frac{n\,\mathrm{Vol}\,C}{\mathrm{Vol}\,K} = \frac{n}{s^d},$$

which yields

$$d((1+\varepsilon)C_1, \mathbb{R}^d) = (1+\varepsilon)^d\,\frac{n}{s^d}.$$

On the other hand, by the selection of the z_j's, the sets $-\varepsilon C + x + z_j$ form a packing of density at most e^{-n/s^d}. Thus,

$$d((1 + \varepsilon)C_2, \mathbb{R}^d) \le \left(\frac{1 + \varepsilon}{\varepsilon}\right)^d e^{-n/s^d}.$$

Putting these bounds together and choosing

$$\varepsilon = 1/(d \ln d) \quad \text{and} \quad n = s^d d(\ln d + \ln \ln d),$$

we obtain

$$d(C, \mathbb{R}^d) = d((1 + \varepsilon)C_1, \mathbb{R}^d) + d((1 + \varepsilon)C_2, \mathbb{R}^d)$$
$$\le (1 + \varepsilon)^d \left(\frac{n}{s^d} + \frac{1}{\varepsilon^d} e^{-n/s^d}\right)$$
$$< \left(1 + \frac{2}{\ln d}\right)(d \ln d + d \ln \ln d + 1)$$
$$< d \ln d + d \ln \ln d + 5d. \qquad \square$$

Erdős and Rogers (1962) managed to refine the above argument to establish the existence of a covering of $\mathbb{R}^d$ with translates of C not exceeding the density bound in Theorem 7.17, and having the additional property that no point is covered too often. In both proofs the first step is that one randomly generates a reasonably thin arrangement of translates covering a very large portion of space.

Considerably improving some earlier results of Rogers (1957, 1959), Gritzmann (1985) has shown that every convex body $C \subseteq \mathbb{R}^d$ with at least $\log_2 \ln d + 4$ orthogonal hyperplanes of symmetry permits a lattice covering of density at most $cd(\ln d)^{1 + \log_2 e} \le cd(\ln d)^{2.443}$. However, for unit balls, Rogers' bound $\vartheta(B^d) \approx cd(\ln d)^{2.047}$ is better. (Here c is some fixed constant.)

All results of this chapter establishing the existence of dense packings and thin coverings were obtained by the *random method* (usually, by a mean value argument). For most of these problems we have very little hope that the corresponding density bounds can be matched by constructive methods.

One of the few exceptions is the sphere packing problem, which has some direct applications in information theory (error-correcting codes). The vigorous research in this field resulted in a large number of surprising and significant discoveries in the last 30 years. Leech (1967) constructed a lattice packing in $\mathbb{R}^{24}$ whose density $(0.00193\ldots)$ is very close to the upper bound of Rogers $(\delta(B^{24}) \le \sigma_{24} = 0.00245\ldots)$ described in Chapter 6. It is widely believed that in $\mathbb{R}^{24}$ there exists no lattice packing of higher density. In the construction of Leech, each ball is touched by 196,560 other balls. In this respect the Leech lattice is known to be optimal. Levenštein (1979) and Odlyzko and Sloane (1979) have proved independently that the maximum number of nonoverlapping unit balls that can touch a given ball in 24-dimensional space is exactly 196,560.

The methods of Leech and Sloane (1971), Rush and Sloane (1987), and Rush (1989) outlined in this chapter suggest that the discovery of some highly symmetric codes may lead to effective constructions of sphere packings, matching the density bounds obtained by random methods.

EXERCISES

7.1 Given any centrally symmetric convex body $C \subseteq \mathbb{R}^d$, let $\delta_T(C)$ (and $\vartheta_T(C)$) denote the maximal (resp. minimal) density of a packing (resp. covering) of $\mathbb{R}^d$ with translates of C. Show that

$$\delta_T(C) \geq \frac{\vartheta_T(C)}{2^d}.$$

7.2 (Mahler, 1946b) Let $\Lambda_1, \Lambda_2, \ldots$ be an infinite sequence of lattices in $\mathbb{R}^d$ with the property that there are constants $\alpha, \beta > 0$ such that, for every i,

(i) Λ_i is admissible for αB^d (see Definition 7.2),

(ii) $\det \Lambda_i \leq \beta$.

Prove that we can select a convergent subsequence $\Lambda_{i_1}, \Lambda_{i_2}, \ldots \to \Lambda$. That is, we can find suitable bases $(u_{i_j, 1}, \ldots, u_{i_j, d})$ for every Λ_{i_j} and a lattice Λ with basis $(u_1, \ldots, u_d)$ such that

$$\lim_{j \to \infty} u_{i_j, k} = u_k \quad (1 \leq k \leq d).$$

7.3 Let $\mu(i)$ denote the Möbius function, i.e.,

$$\mu(i) = \begin{cases} 1 & \text{if } i = 1, \\ 0 & \text{if } p^2 | i \text{ for some prime } p, \\ (-1)^k & \text{if } i = p_1 p_2 \cdots p_k \text{ where the } p_j\text{'s are distinct primes.} \end{cases}$$

Prove that

(i) μ is a multiplicative function; i.e., $\mu(ij) = \mu(i)\mu(j)$ for every pair of relatively prime positive integers (i, j).

(ii) For any multiplicative function m, the function defined by

$$M(k) = \sum_{i \mid k} m(i)$$

is also multiplicative; here the sum is taken over all positive divisors of k.

(iii) $\displaystyle\sum_{i \mid k} \mu(i) = \begin{cases} 1 & \text{if } k = 1, \\ 0 & \text{if } k > 1. \end{cases}$

(iv) (Möbius Inversion Formula) For any real (or complex)-valued function f defined on the set of positive integers, let

$$F(k) = \sum_{i \mid k} f(i).$$

Then

$$f(k) = \sum_{i \mid k} \mu(i) F\left(\frac{k}{i}\right).$$

(v) If

$$f(k) = \sum_{i \mid k} \mu(i) F\left(\frac{k}{i}\right)$$

is true for some pair of real (or complex)-valued functions f and F defined on the set of positive integers, then

$$F(k) = \sum_{i \mid k} f(i).$$

(vi) $\displaystyle \sum_{i=1}^{\infty} \mu(i)\frac{1}{i^d} = \frac{1}{\zeta(d)} = \frac{1}{1 + \dfrac{1}{2^d} + \dfrac{1}{3^d} + \cdots},$

provided that $d > 1$.

7.4 Prove that

(i) $\Delta(C) \leq \mathrm{Vol}\, C$ for any Jordan-measurable set $C \subseteq \mathbb{R}^d$;

(ii) $\Delta(C) \leq \mathrm{Vol}\, C / 2\zeta(d)$ for any unbounded star body having finite volume.

7.5 Show that

$$\int_0^1 d(1-r)^{d-1} r^d \, dr = \frac{(d!)^2}{(2d)!}.$$

7.6 Using (7.1) for centrally symmetric convex bodies, prove that

$$\delta_L(C) > \frac{cd^{3/2}}{4^d}$$

holds for any convex body $C \subseteq \mathbb{R}^d$, where $c > 0$ is a constant, independent of C and d.

7.7 Let $C \subseteq \mathbb{R}^d$ be a compact convex body which is symmetric through each of the coordinate hyperplanes, and let $\|\cdot\|_C$ denote the corresponding

Minkowski metric. Show that

$$\|x\|_C = \|(x_1, \ldots, x_d)\|_C$$

is a nondecreasing function of $|x_i|$, $1 \le i \le d$.

7.8 (Katona et al., 1993) Let $C \subseteq \mathbb{R}^2$ be a centrally symmetric compact convex body, and let $\| \cdot \|_C$ denote the corresponding Minkowski metric. Show that for any $x_1, x_2, x_3 \in \mathbb{R}^d$ with $\|x_i\|_C \ge 1$ $(i = 1, 2, 3)$, one can find two distinct indices $1 \le i \ne j \le 3$ such that $\|x_i + x_j\|_C \ge 1$.

7.9 Let $C \subseteq \mathbb{R}^d$ be a compact convex body which is centrally symmetric about $\mathbf{0}$, and let p be an odd prime. Show that

$$\|x + y\|_{C,p} \le \|x\|_{C,p} + \|y\|_{C,p} \qquad \text{for all } x, y \in \mathbb{Z}_p^d,$$

where $\| \cdot \|$ is the Rush metric (cf. Definition 7.13).

7.10 (Leech and Sloane, 1971) Let p be a prime, and let $L \subseteq \mathbb{Z}_p^d$ be a k-dimensional linear code. Show that

$$\Lambda = L + p\mathbb{Z}^d = \{x + pz \,|\, x \in L, \, p \in \mathbb{Z}^d\}$$

is a lattice in $\mathbb{R}^d$.

7.11 Let H_n be a $2^n \times 2^n$ matrix with entries ± 1 $(n = 0, 1, 2, \ldots)$ defined as follows. Set $H_0 = (1)$, and let

$$H_n = \begin{pmatrix} H_{n-1} & H_{n-1} \\ H_{n-1} & -H_{n-1} \end{pmatrix} \qquad \text{for all } n \ge 1.$$

(i) Show that H_n is a symmetric matrix; i.e., its transpose $H_n^T = H_n$.

(ii) Show that H_n is an *Hadamard matrix*; i.e, $H_n^T H_n = 2^n I$, where I denotes the $2^n \times 2^n$ matrix all of whose entries are 0, except that the entries on its main diagonal are equal to 1.

(iii) Let v_i (resp. v_i') denote 0–1 sequences obtained from the ith row of H_n (resp. from the ith row of $-H_n$) upon replacing all -1's by 0's. Show that

$$L_n = \{v_1, \ldots, v_{2^n}, v_1', \ldots, v_{2^n}'\}$$

is a linear code in $\{0, 1\}^{2^n}$; i.e., the sum of any two elements of L_n taken mod 2 is also in L_n.

(iv) Show that for $n \ge 1$ the minimum Hamming distance between two distinct elements of L_n is 2^{n-1}. Consequently, the minimum Euclidean distance between two distinct elements of $L_n \subseteq \mathbb{R}^{2^n}$ is $2^{(n-1)/2}$.

(v) Show that

$$\Lambda_n = L_n + 2\mathbb{Z}^{2^n} = \{v + 2z | v \in L_n, z \in \mathbb{Z}^{2^n}\}$$

is a lattice, and that the balls of radius min $\{2^{(n-3)/2}, 1\}$ centered at the elements of Λ_n form a packing $(n \geq 1)$. Determine the density of this packing, for $n = 1, 2, 3$. Determine the number of balls tangent to a fixed ball of this packing, for $n = 1, 2, 3$.

7.12 Let C be a finite circular cylinder. Show that the density of the densest lattice packing of C, $\delta_L(C) = \pi/\sqrt{12}$.

7.13 (A. Bezdek and W. Kuperberg, 1991)

(i) Prove that for any two ellipsoids in $\mathbb{R}^d$ $(d \geq 2)$, whose volumes are the same, there is an affine transformation which carries them into congruent bodies.

(ii) Prove that there exists an ellipsoid $C \subseteq \mathbb{R}^3$ for which $\delta(C) \neq \delta_L(C)$.

7.14 (G. Fejes Tóth and W. Kuperberg, 1993c)   Show that there exists an ellipsoid $C \subseteq \mathbb{R}^3$ for which $\vartheta(C) \neq \vartheta_L(C)$.

8

Circle Packings and Planar Graphs

Given a convex body C, what is the maximum number of "neighbors" that can touch an element in a packing of congruent copies of C? The special case of this question when $C = B^3$ (the three-dimensional ball) was the subject of a famous dispute between Gregory and Newton, and it was perhaps the first genuine problem in discrete geometry. In fact, one can set a more ambitious task. Given a packing C of convex bodies, let the *contact graph* of C be defined as a graph whose vertices are associated with the members of C, two vertices being connected by an edge if and only if the corresponding members touch each other. Characterize the class of graphs that can be obtained as the contact graph of a packing of congruent copies (similar copies, translates, etc.) of a fixed convex body C (cf. Jackson and Ringel, 1984; Moser and Pach, 1993; and Pach, 1980). The first nontrivial instance of such a result is Koebe's celebrated representation theorem, which states that a finite graph is a contact graph of a circle packing in the plane if and only if it is planar (see Figure 8.1). This result has many interesting consequences. In particular, it allows us to give a purely geometric proof of Lipton–Tarjan separator theorem for planar graphs.

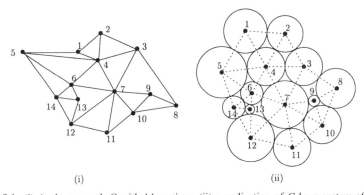

(i)   (ii)

Figure 8.1. (i) A planar graph G with 14 vertices; (ii) a realization of G by a system of circles.

KOEBE REPRESENTATION THEOREM

Koebe (1936) proved the following remarkable theorem, which was rediscovered by Andreev (1970a, 1970b) and Thurston (1985a); see also Brightwell and Scheinerman (1993), and Sachs (1993).

Theorem 8.1 (Koebe). *Given any planar graph G with vertex set $V(G)$ = $\{v_1, \ldots, v_n\}$ and edge set $E(G)$, we can find a packing of n (not necessarily congruent) circular discs $C = \{C_1, \ldots, C_n\}$ in the plane with the property that C_i and C_j touch each other if and only if $v_i v_j \in E(G)$ for $1 \le i \le n$.*

Of course, if we have one such representation of G by *kissing* (i.e., touching) discs, then we can easily construct many others by performing an inversion of the plane with respect to any circle whose center is not covered by the discs. Thurston has also proved that apart from this transformation, the representation of any *maximal* (i.e., *triangulated*) planar graph by kissing circles is unique.

The proof of Theorem 8.1 presented below was given by Colin de Verdière (1989) and Marden and Rodin (1990). It is a straightforward adaptation of Thurston's argument, which was formulated in a slightly more general context.

Proof of Theorem 8.1.   Notice that it is sufficient to prove the result for maximal planar graphs G. Indeed, if G has some nontriangular faces, then put an extra vertex in each of them and connect it to all vertices of the corresponding face. Assume that the resulting graph can be represented by circular discs satisfying the conditions in the theorem. Then, erasing the discs corresponding to the new vertices, we obtain a good representation of the original graph.

Now let G be a fixed triangulated graph with vertex set $V = \{v_1, \ldots, v_n\}$, edge set E, and face set F (including the exterior triangle). By *Euler's polyhedral formula*

$$|V| - |E| + |F| = 2.$$

Since $3|F| = 2|E|$, we have

$$|F| = 2|V| - 4 = 2n - 4. \tag{8.1}$$

Let $r = (r_1, \ldots, r_n)$ be any vector of n positive reals with $r_1 + \cdots + r_n = 1$. For each face $v_i v_j v_k$ of G, consider a triangle cut out of cardboard, whose vertices are the centers of three mutually tangent discs of radii r_i, r_j and r_k. For simplicity, the vertices of the cardboard triangle will also be denoted by v_i, v_j, and v_k, respectively. Try to glue together the cardboard triangles along their edges in the same way as they are connected in G. For any vertex v_i, let $\sigma_r(v_i)$ denote the sum of the angles at v_i of all cardboard triangles that have one of their vertices. If we are lucky, then $\sigma_r(v_i) = 2\pi$ for every vertex of G not belonging to the exterior triangle. In this case, the cardboard triangles will perfectly fit together in the plane, and we obtain a good representation of G by discs of radii $r_1, \ldots, r_n$. (*Warning:* This statement is not as trivial as it seems.)

Suppose that we are not so lucky. In any case, (8.1) implies that

$$\sum_{i=1}^{m} \sigma_r(v_i) = |F|\pi = (2n - 4)\pi.$$

Let $S \subseteq \mathbb{R}^n$ denote the $(n - 1)$-dimensional simplex defined by

$$S = \left\{ r = (r_1, \ldots, r_n) \,\middle|\, r_i > 0 \quad \text{for all } i, \quad \text{and} \quad \sum_{i=1}^{n} r_i = 1 \right\},$$

and let

$$H = \left\{ x = (x_1, \ldots, x_n) \,\middle|\, \sum_{i=1}^{n} x_i = (2n - 4)\pi \right\}.$$

Consider the continuous mapping $f \colon S \to H$, where

$$f(r) = (\sigma_r(v_1), \ldots, \sigma_r(v_n)).$$

Assume without loss of generality that v_1, v_2, and v_3 are the vertices of the exterior face of G. As we have seen before, it is sufficient to show that, for example,

$$x^* = \left(\frac{2\pi}{3}, \frac{2\pi}{3}, \frac{2\pi}{3}, 2\pi, 2\pi, \ldots, 2\pi \right) \tag{8.2}$$

lies in the image of the map f.

Claim A: $f \colon S \to H$ is a one-to-one mapping.

To see this, pick any two distinct points $r, r' \in S$, and let I denote the set of those indices i for which $r_i < r_i'$. Obviously, $I \neq \varnothing$ and $I \neq \{1, \ldots, n\}$.

Consider a cardboard triangle $v_i v_j v_k$ determined by the centers of three mutually touching discs of radii r_i, r_j, and r_k. If we increase r_i but decrease or leave unchanged the other two radii so that the three discs remain tangent, then the angle of the triangle at v_i will decrease (see Exercise 8.1). Similarly, if r_i and r_j grow while r_k decreases or remains the same, then the sum of the angles at v_i and v_j will decrease. Thus,

$$\sum_{i \in I} \sigma_r(v_i) > \sum_{i \in I} \sigma_{r'}(v_i), \tag{8.3}$$

which yields $f(r) \neq f(r')$. This proves Claim A.

Let $s = (s_1, \ldots, s_n)$ be a boundary point of the simplex S, and now let I denote the set of all indices for which $s_i = 0$. Observe that if r tends to s, then in each cardboard triangle which has at least one vertex belonging to $\{v_i \mid i \in I\}$, the sum of the angles at these vertices tends to π. Hence,

$$\lim_{r \to s} \sum_{i \in I} \sigma_r(v_i) = |F(I)|\pi, \tag{8.4}$$

where $F(I)$ denotes the set of faces of G with at least one vertex in $\{v_i | i \in I\}$.

For any fixed $r \in S$ and for any nonempty proper subset $I \subset \{1, \ldots, n\}$, there exists a point $s \in \text{Bd}\, S$ such that $s_i = 0$ for all $i \in I$ and $s_i > r_i$ for all $i \notin I$. If we move r toward s along a line, then, by (8.3), $\sum_{i \in I} \sigma_r(v_i)$ will increase. Taking (8.4) into account, we obtain that

$$\sum_{i \in I} \sigma_r(v_i) < |F(I)|\pi.$$

In other words, the image of the map $f : S \to H$ lies in the $(n - 1)$-dimensional (bounded) convex polytope P^* determined by the relations

$$\sum_{i=1}^{n} x_i = (2n - 4)\pi, \quad \sum_{i \in I} x_i < |F(I)|\pi, \tag{8.5}$$

where I varies over all nonempty proper subsets of $\{1, \ldots, n\}$.

Using the fact that $f : S \to P^*$ is a one-to-one mapping and all accumulation points of $f(r)$, as r tends to $\text{Bd}\, S$, lie on Bd P^* (see (8.4)), we immediately obtain the following.

Claim B: $f : S \to P^*$ is a *surjective* mapping; that is, $f(S) = P^*$.
(See Exercise 8.2.)

To complete the proof of Theorem 8.1, we still have to show that the point $x^* = (x_1^*, \ldots, x_n^*)$, defined in (8.2), belongs to P^*. Clearly,

$$\sum_{i=1}^{n} x_i^* = (2n - 4)\pi.$$

Let I be a nonempty proper subset of $\{1, \ldots, n\}$. If $|I| = n - 1$ or $n - 2$, then all faces of G have at least one vertex belonging to $\{v_i | i \in I\}$, i.e.,

$$|F(I)| = |F| = 2n - 4.$$

Therefore, in these cases,

$$\sum_{i \in I} x_i^* \le 2\pi(n - 3) + \frac{4\pi}{3} < (2n - 4)\pi = |F(I)|\pi.$$

The proof of the following statement is left as an exercise (Exercise 8.3).

Claim C: For any subset $I \subseteq \{1,\ldots,n\}$ with $1 \le |I| \le n-3$, G has more than $2|I|$ faces that have at least one vertex belonging to $\{v_i | i \in I\}$.

For any such I,

$$\sum_{i \in I} x_i^* \le 2\pi|I| < |F(I)|\pi.$$

Thus, x^* satisfies all the relations in (8.5) defining P^*, that is, $x^* \in P^* = f(S)$, as required. $\qquad\square$

Let G be any maximal planar graph with a triangular face $v_1 v_2 v_3$. Notice that by a suitable inversion, any representation of G by discs can be carried into another representation with the property that the discs corresponding to v_1, v_2, and v_3 are of the same size, and all other discs lie in the region enclosed by them. Hence, it follows at once from Claim A that the representation of G is uniquely determined up to inversions and rigid motions of the plane.

Note that Theorem 8.1 immediately implies the following well-known result of Fáry (1948). (See also Tutte, 1963.)

Theorem 8.2 (Fáry). *Any finite planar graph can be drawn in the plane so that its edges are noncrossing line segments.*

LIPTON–TARJAN SEPARATOR THEOREM

Perhaps the most efficient approach to the solution of many algorithmic problems is the so called *divide-and-conquer* paradigm (see Aho et al., 1974; Coreman et al., 1990; Ullman, 1984). The idea is to divide a problem into two smaller subproblems of the same type, which can be solved recursively, and to combine the results of the subproblems to obtain a solution of the original problem. Many applications of this method to planar graphs are based on the following important result of Lipton and Tarjan (1979), which shows that any planar graph can be separated into two much smaller components by the removal of relatively few vertices.

Theorem 8.3 (Lipton–Tarjan). *Let G be a planar graph with n vertices. Then the vertex set of G can be partitioned into three parts, A, B and C, such that $|A|, |B| \le \frac{3}{4}n$, $|C| < 2\sqrt{n}$, and no vertex in A is adjacent to any vertex in B.*

Actually, Lipton and Tarjan established a somewhat stronger result, according to which the sizes of A and B can be bounded from above by $\frac{2}{3}n$ (see Exercise 8.4). The theorem has found many applications in graph theory (Lipton and Tarjan, 1980; Fredrickson, 1987), in VLSI (Leiserson, 1983; Leighton, 1983; Ullman, 1984), in numerical analysis (Gilbert, 1980; Gilbert and Tarjan,

1987; Lipton et al., 1979), in complexity theory (Lipton and Tarjan, 1980), and so on. For many extensions, generalizations, and other applications of this theorem, see Alon et al. (1990, 1994), Chung (1990), Djidjev (1982), Gazit and Miller (1990), Miller (1986), Miller at al. (1991), and Teng (1991).

We shall present a proof of Theorem 8.3, due to Miller and Thurston (1990), which is based on Koebe's representation theorem. We will use the following well-known fact.

Lemma 8.4. *For any n-element set $P \subseteq \mathbb{R}^d$, there exists a point $q \in \mathbb{R}^d$ with the property that any half-space that does not contain q covers at most $dn/(d + 1)$ elements of P. (Such a point q is called a* centerpoint *of P).*

Proof. Let $\mathcal{H}$ be the family of all half-spaces that cover more than $\lfloor dn/(d+1) \rfloor$ elements of P. We have to show that the intersection of all members of $\mathcal{H}$ is nonempty. By Helly's theorem (1923) (see, e.g., Eckhoff, 1993), it is sufficient to prove that any $d + 1$ half-spaces $H_1, \ldots, H_{d+1} \in \mathcal{H}$ have a point in common. Indeed, otherwise we would obtain that

$$|P| \le \sum_{i=1}^{d+1} |(\mathbb{R}^d - H_i) \cap P|$$
$$\le (d + 1)\left(n - \lfloor dn/(d + 1) \rfloor - 1\right)$$
$$< n,$$

a contradiction.   □

Proof of Theorem 8.3. Let S denote the unit sphere in $\mathbb{R}^3$ centered at $(0, 0, 1)$, whose north and south poles are $N = (0, 0, 2)$ and $(0, 0, 0)$, respectively. The *stereographic projection* $\pi: \mathbb{R}^2 \to S$ maps any point p of the xy-plane into the intersection of the segment Np with S. The image of any circular disc under π will be a spherical cap. Conversely, the preimage of any spherical cap S, which does not contain N in its interior, is a circular disc or a half-plane in the xy-plane. (See Figure 8.2.)

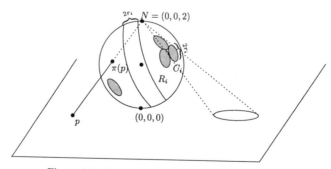

Figure 8.2. Illustration for the proof of Theorem 8.3.

Theorem 8.1 implies that the vertices of G can be represented by nonoverlapping spherical caps $C_1, \ldots, C_n \subseteq S$ such that two of them touch each other if and only if the corresponding vertices are adjacent. Pick a point p_i in the interior of C_i, and set $P = \{p_1, \ldots, p_n\}$. Let $q \in \mathbb{R}^3$ be the centerpoint of P, satisfying the condition in Lemma 8.4. By symmetry, we can assume that q belongs to the segment connecting the center and the south pole of S.

Suppose that q does not coincide with $(0, 0, 1)$, the center of S. Then

$$d = |q - (0, 0, 1)| \neq 0.$$

Let $\sigma : S \to S$ be a mapping defined by

$$\sigma(p) = \begin{cases} \pi\left(\sqrt{\dfrac{1+d}{1-d}} \, \pi^{-1}(p) \right) & \text{if } p \neq N, \\[2mm] N & \text{if } p = N. \end{cases}$$

Since σ is the composition of an inverse stereographic projection, a dilatation of the xy-plane (by a factor of $\sqrt{\frac{1+d}{1-d}}$) and a stereographic projection, it carries spherical caps into spherical caps, and it also preserves the incidences between them. Obviously, $\sigma(p_i)$ will lie in the interior of the cap $\sigma(C_i)$, and it can be shown that $(0, 0, 1)$ will be a centerpoint of $\sigma(P)$.

Thus, we can assume without loss of generality, that $(0, 0, 1)$ is a centerpoint of P. This means that any plane passing through $(0, 0, 1)$ has at most $\frac{3}{4} n$ elements of P strictly on its right side (and left side).

Let r_i denote the (spherical) radius of the cap C_i. Since $C_1, \ldots, C_n$ form a packing, and the area of C_i is at most πr_i^2, we have

$$\sum_{i=1}^{n} \pi r_i^2 < 4\pi.$$

By Jensen's inequality, this yields that

$$4 > \sum_{i=1}^{n} r_i^2 \geq n \left(\frac{\sum_{i=1}^{n} r_i}{n} \right)^2.$$

Therefore,

$$\sum_{i=1}^{n} r_i < 2\sqrt{n}.$$

Consider now any plane passing through $(0, 0, 1)$, and let $u(H)$ denote the unit normal vector sitting at $(0, 0, 1)$. The locus of the endpoint of $u(H)$, as H varies over all planes through $(0, 0, 1)$ that intersect a fixed spherical cap C_i, will be a ringlike region R_i symmetric about a great circle of S (see Figure 8.2).

Clearly, the area of R_i satisfies

$$A(R_i) < (2r_i)(2\pi) = 4\pi r_i.$$

Hence, there exists a point of S which is covered by at most

$$\frac{\sum_{i=1}^{n} A(R_i)}{A(S)} < \frac{\sum_{i=1}^{n} 4\pi r_i}{4\pi} = \sum_{i=1}^{n} r_i < 2\sqrt{n}$$

regions R_i. Equivalently, there is a plane H_0 intersecting fewer than $2\sqrt{n}$ caps of C_i.

Let A, B, and C denote the set of those vertices of G, for which the corresponding spherical caps C_i lie entirely on one side of H_0, on the other side of H_0, or meet H_0, respectively. It follows directly from the above properties of H_0 that $|A|, |B| \leq \frac{3}{4} n$ and $|C| < 2\sqrt{n}$. Clearly, no vertex in A can be adjacent to any vertex in B, because two closed caps lying in two complementary half-spaces can never touch each other. $\square$

Notice that the above argument can easily be modified to give a stronger result.

Theorem 8.5. *Let G be a graph whose vertices can be represented by a system of d-dimensional balls $\mathcal{B} = \{B_1^d, \ldots, B_n^d\}$ such that two balls have a point in common if and only if the corresponding vertices are adjacent. Assume further that no point of $\mathbb{R}^d$ belongs to the interior of more than k members of $\mathcal{B}$. Then the vertex set of G can be partitioned into three subsets, A, B, and C, such that*

$$|A|, |B| \leq \frac{d+1}{d+2}\, n, \quad |C| \leq c_d k^{1/d} n^{1-(1/d)},$$

and no vertex in A is adjacent to any vertex in B (c_d is a constant depending only on d).

DISCRETE CONVEX FUNCTIONS

Koebe's original proof of Theorem 8.1 is based on some deep results from the theory of conformal mappings. Thurston's proof, presented at the beginning of this chapter, uses discrete methods. In fact, his arguments can be easily translated into an algorithm to construct an "approximate" circle representation of a planar graph with arbitrary precision. Extending this approach, Thurston (1985b) suggested that many central results in the theory of conformal mappings (e.g., Schwarz lemma, Riemann mapping theorem, Koebe uniformization) can be "discretized" and proved constructively in a similar manner. (See Doyle et al., 1994; Rodin, 1987, 1989; Rodin and Sullivan, 1987; Stephenson, 1990; and Schramm, 1991). A crucial element of many arguments of this type is to establish certain rigidity properties for infinite circle packings, in which each circle touches six other circles.

In this section we prove such a result.

Definition 8.6. A packing of convex sets $C = \{C_1, C_2, \ldots, C_n\}$ is called *k-neighbored* if every C_i is tangent to at least k other elements of C.

For instance, the densest lattice packing of unit circles (see Figure 3.1) is six-neighbored. On the other hand, it is easy to see that every six-neighbored circle packing of not necessarily equal circles must be infinite (Exercise 8.8).

L. Fejes Tóth (1977) conjectured that the regular circle packing has the interesting extremal property that if we "perturb" it, then we shall necessarily create either arbitrarily large or arbitrarily small circles. More precisely, the following "zero-or-one law" holds. If C is a six-neighbored circle packing, then

$$\frac{\inf_{C \in C} r(C)}{\sup_{C \in C} r(C)} = \begin{cases} 1 & \text{if } C \text{ is regular,} \\ 0 & \text{otherwise,} \end{cases}$$

where $r(C)$ denotes the radius of C.

Bárány, Füredi, and Pach (1984) proved this conjecture in the following stronger form.

Theorem 8.7 (Bárány et al.). *If C is a six-neighbored circle packing containing at least two circles of different radii, then $\inf_{C \in C} r(C) = 0$.*

One can be tempted to conjecture that for any six-neighbored circle packing C, there is a bounded region in the plane that contains infinitely many members of C. However, this is not the case. The circle packing depicted in Figure 8.3 (which was discovered by K. Stephenson) has no finite point of accumulation.

To establish Theorem 8.7, we need some preparation.

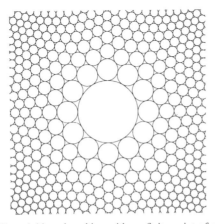

Figure 8.3. Six-neighbored packing with no finite point of accumulation.

Definition 8.8. A graph $G = (V, E)$ is called *locally finite* if every vertex is adjacent to only finitely many other vertices. For any $x \in V$, let $N_G(x)$ denote the set of all neighbors of x.

A function $f: V \to \mathbb{R}$ is called *convex* (or *subharmonic*) on G if

$$\frac{1}{|N_G(x)|} \sum_{y \in N_G(x)} f(y) \geq f(x) \quad \text{for every } x \in V.$$

Obviously, any convex function defined on a connected finite graph is constant.

Theorem 8.9 (Bárány et al., 1984). *Let $G = (V, E)$ be a locally finite graph, $f: V \to \mathbb{R}$ a convex function on G. Assume further that there is an edge $x_0 y_0 \in E$ with $f(x_0) = 0$, $f(y_0) = 1$, and let V_i denote the set of those points of G that can be reached from y_0 by a monotone increasing path of length at most i, i.e.,*

$$V_i = \left\{ y \in V \,\middle|\, \begin{array}{l} \exists y_1, \ldots, y_{j-1} \in V \text{ s.t. } j \leq i, \ y_0 y_1, y_1 y_2, \ldots, y_{j-1} y \in E, \\ \text{and} \quad f(y_0) \leq f(y_1) \leq \cdots \leq f(y) \end{array} \right\}.$$

Then, for every natural number n,

$$\max_{x \in V_n} f(x) \geq 1 + \sum_{i=1}^{n} \frac{1}{|E_i|},$$

where

$$E_i = \{uv \in E \mid u \in V_{i-1}, \ v \in V_i - V_{i-1}, \ f(u) \leq f(v)\}.$$

Proof. Throughout the proof n will be kept fixed, and $M = \max_{x \in V_n} f(x)$. Define a subgraph $G' = (V', E') \subseteq G$ by

$$V' = V_n \cup \{x_0\},$$
$$E' = \{uv \in E \mid u, v \in V_n\} \cup \{x_0 y_0\}.$$

We turn G' into a *directed graph* $\vec{G}' = (V', \vec{E}')$ by orienting its edges uv according to the following rules.

(a) If $\vec{uv} \in E'$ and $f(u) < f(v)$, then let $\vec{uv} \in \vec{E}'$.

(b) If $uv \in E_i$ ($\subseteq E'$) for some $u \in V_{i-1}$, $v \in V_i - V_{i-1}$, then let $\vec{uv} \in \vec{E}'$ ($1 \leq i \leq n$).

(c) All other edges $uv \in E'$ [with $f(u) = f(v)$] can be oriented arbitrarily.

Observe that if $uv \in E'$ for some $u \in V_{i-2}$, $v \in V_i - V_{i-1}$, then $f(u)$ is necessarily larger than $f(v)$; consequently, $\vec{vu} \in \vec{E}'$.

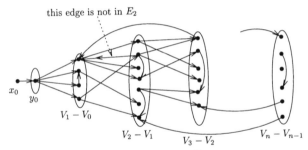

Figure 8.4. Graph $\vec{G}'$.

Let Φ denote the class of all functions $\varphi : V' \to \mathbb{R}$ satisfying the following properties:

(i) $\varphi(x_0) = 0$,

(ii) $f(x) \le \varphi(x) \le M$ for every $x \in V'$;

(iii) there exists a directed subgraph $\vec{G}'' = (V', \vec{E}'') \subseteq \vec{G}'$ such that $\overrightarrow{x_0 y_0} \in \vec{E}'' \subseteq \vec{E}'$, $\varphi(u) \le \varphi(v)$ for every $\overrightarrow{uv} \in \vec{E}''$, and

$$\frac{1}{|N_{G''}(x)|} \sum_{y \in N_{G''}(x)} \varphi(y) \ge \varphi(x) \qquad \text{for every } x \in V_{n-1}.$$

Here G'' denotes the graph obtained from $\vec{G}''$ by ignoring the orientation of the edges.

Clearly, $f \in \Phi$. To see this, it suffices to show that $\varphi = f$ satisfies condition (iii) with $\vec{G}'' = \vec{G}'$. But this follows immediately from the fact that if $x \in V_{n-1}$, then $f(y) < f(x)$ for every $y \in N_G(x) - N_{G'}(x)$.

Given any $\varphi_1, \varphi_2 \in \Phi$, we say that $\varphi_1 \le \varphi_2$ if $\varphi_1(x) \le \varphi_2(x)$ for every $x \in V'$. In the sequel, let us fix a *maximal* element φ of Φ in this partial ordering. (The existence of such a maximal element follows by an easy compactness argument.) Let $\vec{G}''$ denote a corresponding graph in (iii).

We claim that φ satisfies

$$\frac{1}{|N_{G''}(x)|} \sum_{y \in N_{G''}(x)} \varphi(y) = \varphi(x)$$

for every $x \in V_{n-1}$, which is not an isolated vertex in G''. To obtain a contradiction, assume that the left-hand side is strictly larger than the right-hand side for some $x \in V_{n-1}$. Then, for a sufficiently small $\varepsilon > 0$,

$$\varphi_\varepsilon(z) = \begin{cases} \varphi(z), & \text{if } z \ne x, \\ \varphi(x) + \varepsilon & \text{if } z = x, \end{cases}$$

and $\vec{G}''_{\varepsilon} = (V', \vec{E}'' - \{\vec{xz} | \varphi(x) = \varphi(z)\})$ would meet all the conditions (i), (ii), and (iii), contradicting the maximality of φ.

In other words, letting $\varphi(\vec{xy}) = \varphi(y) - \varphi(x) \geq 0$ for every $\vec{xy} \in \vec{E}''$, we have

$$\sum_{\vec{xy} \in \vec{E}''} \varphi(\vec{xy}) = \sum_{\vec{zx} \in \vec{E}''} \varphi(\vec{zx}) \qquad \text{for every } x \in V_{n-1}. \qquad (8.6)$$

Using graph-theoretic terminology, we can say that $\vec{G}''$ is a *network* with one *source* x_0 and some *sinks* in $V_n - V_{n-1}$, and assigning the value $\varphi(\vec{xy}) \geq 0$ to each $\vec{xy} \in \vec{E}''$, we obtain a *flow*. Notice that by the maximality of φ, $\varphi(x) = M$ for every $x \in V_n - V_{n-1}$, so there can be no source in $V_n - V_{n-1}$.

Since the flow value disappearing at the sinks is equal to the values entering our network, we have

$$\sum_{y \in V_n - V_{n-1}} \sum_{\vec{xy} \in \vec{E}''} \varphi(\vec{xy}) = \varphi(\vec{x_0 y_0}) = \varphi(y_0). \qquad (8.7)$$

It is clear that every edge of positive flow value leaving V_{i-1} belongs to E_i; thus

$$\sum_{\substack{\vec{xy} \in \vec{E}'' \\ xy \in E_i}} \varphi(\vec{xy}) \geq \varphi(\vec{x_0 y_0}) = \varphi(y_0) \qquad (1 \leq i \leq n). \qquad (8.8)$$

Hence, we obtain by (8.6) and (8.7) that

$$\sum_{xy \in E''} (\varphi(y) - \varphi(x))^2 = \sum_{y \in V'} \varphi(y) \sum_{x \in N_{G''}(y)} (\varphi(y) - \varphi(x))$$

$$= \sum_{y \in V_n - V_{n-1}} \varphi(y) \sum_{x \in N_{G''}(y)} (\varphi(y) - \varphi(x))$$

$$= \sum_{y \in V_n - V_{n-1}} M \sum_{\vec{xy} \in \vec{E}''} \varphi(\vec{xy})$$

$$= M\varphi(y_0).$$

On the other hand, using (8.8) and Jensen's inequality, we get

$$\sum_{xy \in E''} (\varphi(y) - \varphi(x))^2 \geq (\varphi(y_0) - \varphi(x_0))^2 + \sum_{i=1}^{n} \sum_{\substack{x\vec{y} \in \vec{E}'' \\ xy \in E_i}} \varphi^2(\vec{xy})$$

$$\geq \varphi^2(y_0) + \sum_{i=1}^{n} \frac{\varphi^2(y_0)}{|E'' \cap E_i|}$$

$$\geq \varphi^2(y_0) \left(1 + \sum_{i=1}^{n} \frac{1}{|E_i|} \right).$$

Comparing the last two relations and noticing that $\varphi(y_0) \geq f(y_0) = 1$, the theorem follows immediately. □

Theorem 8.9 is an extension of a result by Nash-Williams (1959) on infinite percolations. For the further properties of discrete convex functions, consult Doyle and Snell (1984).

We shall also make use of the following inequality for the radii of some circles surrounding a fixed circle.

Theorem 8.10 (Bárány et al., 1984). *Let C be a (circular) disc of radius $r(C)$ which is tangent to at least six nonoverlapping discs $C_1, \ldots, C_k$ ($k \geq 6$). Then*

$$\frac{1}{k} \sum_{i=1}^{k} \frac{1}{r(C_i)} \geq \frac{1}{r(C)},$$

with equality if and only if $k = 6$, $r(C) = r(C_1) = \cdots = r(C_6)$.

The following elegant proof is due to I. Vincze.

Proof. We may assume without loss of generality that C is *exactly* surrounded by $C_1, \ldots, C_k$, i.e., each C_i is tangent to C_{i-1}, C_{i+1}, and C (where the subscripts are taken modulo k). Let O and O_i denote the centers of C and C_i, respectively, and let $\angle O_i O O_{i+1} = \varphi_i$.

Let us have a closer look at the discs C, C_1, and C_2. Let P denote the center of the circle D inscribed in $\triangle O O_1 O_2$, and set $\angle O O_1 P = \alpha$, $\angle O O_2 P = \beta$.

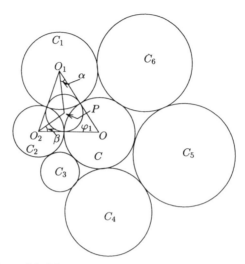

Figure 8.5. Discs $C_1, \ldots, C_k$ exactly surround the disc C.

Then we have

$$\frac{r(D)}{r(C_1)} + \frac{r(D)}{r(C_2)} = \tan \alpha + \tan \beta$$

$$\geq 2 \tan\left(\frac{\alpha + \beta}{2}\right)$$

$$= 2 \tan \frac{\pi - \varphi_1}{4}$$

$$= \tan \frac{\varphi_1}{2} \frac{\left(1 - \tan \frac{\varphi_1}{4}\right)^2}{\tan \frac{\varphi_1}{4}}$$

$$= \frac{r(D)}{r(C)} \left(\tan \frac{\varphi_1}{4} + \cot \frac{\varphi_1}{4} - 2\right).$$

Similarly, we obtain

$$\frac{1}{r(C_i)} + \frac{1}{r(C_{i+1})} \geq \frac{1}{r(C)} \left(\tan \frac{\varphi_i}{4} + \cot \frac{\varphi_i}{4} - 2\right)$$

for every $1 \leq i \leq k$.

Summing up these inequalities and using Jensen's inequality, we get

$$2 \sum_{i=1}^{k} \frac{1}{r(C_i)} \geq \frac{1}{r(C)} \sum_{i=1}^{k} \left(\tan \frac{\varphi_i}{4} + \cot \frac{\varphi_i}{4} - 2 \right)$$

$$\geq \frac{k}{r(C)} \left(\tan \frac{\pi}{2k} + \cot \frac{\pi}{2k} - 2 \right)$$

$$\geq \frac{2k}{r(C)},$$

provided that $k \geq 6$. $\qquad\qquad\square$

Proof of Theorem 8.7. Let C be a six-neighbored circle packing. Construct a planar graph G by assigning a vertex v_i to each $C_i \in C$, and by letting v_i and v_j be adjacent if and only if C_i and C_j touch each other.

If G is not locally finite or C is a regular lattice packing of unit discs, then there is nothing to prove. Thus, we can assume without loss of generality that there are two elements, say $C_1, C_2 \in C$, touching each other and $r(C_1) > r(C_2)$. Theorem 8.10 implies that

$$f(v_i) = \left(\frac{1}{r(C_i)} - \frac{1}{r(C_1)} \right) \Big/ \left(\frac{1}{r(C_2)} - \frac{1}{r(C_1)} \right), \qquad \text{for every } C_i \in C$$

is a convex function on G. Furthermore, $f(v_1) = 0$ and $f(v_2) = 1$. Hence, f meets the requirements of Theorem 8.9 with $x_0 = v_1$, $y_0 = v_2$.

Suppose now, for contradiction, that

$$\inf_{C_i \in C} r(C_i) = \varepsilon > 0.$$

Then all elements of the set V_i defined in Theorem 8.9 correspond to circles contained in disc of radius $2r(C_2)i$ around the center of C_2. Consequently,

$$|V_i| \leq \left(\frac{2r(C_2)}{\varepsilon} \right)^2 i^2 \qquad \text{for every } i \geq 1.$$

On the other hand, since G is a planar graph, the subgraph of G induced by $V_i \subseteq V(G)$ has at most $3|V_i| - 6$ edges. Thus,

$$\sum_{j=1}^{i} |E_j| < 3|V_i| \leq 3 \left(\frac{2r(C_2)}{\varepsilon} \right)^2 i^2 \qquad \text{for every } i \geq 1,$$

where E_j denotes the same as in Theorem 8.9. However, this yields

$$\sum_{j=1}^{n} \frac{1}{|E_j|} \geq \frac{\varepsilon^2}{24r^2(C_2)} \ln n \qquad \text{for every } n \geq 1$$

(see Exercise 8.14).

Thus, by Theorem 8.9,

$$\lim_{n \to \infty} \max_{x \in V_n} f(x) = \infty,$$

contradicting our assumption that

$$f(v_i) = \frac{\dfrac{1}{r(C_i)} - \dfrac{1}{r(C_1)}}{\dfrac{1}{r(C_2)} - \dfrac{1}{r(C_1)}} < \frac{\dfrac{1}{\varepsilon} - \dfrac{1}{r(C_1)}}{\dfrac{1}{r(C_2)} - \dfrac{1}{r(C_1)}}$$

for every $C_i \in C$. $\qquad\qquad\qquad\qquad\qquad\qquad\qquad\qquad\qquad\qquad\qquad\square$

EXERCISES

8.1 Let v_i, v_j, and v_k be the centers of three mutually tangent discs on the plane with radii r_i, r_j, and r_k, respectively. Prove that if we increase r_i and decrease r_j and r_k so that the discs remain tangent, then $\angle v_j v_i v_k$ will decrease.

8.2 Let f be a continuous one-to-one mapping from the interior of the d-dimensional ball B^d ($d \geq 1$) to itself. Prove that if all accumulation points of $f(p)$, as p tends to some boundary point of B^d, lie on Bd B^d, then $f(B^d) = B^d$.

8.3 Let G be a triangulated planar graph with vertex set $V(G) = \{v_1, \ldots, v_n\}$, $n > 3$. For any $U \subseteq V(G)$, let $F(U)$ denote the set of all faces (including the exterior face) of G having at least one vertex that belongs to U. Prove that $|F(U)| > 2|U|$ for any U with $1 \leq |U| \leq n - 3$.

8.4 (Alon, Seymour, and Thomas, 1994) Let G be a planar graph of n vertices, all of whose faces are triangles. For any cycle C of G, let $A(C)$ and $B(C)$ denote the set of vertices drawn strictly inside C and strictly outside C, respectively. Let C be a cycle such that $|C| \leq 2\lfloor \sqrt{2n} \rfloor$ vertices, $|B(C)| < \frac{2}{3}n$, and $|A(C)| - |B(C)|$ is minimal subject to these conditions. Prove that $|A(C)| < \frac{2}{3}n$, and that no vertex in $A(C)$ is adjacent to any vertex in $B(C)$.

8.5 Let G be a planar graph whose vertices are $v_1 \ldots, v_n$, and let d_i denote the degree of v_i. Suppose that each vertex v_i has a nonnegative weight $w(v_i)$, and $\sum_{i=1}^{n} w(v_i) = 1$.

    (i) Prove that the vertex set of G can be partitioned into three parts, A, B, and C, such that

$$\sum_{v_i \in A} w(v_i), \sum_{v_i \in B} w(v_i) \leq \frac{3}{4}, \qquad |C| < 2\sqrt{n},$$

and no vertex in A is adjacent to any vertex in B.

(ii) Prove that for every $\varepsilon > 0$, there exists a constant c_ε (independent of G) such that the vertex set can be partitioned into two parts, A and B, such that

$$\sum_{v_i \in A} w(v_i), \sum_{v_i \in B} w(v_i) \leq \frac{3}{4} + \varepsilon,$$

and the number of edges between A and B is at most $c_\varepsilon(\sum_{i=1}^{n} d_i^2)^{1/2}$.

(iii)⋆ (Gazit and Miller, 1990) Prove that the vertex set of G can be partitioned into two parts A and B, such that

$$\sum_{v_i \in A} w(v_i), \sum_{v_i \in B} w(v_i) \leq \frac{2}{3},$$

and the number of edges of G between A and B is at most $1.58(\sum_{i=1}^{n} d_i^2)^{1/2}$.

8.6 (de Fraysseix et al., 1994) Prove that the vertices of any planar graph can be represented by nonoverlapping triangles in the plane such that two triangles touch each other if and only if the corresponding vertices are adjacent.

8.7⋆ (Rodin and Sullivan, 1987) Given a circle packing $C = \{C_1, C_2, \ldots\}$ in the plane, let $C(C)$ denote the cell decomposition of $\mathbb{R}^2$ determined by the set of segments connecting the centers of those pairs of circles that touch each other. Furthermore, let C_0 denote the regular hexagonal packing of unit circles. Prove that $C(C)$ is topologically isomorphic to $C(C_0)$ if and only if C is similar to C_0.

8.8 Prove that every six-neighbored circle packing in the plane consists of infinitely many circular discs.

8.9 (Grünbaum, 1961; Hadwiger, 1969) Given a convex body $C \subseteq \mathbb{R}^d$, let $H(C)$ be defined as the maximum number of nonoverlapping translates of C that can touch C. $H(C)$ is called the *Hadwiger number* of C. Prove that $H(C) \leq 3^d - 1$, with equality if C is a parallelepiped.

8.10 (Danzer and Grünbaum, 1962) Let $C \subseteq \mathbb{R}^d$ be a convex body. Prove that the maximum number of nonoverlapping translates of C that can pairwise touch each other is 2^d, with equality if C is a parallelepiped.

8.11 (Malitz and Papakostas, 1992) Prove that for any d there exists $\varepsilon(d) > 0$ with the property that every planar graph, all of whose vertices are of degree at most d, can be drawn in the plane by noncrossing line segments so that the angle between any two edges meeting at a vertex is at least $\varepsilon(d)$.

8.12 (Formann et al., 1993) Prove that there is a constant $c > 0$ with the property that any graph, all of whose vertices are of degree at most d, can be drawn in the plane by (possibly crossing) line segments so that the angle between any two edges meeting at a vertex is at least c/d^2.

8.13 (L. Fejes Tóth, 1978–1982)   Let $C_1, \ldots, C_k$ $(k \geq 3)$ be nonoverlapping circular discs in the plane exactly surrounding a disc C; i.e., each C_i is tangent to C_{i-1}, C_{i+1}, and C. Prove that

$$\frac{1}{k} \sum_{i=1}^{k} r^n(C_i) \geq \left(\frac{\sin(\pi/k)}{1 - \sin(\pi/k)} r(C) \right)^n$$

holds for every $n \geq 1$, with equality if and only if

$$r(C_1) = r(C_2) = \cdots = r(C_k).$$

8.14 Let $e_1, \ldots, e_n$ and γ be positive numbers satisfying

$$\sum_{j=1}^{i} e_j \leq \gamma i^2 \qquad (1 \leq i \leq n).$$

Then

$$\sum_{j=1}^{n} \frac{1}{e_j} > \frac{\ln n}{2\gamma}.$$

[*Note:* This is a simple special case of Karamata's inequality (1932).]

8.15 Let G be a graph whose vertex set is the integer lattice in the plane, i.e.,

$$V(G) = \{(m_1, m_2) \,|\, m_1, m_2 \in \mathbb{Z}\}$$

and two points $(m_1, m_2), (m_1', m_2')$ are connected with an edge if and only if their Hamming distance is 1, i.e.,

$$|m_1 - m_1'| + |m_2 - m_2'| = 1.$$

Prove that if f is a convex function on G, which is bounded from above, then f is constant.

8.16 (Österreicher and Linhart, 1981)   Prove that

(i) there does not exist a four-neighbored packing of finitely many equal circles in the plane;

(ii) there exists a five-neighbored circle packing in the plane consisting of a finite number of (not necessarily equal) circles (cf. Definition 8.6).

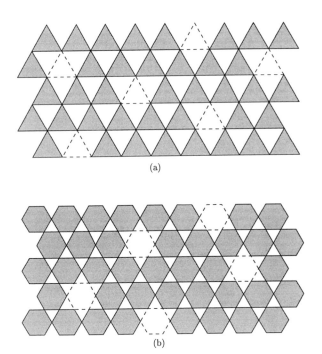

(a)

(b)

Figure 8.6. Five-neighbored packings with (a) congruent convex discs, and (b) centrally symmetric congruent discs.

8.17 (Makai, 1987) Prove that

    (i) there exists a five-neighbored packing C in $\mathbb{R}^2$ with translates of a parallelogram, whose density $d(C, \mathbb{R}^2) = 0$;

   (ii)⋆ for any five-neighbored packing C with translates of a convex disc C which is not a parallelogram, $d(C, \mathbb{R}^2) \geq 3/7$, where equality can occur only if C is a triangle (see Figure 8.6a);

 (iii)⋆ for any five-neighbored packing C in $\mathbb{R}^2$ with translates of a centrally symmetric convex disc C which is not a parallelogram $d(C, \mathbb{R}^2) \geq 9/14$, where equality can occur only if C is an affinely regular hexagon (see Figure 8.6b).

8.18 Is the following generalization of Theorem 8.1 true? If C is a 12-neighbored packing of balls in $\mathbb{R}^3$ containing at least two balls of different radii, then

$$\inf r(B) = 0,$$

where $r(B)$ denotes the radius of B, and the infimum is taken over all elements of C.

Part II

Arrangements of Points and Lines

9

Extremal Graph Theory

Extremal graph theory was started by a group of enthusiastic young students attending D. König's graph theory classes at the Budapest Polytechnic in the early 1930s. P. Erdős, T. Gallai, E. Klein, G. Szekeres, and P. Turán established a number of fundamental theorems in this field, which turned out to have important geometric applications. Some of their early results are mentioned in the classical book of König (1936).

Let G be a *simple* graph with vertex set $V(G)$ and edge set $E(G)$; that is, G has no loops or parallel edges. One of the basic problems of extremal graph theory is to determine the maximum number of edges in a graph G with n vertices under the assumption that it satisfies a certain property $\mathcal{P}$. In most applications the property $\mathcal{P}$ is that "G does not contain a subgraph isomorphic to some fixed graph H." H is usually called the *forbidden subgraph*. In this chapter we present some basic results of this kind. Many other problems and results can be found in B. Bollobás's excellent monographs (1978, 1979).

FORBIDDEN PATHS AND CYCLES

H is said to be a *subgraph* (an *induced subgraph*) of G if it can be obtained from G by deleting some of its edges and vertices (resp. some of its vertices and only those edges which are incident to them). In other words, an induced subgraph of G can be regarded as the restriction of G to a subset of $V(G)$. The *degree* of a vertex $x \in V(G)$ is defined as the number of vertices in G adjacent to x, and it is denoted by $d(x) = d_G(x)$. Let P_r (resp. C_r) denote a path (resp. cycle) of length r, and let K_r denote a complete graph on r vertices. That is,

$$V(P_r) = \{x_0, x_1, \ldots, x_r\}, \quad E(P_r) = \{x_0x_1, x_1x_2, \ldots, x_{r-1}x_r\};$$

$$V(C_r) = \{x_1, x_2, \ldots, x_r\}, \quad E(C_r) = \{x_1x_2, \ldots, x_{r-1}x_r, x_rx_1\};$$

$$V(K_r) = \{x_1, x_2, \ldots, x_r\}, \quad E(K_r) = \{x_ix_j \mid 1 \leq i < j \leq r\}.$$

A graph G is called *connected* if for any $x \neq x' \in V(G)$ there is a path $P_r \subseteq G$ connecting x to x' (i.e., $x_0 = x$, $x_r = x'$). A subgraph $C_r \subseteq G$ is said to be a

Hamiltonian cycle if C_r is a cycle of length $r = |V(G)|$; that is, it passes through every vertex of G. Clearly, any graph G with a Hamiltonian cycle is connected.

To determine the largest number of edges in a graph with n vertices containing no P_r as a subgraph, we need a technical result due to G. Dirac (1952). The elegant proof presented below was found by the 13-year-old L. Pósa (1962).

Theorem 9.1 (Dirac).   *Let G be a connected graph with $n \geq 3$ vertices such that*

$$d(x) + d(y) \geq r$$

for every pair of nonadjacent vertices $x, y \in V(G)$.

(i) *If $r = n$, then G has a Hamiltonian cycle;*
(ii) *If $r < n$, then $G \supseteq P_r$ and $G \supseteq C_{\lceil (r+2)/2 \rceil}$.*

Proof.   Let $P = x_1 x_2 \cdots x_m$ be a path of maximum length in G. In particular, all neighbors of x_1 and x_m are elements of this path.

Assume first that G contains a cycle C_m of length m. Then $m = n$, otherwise, by the connectivity of G, we could find a vertex of C_m adjacent to some $y \notin V(C_m)$, and y would be the endpoint of a path with $m + 1$ vertices, contradicting the maximality of m. Thus, in this case G has a Hamiltonian cycle, and the result is true.

Assume next that G has no cycle of length m. Let

$$N(x_1) = \{x_i \in V(P) \,|\, x_1 x_i \in E(G)\},$$
$$N^+(x_m) = \{x_i \in V(P) \,|\, x_{i-1} x_m \in E(G)\}.$$

Evidently, x_1 and x_m are not adjacent, and $N(x_1) \cap N^+(x_m) = \varnothing$; otherwise, G would contain a cycle of length m (see Figure 9.1). Hence, we can apply our condition to the degrees of x_1 and x_m to obtain

$$r \leq d(x_1) + d(x_m) = |N(x_1)| + |N^+(x_m)| < m.$$

Then $r < n$, and therefore $G \supseteq P_r$, because the length of P is $m - 1 \geq r$. To prove the second part of (ii), observe that one of $d(x_1)$ and $d(x_m)$, say $d(x_1)$, is at least $\lceil r/2 \rceil$. This implies that there is $x_i \in N(x_1)$ with $i \geq \lceil r/2 \rceil + 1$, that is, $x_1, x_2, \ldots, x_i$ form a cycle of length at least $\lceil (r+2)/2 \rceil$, as desired.   □

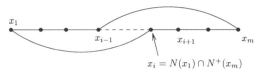

$$x_i = N(x_1) \cap N^+(x_m)$$

Figure 9.1. Illustration for the proof of Theorem 9.1.

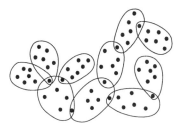

Figure 9.2. Lower-bound construction for C_r-free graph $(r = 8)$.

Corollary 9.2. *Let G be a graph with n vertices containing no path of length r. Then*

$$|E(G)| \le \frac{r-1}{2} n,$$

with equality if and only if all connected components of G are complete graphs on r vertices.

Proof. We prove the upper bound by induction on n. For $n \le r$ the assertion is trivial. Let $n > r$, and assume that we have already proved the result for every graph G with fewer than n vertices. If G is disconnected, then applying the induction hypothesis to each of its connected components, the result follows.

If G is connected, then by Theorem 9.1(ii), it must have at least one vertex x such that $d(x) < r/2$, because otherwise, $G \supseteq P_r$. Let $G - x$ denote the graph obtained from G by deleting x (and all edges incident with x). Obviously, $G - x$ is not the union of disjoint complete subgraphs of r vertices; therefore, by the induction hypothesis, $|E(G - x)| < (r-1)(n-1)/2$ with strict inequality. Hence,

$$\begin{aligned}
|E(G)| &= |E(G - x)| + d(x) \\
&< \frac{(r-1)(n-1)}{2} + \frac{r-1}{2} \\
&= \frac{r-1}{2} n.
\end{aligned}$$ $\square$

According to a similar result of Erdős and Gallai (1959), every graph with n vertices containing no cycles of length at least r has at most $(r-1)(n-1)/2$ edges. Equality holds if and only if G consists of $(n-1)/(r-2)$ "leaves" connected in a "cactus-like" structure, and each "leaf" is a complete subgraph on $r-1$ vertices (see Figure 9.2).

FORBIDDEN COMPLETE SUBGRAPHS

In this section we present a short proof of Turán's classical theorem (1941, 1954), due to Erdős (1970).

A graph G of n vertices is said to be *s-partite* $(1 \leq s \leq n)$ if its vertex set can be decomposed into s groups $V_1 \cup V_2 \cup \cdots \cup V_s$ such that no two vertices within the same group are connected by an edge. If, in addition,

   (i) $|V_i| = \lfloor n/s \rfloor$ or $\lceil n/s \rceil$ for every i, and
   (ii) any two vertices belonging to different groups V_i and V_j are connected by an edge of G (for all $1 \leq i < j \leq s$),

then G is called a *balanced complete s-partite graph*, and is denoted by $T_s(n)$. Evidently, $K_r \not\subseteq T_{r-1}(n)$; in other words, $T_{r-1}(n)$ is K_r-free.

Theorem 9.3 (Turán).   *Let G be a K_r-free graph with n vertices. Then*

$$|E(G)| \leq |E(T_{r-1}(n))|,$$

with equality if and only if G is isomorphic to $T_{r-1}(n)$.

The following result immediately implies Turán's theorem. One has only to check that in the class of all $(r-1)$-partite graphs with n vertices, $T_{r-1}(n)$ has the maximum number of edges (see Exercise 9.5).

Theorem 9.4 (Erdős).   *Given any K_r-free graph G with n vertices, one can always construct an $(r-1)$-partite graph H on the same vertex set such that*

$$d_H(x) \geq d_G(x) \qquad \text{for every } x \in V(G) = V(H).$$

Moreover, if G is not a complete $(r-1)$-partite graph, then we can also assume that $d_H(x) > d_G(x)$ for at least one vertex x.

Proof.   The proof proceeds by induction on r. For $r = 2$, the assertion is trivially true. Let $r > 2$, and assume that we have already proved the result for every integer smaller than r.

Pick a vertex $v \in V(G)$ whose degree in G is maximum. Let $N_G(v)$ denote

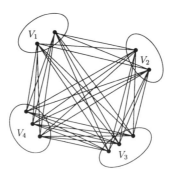

Figure 9.3. $T_4(10)$—a balanced complete four-partite graph with 10 vertices.

the set of all neighbors of v in G. Then $G[N_G(v)]$, the subgraph of G *induced* by $N_G(v)$, is obviously K_{r-1}-free. Hence, by the induction hypothesis, one can find an $(r-2)$-partite graph H' on $N_G(v)$ so that the degree of every vertex in H' is at least as large as in $G[N_G(v)]$. Let H be defined as the graph obtained from H' by connecting every vertex of $V(G) - N_G(v)$ to every vertex of $N_G(v)$. No two vertices in $V(G) - N_G(v)$ will be adjacent in H.

H is clearly $(r-1)$-partite and, by the maximality of $d_G(v)$,

$$d_H(x) = d_G(v) \geq d_G(x) \qquad \text{for all } x \in V(G) - N_G(v).$$

On the other hand, for every $x \in N_G(v)$,

$$d_H(x) = d_{H'}(x) + (n - d_G(v))$$
$$\geq d_{G[N_G(v)]}(x) + (n - d_G(v))$$
$$\geq d_G(x).$$

If $d_H(x) = d_G(x)$ for all x, then $G[N_G(v)]$ must be a complete $(r-2)$-partite graph, and every vertex of $N_G(v)$ must be adjacent to all vertices in $V(G) - N_G(v)$. Thus, G is a complete $(r-1)$-partite graph. $\qquad \square$

The next theorem can be considered as an analogue of Turán's theorem for bipartite graphs. It was proved in a slightly different setting by Kővári, Sós, and Turán (1954) and independently by Erdős (unpublished). The problem is usually referred to in the literature as the problem of Zarankiewicz (1951). Let $K_{r,s}$ denote the complete bipartite graph with r and s vertices in its first and second classes, respectively.

Theorem 9.5 (Kővári et al.). *Let $G_{m,n}$ be a bipartite (two-partite) graph with m vertices in its first class and n in the second. Assume that $G_{m,n}$ does not contain $K_{r,s}$ as a subgraph so that the first and second classes of $K_{r,s}$ are contained in the first and second classes of $G_{m,n}$, respectively. Then*

$$|E(G_{m,n})| \leq c_{r,s}(mn^{1 - 1/r} + n),$$

where $c_{r,s}$ is a constant depending only on r and s.

Proof. Let V_1 and V_2 denote the classes of $G_{m,n}$, $|V_1| = m$, $|V_2| = n$. Using the fact that every r-tuple W of V_1 can be completely connected to at most $s-1$ vertices $x \in V_2$, we obtain that the number of such pairs (W, x) satisfies

$$\sum_{x \in V_2} \binom{d(x)}{r} = \text{number of completely connected pairs } (W, x)$$

$$\leq (s-1)\binom{m}{r}. \tag{9.1}$$

Observe that

$$f(z) = \begin{cases} \dfrac{z(z-1)\cdots(z-r+1)}{r!} & \text{if } z \le r-1 \\ 0 & \text{if } z \ge r-1 \end{cases}$$

is a convex function. Hence, we can apply Jensen's inequality to obtain the result that

$$\sum_{x \in V_2} \binom{d(x)}{r} = \sum_{x \in V_2} f(d(x))$$

$$\ge nf\left(\sum_{x \in V_2} d(x)/n\right)$$

$$= nf(|E(G_{m,n})|/n). \tag{9.2}$$

If $|E(G_{m,n})|/n \le r-1$, then the theorem is true. In the opposite case, putting (9.1) and (9.2) together, we get

$$(s-1)\binom{m}{r} \ge n\binom{|E(G_{m,n})|/n}{r}$$

$$\Rightarrow \qquad (s-1)m^r \ge n\left(\frac{|E(G_{m,n})|}{n} - r + 1\right)^r$$

$$\Rightarrow \quad (s-1)^{1/r}mn^{1-1/r} + (r-1)n \ge |E(G_{m,n})|. \qquad \square$$

The proof of the following simple but extremely useful lemma is left as an exercise (see Exercise 9.16).

Lemma 9.6. *The vertex set of any graph G can be decomposed into two disjoint parts V_1 and V_2 such that $||V_1| - |V_2|| \le 1$, and the number of edges $xy \in E(G)$ with $x \in V_1$, $y \in V_2$ is at least $|E(G)|/2$.*

Corollary 9.7. *Let $r \le s$ be fixed. Then any $K_{r,s}$-free graph G with n vertices has at most $c_s n^{2-1/r}$ edges, where c_s is a constant depending only on s.*

Proof. By Lemma 9.6, there is a bipartite subgraph $G_{\lfloor n/2 \rfloor, \lceil n/2 \rceil} \subseteq G$, satisfying

$$|E(G)| \le 2|E(G_{\lfloor n/2 \rfloor, \lceil n/2 \rceil})|.$$

Applying Theorem 9.5 to $G_{\lfloor n/2 \rfloor, \lceil n/2 \rceil}$, the result follows. $\qquad \square$

A well-known construction of Reiman (1958) shows that the bound in the above corollary is asymptotically tight for $r = 2$ (see Erdős and Rényi, 1962; Mörs, 1981; and Figure 9.4). Brown (1966) used an ingenious algebraic

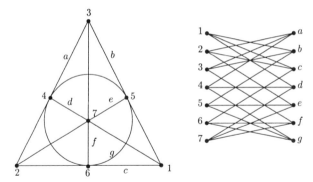

Figure 9.4. Fano plane, **PG**(2, 2), and its incidence graph.

construction to prove that the exponent cannot be improved for $r = 3$ either. To decide whether the same is true for any $r > 3$ is one of the most challenging unsolved problems in this area.

Theorem 9.8. *Let $r \geq 2$ be fixed. For every n, there exists a $K_{r,r}$-free graph G with n vertices such that*

(i) $|E(G)| \geq c_2 n^{3/2}$ *if $r = 2$,*
(ii) $|E(G)| \geq c_r n^{2 - 2/r}$ *if $r > 2$,*

where c_r is a positive constant depending only on r.

Proof.   (i) Assume for simplicity that $n = p^2 + p + 1$ for some prime p. Then one can construct a *projective plane* **PG**$(2,p)$, *of order p*, as follows.

The *points* of **PG**$(2,p)$ are represented by ordered triples (a,b,c), where $a, b, c \in \mathbb{Z}_p$ (not necessarily distinct), and not all of them are equal to 0. Two triples (a,b,c) and (a',b',c') represent the same point if and only if $(a',b',c') = (\lambda a, \lambda b, \lambda c)$ for some $\lambda \neq 0$.
A *line* in **PG**$(2,p)$ consists of all triples (x,y,z) satisfying

$$ax + by + cz \equiv 0 \pmod{p}$$

for a fixed triple (a,b,c). Obviously, the number of points in **PG**$(2,p)$, as well as the number of lines, is equal to

$$\frac{p^3 - 1}{p - 1} = p^2 + p + 1 = n.$$

Every line in **PG**$(2,p)$ contains exactly $p + 1$ points, and the number of lines passing through any point is also $p + 1$.

Let G be a graph whose vertices are the points of $\mathbf{PG}(2,p)$, two distinct points (a, b, c) and (a^*, b^*, c^*) being joined by an edge if and only if

$$aa^* + bb^* + cc^* \equiv 0 \pmod{p}.$$

The neighbors of (a, b, c) in G obviously form a line in $\mathbf{PG}(2,p)$, which may contain (a, b, c) itself. Depending on this, every vertex of G has degree p or $p + 1$. Finally, G is clearly $K_{2,2}$-free, and

$$|E(G)| \geq \frac{(p^2 + p + 1)p}{2} > c_2 n^{3/2}$$

for a suitable constant $c_2 > 0$.

(ii)   We shall apply the "random method"; that is, we are going to prove that for a proper choice of c_r, at least half of the graphs with n vertices and $c_r n^{2 - 2/r}$ edges contain no $K_{r,r}$ as a subgraph, while $n \to \infty$.

The total number of graphs with n vertices and e edges is

$$\left(\binom{n}{2} \atop e \right).$$

The number of those graphs G among them, which contain $K_{r,r}$ is at most

$$\frac{1}{2} \cdot \binom{n}{2r} \cdot \binom{2r}{r} \cdot \left(\binom{n}{2} - r^2 \atop e - r^2 \right),$$

because the two vertex classes of $K_{r,r}$ can be chosen in

$$\frac{1}{2} \binom{n}{2r} \binom{2r}{r}$$

different ways, and in each case we can choose the other $e - r^2$ edges of G arbitrarily from the remaining

$$\binom{n}{2} - r^2$$

possibilities. The ratio of these two numbers is

$$
\frac{\dfrac{1}{2}\cdot\binom{n}{2r}\cdot\binom{2r}{r}\cdot\binom{\binom{n}{2}-r^2}{e-r^2}}{\binom{\binom{n}{2}}{e}} < c_r' n^{2r}\frac{\left(\binom{n}{2}-r^2\right)!}{\binom{n}{2}!}\cdot\frac{e!}{(e-r^2)!}
$$

$$
< c_r' n^{2r}\left(e\Big/\binom{n}{2}\right)^{r^2}
$$

$$
< \frac{1}{2},
$$

for $e = c_r n^{2-2/r}$, provided that $c_r > 0$ is a sufficiently small constant.   □

We mention without proof another important result of Erdős (unpublished) and Bondy and Simonovits (1974), generalizing the $r = 2$ special case of Corollary 9.7.

Theorem 9.9 (Bondy and Simonovits).   *Let $r \geq 4$ be a fixed even number. Then any C_r-free graph G with n vertices has at most $c_r n^{1+2/r}$ edges, where c_r is an appropriate constant.*

This result is known to be asymptotically sharp for $r = 4, 6$, and 10 (see Brown, 1966; Benson, 1966; Singleton, 1966; Bondy, 1971; and Wenger, 1991). For other values of r, Margulis (1982), Imrich (1984), and Lubotzky et al. (1988) found some algebraic constructions of C_r-free graphs with n vertices and $c_r' n^{1+4/(3r+1)}$ edges.

ERDŐS–STONE THEOREM

The next theorem of Erdős and Stone (1946) is perhaps the most useful extension of Turán's theorem (Theorem 9.3). Turán's theorem states that if a graph G with n vertices has more than $|E(T_{r-1}(n))| \approx (n^2/2)(1-1/(r-1))$ edges, then G contains K_r as a subgraph. The Erdős–Stone theorem adds to this that if the number of edges in G is at least $|E(T_{r-1}(n)| + \varepsilon n^2$ and n is sufficiently large, then we can also guarantee the existence of a complete r-partite subgraph, whose classes are large. More precisely, let $K_{\underbrace{t,t,\ldots,t}_{r\ \text{times}}}$ denote a complete r-partite graph with t vertices in each class. [Recall that using our previous notation, $K_{t,t,\ldots,t} = T_r(rt)$.]

Theorem 9.10 (Erdős and Stone).   *Let $r \geq 2$, t be fixed natural numbers, $\varepsilon > 0$. Then there exists an integer $n_0 = n_0(r, t, \varepsilon)$ such that any graph G with $n \geq n_0$ vertices and at least $\frac{n^2}{2}(1 - \frac{1}{r-1} + \varepsilon)$ edges has a complete r-partite subgraph $K_{t,t,\ldots,t}$ whose classes are of size t.*

Note that for $r = 2$ this theorem states that if n is sufficiently large, then any graph with n vertices and at least εn^2 edges contains a $K_{t,t}$. However, this is an immediate consequence of Corollary 9.7.

We shall prove Theorem 9.10 in two steps. First we settle the special case when every vertex has large degree.

Lemma 9.11.   *There exists $n_1 = n_1(r, t, \varepsilon)$ such that if G has $n \geq n_1$ vertices and*

$$d(x) \geq n\left(1 - \frac{1}{r-1} + \varepsilon\right) \qquad \text{for every } x \in V(G),$$

then $G \supseteq \underbrace{K_{t,t,\ldots,t}}_{r \text{ times}}.$

Proof.   We prove the claim by induction on r. As we have already noted, the claim is true for $r = 2$. Assume that we have proved it for some $r \geq 2$, and we want to show that it also holds for $r + 1$.

Set $T = \lceil t/\varepsilon \rceil$. If $n \geq n_1(r, T, \varepsilon)$, then, by the inductive hypothesis, $G \supseteq \underbrace{K_{T,T,\ldots,T}}_{r \text{ times}}$. Let V_i $(1 \leq i \leq r)$ denote the classes of $K_{T,T,\ldots,T}$, $|V_i| = T$. A vertex $x \in V(G) - \bigcup_{i=1}^{r} V_i$ is called *regular* if it has at least t neighbors in each V_i $(1 \leq i \leq r)$. The number of regular vertices in G will be denoted by R. Let m denote the number of "missing" edges between $V(G) - \bigcup_{i=1}^{r} V_i$ and $\bigcup_{i=1}^{r} V_i$, i.e.,

$$m = \left|\left\{ xy \notin E(G) \,\middle|\, x \in V(G) - \bigcup_{i=1}^{r} V_i \ \text{ and } \ y \in \bigcup_{i=1}^{r} V_i \right\}\right|.$$

Using the assumption

$$d(y) \geq n\left(1 - \frac{1}{r} + \varepsilon\right) \qquad \text{for every } y \in \bigcup_{i=1}^{r} V_i,$$

we obtain the inequality

$$m < \left|\bigcup_{i=1}^{r} V_i\right| n\left(\frac{1}{r} - \varepsilon\right) = Tn(1 - \varepsilon r). \tag{9.3}$$

On the other hand, any nonregular vertex of $V(G) - \bigcup_{i=1}^{r} V_i$ is nonadjacent to more than $T - t$ vertices in at least one V_i. Hence,

$$m > (n - rT - R)(T - t) \geq (n - rT - R)T(1 - \varepsilon). \tag{9.4}$$

Comparing (9.3) and (9.4), we get

$$R > n\frac{\varepsilon(r - 1)}{1 - \varepsilon} - rT.$$

Hence, by choosing n sufficiently large, we can ensure that the number of regular vertices,

$$R > \left(\frac{T}{t}\right)^r (t - 1).$$

However, in this case, by the pigeonhole principle, we can find t regular vertices $v_1, \ldots, v_t \in V(G) - \bigcup_{i=1}^{r} V_i$ and t-element subsets $V_i' \subseteq V_i$ ($1 \leq i \leq r$) such that every v_j is connected to all elements of $\bigcup_{i=1}^{r} V_i'$. This means that the classes $\{v_1, \ldots, v_t\}$ and V_i' ($1 \leq i \leq r$) determine an $(r + 1)$-partite complete subgraph $K_{\underbrace{t,t,\ldots,t}_{r+1 \text{ times}}} \subseteq G$. This proves the claim for $r + 1$. $\qquad\square$

Proof of Theorem 9.10. Let $|E(G)| \geq \dfrac{n^2}{2}\left(1 - \dfrac{1}{r-1} + \varepsilon\right)$.

If $d(x) \geq n\left(1 - \dfrac{1}{r-1} + \dfrac{\varepsilon}{2}\right)$ for all $x \in V(G)$, then we are done by Lemma 9.11. If not, then pick a vertex x_1, violating this condition, and let G_1 be the graph obtained from $G = G_0$ by deleting x_1 and all edges incident with it. If $x_1, x_2, \ldots, x_i$, $G_1, G_2, \ldots, G_i$ have already been defined and $|V(G_i)| = n - i \geq n_1(r, t, \varepsilon/2)$, then there are two possibilities:

(i) $d_{G_i}(x) \geq (n - i)\left(1 - \dfrac{1}{r-1} + \dfrac{\varepsilon}{2}\right)$ for every $x \in V(G_i)$, and in this case we are done by Step 1. Or

(ii) one can pick $x_{i+1} \in V(G_i)$ with

$$d_{G_i}(x_{i+1}) < (n - i)\left(1 - \frac{1}{r-1} + \frac{\varepsilon}{2}\right).$$

Set $G_{i+1} = G_i - x_{i+1}$, and repeat this procedure until we get stuck.

We are going to prove that if we get stuck after the ith step, then $|V(G_i)| \geq n_1(r, t, \varepsilon/2)$, hence (i) holds, and $G \supseteq K_{t,t,\ldots,t}$, as desired. Obviously,

$$|E(G)| = \sum_{j=1}^{i} d_{G_{j-1}}(x_j) + |E(G_i)|$$

$$< \sum_{j=1}^{i} (n - j + 1)\left(1 - \frac{1}{r - 1} + \frac{\varepsilon}{2}\right) + \binom{n - i}{2}$$

$$= \left(\binom{n}{2} - \binom{n - i}{2} + i\right)\left(1 - \frac{1}{r - 1} + \frac{\varepsilon}{2}\right) + \binom{n - i}{2}.$$

On the other hand, according to our assumptions,

$$|E(G)| \geq \binom{n}{2}\left(1 - \frac{1}{r - 1} + \varepsilon\right).$$

Combining the last inequalities, we can conclude that

$$\binom{n - i}{2}\left(\frac{1}{r - 1} - \frac{\varepsilon}{2}\right) + i\left(1 - \frac{1}{r - 1} + \frac{\varepsilon}{2}\right) \geq \frac{\varepsilon}{2}\binom{n}{2},$$

or

$$\binom{n - i}{2} \geq \left(\frac{\varepsilon}{4}n^2 - 2n\right)\bigg/\left(\frac{1}{r - 1} - \frac{\varepsilon}{2}\right) \geq \binom{n_1(r, t, \varepsilon/2)}{2},$$

provided that n is sufficiently large. Thus, $|V(G_i)| = n - i \geq n_1(r, t, \varepsilon/2)$, completing the proof. $\qquad \square$

Given a graph H, let $\chi(H)$ denote, as usual, its *chromatic number*, that is, the minimum number of colors needed to color the vertices so that no two vertices of the same color are adjacent.

We mentioned at the beginning of this chapter that one of the basic problems in extremal graph theory is the "forbidden subgraph problem," i.e., to determine or estimate the function $ex(n, H)$, defined as the maximum number of edges an H-free graph with n vertices can have. In fact, almost all problems we have discussed above fall into this category.

Now we are able to give an asymptotically exact answer to this question if $\chi(H) \geq 3$; i.e., H is not a bipartite graph. The following result is a straightforward consequence of the Erdős–Stone theorem.

Corollary 9.12. *Given a nonempty graph H with chromatic number $\chi(H)$,*

$$ex(n, H) = \frac{n^2}{2}\left(1 - \frac{1}{\chi(H) - 1}\right) + o(n^2).$$

THEOREMS OF RAMSEY AND SZEMERÉDI

In the introduction to this chapter we remarked that extremal graph theory was started by a group of young Hungarian mathematicians in the 1930s. However, the first genuine theorem in this field was proved by a brilliant English philosopher, economist, and mathematician, F. P. Ramsey. Ramsey belonged to the Cambridge circle headed by J. Keynes, his father was the President of Magdalene College, and his brother was the Archbishop of Canterbury. His paper (Ramsey, 1930) had remained unnoticed until Erdős and Szekeres (1935) rediscovered its main results and applied it to solve a beautiful geometric problem raised by E. Klein. (Erdős likes to call it the "happy ending" problem, because shortly afterward Klein and Szekeres got married, and they have lived happily together ever since.)

Theorem 9.13 (Ramsey). *For any positive integers $r \geq k$, there exists $n_0 = n_0(r, k)$ satisfying the following condition. For any $n \geq n_0$, if we partition the set of all k-tuples of an n-element set X into two classes, then we can always find an r-element subset $Y \subseteq X$, all of whose k-tuples belong to the same class.*

By iterating this result, one can easily obtain a similar theorem for partitions into more than two classes (Exercise 9.20).

For $k = 2$, the above theorem states that any graph G with sufficiently many vertices contains either a complete subgraph or an empty subgraph with r vertices. (A graph is called *empty*, if it has no edges.)

Theorem 9.14 (Erdős and Szekeres, 1935; Erdős, 1947). *Let $R(r)$ denote the smallest integer n with the property that any graph with n vertices has r pairwise adjacent vertices or r pairwise nonadjacent vertices. Then*

$$\frac{r}{6} 2^{r/2} \leq R(r) \leq \binom{2r - 2}{r - 1} < 2^{2r}.$$

Proof. For any $r, s \geq 2$, let $R(r, s)$ denote the minimum value of n such that every graph with n vertices contains either a complete subgraph of r vertices or an empty subgraph of s vertices. Clearly, $R(r, s) = R(s, r)$ and $R(r, 2) = R(2, r) = r$. Furthermore, for $r, s > 2$, we have

$$R(r, s) \leq R(r - 1, s) + R(r, s - 1). \tag{9.5}$$

Indeed, if G is any graph with $R(r-1, s)+R(r, s-1)$ vertices, then any $x \in V(G)$ is either adjacent to at least $R(r - 1, s)$ vertices, or is nonadjacent to at least $R(r, s - 1)$ vertices. Assume, by symmetry, that the first possibility holds. If there are no s pairwise nonadjacent vertices among the neighbors of x, then x must have at least $r - 1$ pairwise adjacent neighbors. Hence, G contains K_r as a graph, and (9.5) holds.

This yields, by induction on $r + s$, that

$$R(r, s) \leq \binom{r + s - 2}{r - 1}.$$

In particular, for $s = r$, we obtain

$$R(r) = R(r, r) \leq \binom{2r - 2}{r - 1},$$

the desired upper bound.

The lower bound can be established by the *random method*. Let $n = \lfloor r2^{r/2}/6 \rfloor$, and choose randomly a graph G on the vertex set $\{1, \ldots, n\}$. For any r-element subset $X \subseteq \{1, \ldots, n\}$, there are exactly

$$2 \cdot 2^{\binom{n}{2} - \binom{r}{2}}$$

graphs G, in which X induces a complete or an empty subgraph. Since the total number of graphs on $\{1, \ldots, n\}$ is $2^{\binom{n}{2}}$, the probability that a randomly chosen G contains a complete or an empty subgraph with r vertices is at most

$$\frac{\binom{n}{r} \cdot 2 \cdot 2^{\binom{n}{2} - \binom{r}{2}}}{2^{\binom{n}{2}}} \leq \left(\frac{ne}{r}\right)^r \cdot 2^{1 - \binom{r}{2}} < 1.$$

Thus, there exists a graph with n vertices which contains neither a complete nor an empty subgraph with r vertices.   □

It would be of great interest to determine where $\lim_{r \to \infty} (R(r))^{1/r}$ lies in the interval $[\sqrt{2}, 4]$. In fact, it is not even known whether this limit exists.

It is easy to extend the above argument to the general case addressed in Ramsey's theorem. The proof of Theorem 9.13 along the same lines is left to the reader (Exercise 9.24).

Theorem 9.8 implies that the following variant of Ramsey's theorem holds for bipartite graphs.

Corollary 9.15. *For any r, there exists an $n = n(r)$ satisfying the following condition. For any partition of the edge set of $K_{n,n}$ (the complete bipartite graph with n vertices in each of its classes) into two parts, there always exists a subgraph $K_{r,r} \subseteq K_{n,n}$, all of whose edges belong to the same part.*

Surprisingly, the analogous statement for infinite bipartite graphs is false (Exercises 9.21 and 9.22). For many other finite and infinite Ramsey-type results, consult the excellent monographs Graham (1981), Graham et al. (1990), and Erdős, Hajnal et al. (1984).

The general "philosophical" principle behind Ramsey theory is that even the most chaotic systems contain some relatively large subsystems that are highly regular. We close this section with a famous (and very useful) theorem of

Szemerédi (1978), pushing this principle to an extreme. It can be paraphrased, as follows. Even the most chaotic systems can be decomposed into a relatively small number of (almost) regular subsystems.

For a fixed graph G and two disjoint subsets $V_1, V_2 \subseteq V(G)$, let $E(V_1, V_2)$ denote the set of all edges of G with one endpoint in V_1 and the other in V_2. The *edge density* between V_1 and V_2 is

$$\delta(V_1, V_2) = \frac{E(V_1, V_2)|}{|V_1| \cdot |V_2|}.$$

The pair (V_1, V_2) is called ε-*regular* if

$$|\delta(V_1', V_2') - \delta(V_1, V_2)| < \varepsilon$$

holds for every $V_1' \subseteq V_1$ and $V_2' \subseteq V_2$ such that $|V_1'| \geq \varepsilon |V_1|$ and $|V_2'| \geq \varepsilon |V_2|$.

Theorem 9.16 (Szemerédi Regularity Lemma). *Given any $0 < \varepsilon < 1$ and any natural number m, there exist $n_0 = n_0(\varepsilon, m)$ and $M = M(\varepsilon, m)$ with the following property. The vertex set of any graph G with $|V(G)| \geq n_0$ can be partitioned into ℓ classes $V_1, \ldots, V_\ell$ as equally as possible such that $m \leq \ell \leq M$ and all but at most $\varepsilon \ell^2$ pairs (V_i, V_j) are ε-regular.*

It is not hard to show that in the above statement we cannot require all pairs (V_i, V_j) to be ε-regular (Rödl, 1985; Elekes, 1992).

Let $V_1, \ldots, V_r$ be disjoint m-element sets, and let $0 \leq \delta \leq 1$. Define a *random r-partite graph* G on the vertex set $V(G) = V_1 \cup \cdots \cup V_r$. For any pair of vertices $u \in V_i$, $v \in V_j$ ($1 \leq i < j \leq r$), let uv belong to $E(G)$ with probability δ. (The choice is made independently for each pair.) Then, for any $i \neq j$, the expected edge density between V_i and V_j will be δ. Furthermore, the expected number of complete subgraphs of G with r vertices is $\delta^{\binom{r}{2}} m^r$.

Our next lemma shows that a similar statement is true for the graph induced by any r vertex classes of the partition described in Theorem 9.16, provided that any two of them form an ε-regular pair whose edge density exceeds δ. In fact, the wide applicability of Szemerédi's regularity lemma is based on the fact that for most technical purposes, the bipartite graph induced by an ε-regular pair (V_i, V_j) can be regarded as a random graph (see Simonovits and Sós, 1991).

Lemma 9.17. *Let r be a natural number, and let $0 < \varepsilon \leq r^{-r}$, $\varepsilon^{1/r} \leq \delta \leq \frac{1}{2}$. Suppose that a graph G has r pairwise disjoint sets $V_1, \ldots, V_r \subseteq V(G)$, $|V_i| \geq N$ for all $1 \leq i \leq r$, and*

$$\delta(V_i', V_j') \geq \delta$$

for all $V_i' \subseteq V_i, V_j' \subseteq V_j$ ($i \neq j$) such that $|V_i'| \geq \varepsilon |V_i|$, $|V_j'| \geq \varepsilon |V_j|$. Then G

contains at least

$$\delta^{\binom{r}{2}}\left(\frac{N}{2}\right)^r$$

complete subgraphs with r vertices, $v_1, \ldots, v_r$, such that $v_i \in V_i$.

Proof. By induction on r. For $r = 1$ there is nothing to prove. Assume that the assertion is true for $r - 1$, and we want to prove it for r. Let V'_r denote the set of all points in V_r that have at least $\delta|V_i|$ neighbors in V_i, for all $i < r$. Clearly, $|V'_r| \geq |V_r|/2$. Otherwise, for some $i < r$, at least $|V_r|/(2r - 2) > \varepsilon|V_r|$ points of V_r would have fewer than $\delta|V_i|$ neighbors in V_i. Letting V''_r denote the set of these vertices, we would obtain

$$\delta(V''_r, V_i) < \delta,$$

a contradiction.

Fix any vertex $x \in V'_r$, and let V^*_i denote the set of neighbors of x in V_i $(i < r)$. Applying the induction hypothesis to the sets $V^*_1, \ldots, V^*_{r-1}$ (with $\varepsilon^* = \varepsilon/\delta$, $\delta^* = \delta$ and $N^* = \delta N$), it follows that they induce at least

$$\delta^{\binom{r-1}{2}}\left(\frac{\delta N}{2}\right)^{r-1}$$

complete subgraphs. Thus, each $x \in V'_r$ belongs to at least that many K_r's, and the total number of K_r's induced by $V_1, \ldots, V_r$ is at least

$$|V'_r|\delta^{\binom{r-1}{2}}\left(\frac{\delta N}{2}\right)^{r-1} \geq \delta^{\binom{r}{2}}\left(\frac{N}{2}\right)^r,$$

as desired.   □

Corollary 9.18. *For any $\gamma > 0$ and any natural number $r \geq 2$, there exists a constant $\gamma' = \gamma'(\gamma, r)$ such that any graph G with n vertices and at least $\frac{n^2}{2}\left(1 - \frac{1}{r-1} + \gamma\right)$ edges contains at least $\gamma'n^r$ complete subgraphs with r vertices.*

Proof. Consider a partition $V(G) = V_1 \cup \cdots \cup V_\ell$ satisfying the conditions in Theorem 9.16, for some ε and m which will be specified later. Define a graph G^* on the vertex set $V(G^*) = \{V_1, \ldots, V_\ell\}$ by letting

$$V_iV_j \in E(G^*) \Leftrightarrow (V_i, V_j) \quad \text{is } \varepsilon\text{-regular and} \quad \delta(V_i, V_j) \geq \frac{\gamma}{4}.$$

Any edge of G connects

1. two classes (V_i, V_j) such that $V_i V_j \in E(G^*)$,

2. or an ε-irregular pair (V_i, V_j),

3. or a pair (V_i, V_j) with $\delta(V_i, V_j) < \gamma/4$,

4. or two points belonging to the same class V_i.

Let $e_1, \ldots, e_4$ denote the number of edges of G satisfying (1), $\ldots$, (4), respectively. We have

$$\frac{n^2}{2}\left(1 - \frac{1}{r-1} + \gamma\right)$$

$$\leq e_1 + e_2 + e_3 + e_4$$

$$\leq |E(G^*)|\left\lceil \frac{n}{\ell} \right\rceil^2 + \varepsilon \ell^2 \left\lceil \frac{n}{\ell} \right\rceil^2 + \frac{\gamma}{4}\binom{n}{2} + \ell\binom{\lceil n/\ell \rceil}{2}.$$

If $\varepsilon = (\gamma/5)^r$, $\ell \geq m = \lceil 5/\gamma \rceil$, and n is sufficiently large, then each of the last three terms is at most $\gamma n^2/8$, and we obtain

$$|E(G^*)| \geq \frac{n^2}{2\lceil n/\ell \rceil^2}\left(1 - \frac{1}{r-1} + \frac{\gamma}{4}\right)$$

$$\geq \frac{\ell^2}{2}\left(1 - \frac{1}{r-1} + \frac{\gamma}{5}\right).$$

By Turán's theorem (Theorem 9.3 or 9.10), this implies that G^* has a complete subgraph with r vertices, say, $V_1, \ldots, V_r$. Setting $\delta = \gamma/5$, we can apply Lemma 9.17 to conclude that the number of K_r's in G is at least

$$\delta\binom{r}{2}\left(\frac{N}{2}\right)^r \geq \delta\binom{r}{2}\left(\frac{\lfloor n/M(\varepsilon, m)\rfloor}{2}\right)^r. \qquad \square$$

Using completely different techniques, Moon and Moser (1962) have shown that Corollary 9.18 holds with $\gamma' = \gamma/r^r$ (see also Lovász, 1972; Bollobás, 1976; and Exercise 9.25).

TWO GEOMETRIC APPLICATIONS

In this section we present two simple applications of Turán's and Ramsey's theorems (Theorems 9.3 and 9.13).

Suppose that we want to arrange r points on $\mathbb{S}^2$, the boundary of the three-dimensional unit ball, so as to maximize the minimum distance between them. Equivalently, we want to determine the largest value ρ_r such that one can arrange r nonoverlapping spherical caps of radius ρ_r in $\mathbb{S}^2$. In other words, we want to find the maximum density δ_r of a packing of r congruent spherical caps in $\mathbb{S}^2$. For very large values of r, this problem is essentially equivalent to its planar version discussed in Chapter 3. In particular, $\lim_{r \to \infty} \delta_r = \pi/\sqrt{12}$. As far as we know, the exact values of δ_r and ρ_r are known only for $r \leq 12$ and $r = 24$. The best currently known constructions are described in Kottwitz (1991). The question was raised by the botanist Tammes (1930), who studied the distribution of pores on pollen grains.

Let us formalize the above problem in a slightly more general setting. Throughout this section, let C denote a fixed compact subset of S^d, which is equal to the closure of its interior, or let C be the boundary of such a set. ($C = S^2$ satisfies the latter condition.) For any $r \geq 2$, let the rth *packing constant* of C be defined as

$$d_r = \max_{\substack{P \subseteq C \\ |P| = r}} \min_{p \neq q \in P} |p - q|.$$

Obviously, d_r $(r = 1, 2, \ldots)$ is a monotone decreasing sequence, approaching zero as $r \to \infty$. However, it is not necessarily strictly decreasing (e.g., if $C = S^2$, then $d_5 = d_6$).

Generalizing an old observation of Erdős (1955b), Turán (1970a, 1970b) noticed that the packing constants of C can be used to formulate some interesting properties of the distribution of distances determined by any (finite) set of points in C.

Theorem 9.19 (Turán).   *Let $r \geq 2$ be an integer such that $d_{r+1} \neq d_r$. Then, for any set of $n \geq r + 1$ points $p_1, \ldots, p_n \in C$, the number of pairs $p_i p_j$ $(i < j)$ satisfying $|p_i - p_j| \leq d_{r+1}$ is at least*

$$\frac{n^2}{2r} - \frac{n}{2}.$$

Moreover, this bound is tight whenever n is a multiple of r.

Proof.   Define a graph G on the vertex set $\{p_1, \ldots, p_n\}$ by connecting p_i and p_j with an edge if and only if their distance does not exceed d_{r+1}. It follows from the definition of d_{r+1} that G cannot contain $r + 1$ pairwise nonadjacent vertices. Thus, by Turán's theorem (Exercise 9.6), $|E(G)| \geq n^2/2r - n/2$.

To see that this bound cannot be improved, notice that the assumption $d_{r+1} \neq d_r$ implies that one can pick r points of C with all mutual distances larger than d_{r+1}. Replacing each of them by n/r distinct points taken from their very small neighborhoods, we obtain an n-element point set, for which the bound in the theorem is attained.   □

Some further questions and generalizations of this idea can be found in Erdős, Meir et al. (1971, 1972a, 1972b), and Exercises 9.26 and 9.27.

We close this chapter with a real "gem," whose discovery turned out to be the starting point of many new developments in graph theory and combinatorial geometry.

Theorem 9.20 (Erdős and Szekeres, 1935). *For any $r \geq 3$, there exists $n = n(r)$ with the following property. Any set of n points in the plane, no three of which are on a line, contains the vertex set of a convex r-gon.*

Proof. Let $P = \{p_1, \ldots, p_n\}$ be any set of n points in the plane in general position. Partition the set of all triples $p_i p_j p_k$ ($1 \leq i < j < k \leq n$) into two classes according to the orientation of the triangles $p_i p_j p_k$ (clockwise or counter-clockwise). By Theorem 9.13, if $n \geq n_0(r, 3)$, then one can find an r-element subset $Q \subseteq P$, all of whose triples belong to the same class. However, this implies that Q is the vertex set of a convex r-gon (Exercise 9.29). □

EXERCISES

9.1 Prove that the maximum number of edges in a graph of n vertices containing no even cycles C_{2k} ($k = 2, 3 \ldots$) is $\lfloor 3(n-1)/2 \rfloor$.

9.2 (Pósa, 1962) Given a graph G with n vertices such that for any $k < n/2$ there are fewer than k vertices of degree at most k, prove that G has a Hamiltonian cycle.

9.3 (Erdős, 1962b) Given a graph G with n vertices such that every vertex has degree at least $k < n/2$, prove that if

$$|E(G)| \geq 1 + \max_{k \leq t < n/2} \left[\binom{n-t}{2} + t^2 \right],$$

then G has a Hamiltonian cycle. Show that the condition on the number of edges cannot be weakened.

9.4 (Erdős, 1938) Let $1 < n_1 < n_2 < \cdots < n_k \leq n$ be a sequence of natural numbers with the property that no n_i divides the product of any two others $n_g n_h$ (i, g, h are distinct). Show that $k \leq \pi(n) + \lfloor n^{2/3} \rfloor$, where $\pi(n)$ denotes the number of primes not exceeding n.

9.5 Prove that the maximum number of edges in an $(r-1)$-partite graph with n vertices is

$$|E(T_{r-1}(n))| = \frac{1}{2}\left(1 - \frac{1}{r-1}\right)(n^2 - s^2) + \binom{s}{2},$$

where s is the remainder of n upon division by $r - 1$.

9.6 (Turán, 1941) Let G be a graph with n vertices. Show that

  (i) if G contains no complete subgraph with $r + 1$ vertices, then

$$|E(G)| \leq \frac{n^2}{2}\left(1 - \frac{1}{r}\right);$$

(ii) if G contains no $r + 1$ pairwise nonadjacent vertices (i.e., no *empty subgraph* with $r + 1$ vertices), then $|E(G)| \geq n^2/(2r) - n/2$.

9.7 ⋆ (Füredi, 1991b)

(i) Let $H = \{E_1, \ldots, E_m\}$ be a family of subsets of an n-element set, let $s = \sum_{i=1}^{m} |E_i|/m$, and let $k, d \geq 2$ be integers. Show that if

$$s\left(\frac{ms}{n} - k + 1\right) \geq (m - 1)(k - 1)(d - 1),$$

then one can find k members of H with an element in common such that the size of the intersection of any two of them is at least d.

(ii) Let G_k denote the bipartite graph with vertex classes

$$V_1 = \{x_0\} \cup \{x_{ij} | 1 \leq i < j \leq k\}, \quad V_2 = \{y_i | 1 \leq i \leq k\}$$

and edge set

$$E(G_k) = \{x_0 y_i | 1 \leq i \leq k\} \cup \{x_{ij} y_i, x_{ij} y_j | 1 \leq i < j \leq k\}.$$

Show that for every $k \geq 2$ there exists a constant c_k such that any G_k-free graph with n vertices has at most $c_k n^{3/2}$ edges (see Figure 9.5).

9.8 Let $v_1, \ldots, v_n$ be vectors of length at least 1 in d-dimensional Euclidean space. Show that the number of pairs $v_i v_j$ ($1 \leq i < j \leq n$) with $|v_i + v_j| < 1$ is at most $\lfloor n^2/4 \rfloor$.

9.9 (Katona, 1969, 1978) Let $\mathbf{X}$ and $\mathbf{Y}$ be two identically distributed discrete random variables whose (finitely many possible) values are vectors in Euclidean d-space. Show that

$$\Pr\left[|\mathbf{X} + \mathbf{Y}| \geq r\right] \geq \tfrac{1}{2}(\Pr\left[|\mathbf{X}| \geq r\right])^2.$$

9.10 Given a triangle-free (K_3-free) graph G, let $\alpha(G)$ and $\tau(G)$ denote, as usual, the maximum number of pairwise nonadjacent vertices and the

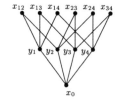

Figure 9.5. The graph G_4.

minimum number of vertices such that every edge of G is incident with at least one of them, respectively. Show that

$$|E(G)| \leq \alpha(G)\tau(G).$$

Deduce from this Turán's theorem in the special case $r = 3$.

9.11 Prove that if a graph G with n vertices and at least $\lfloor n^2/4 \rfloor$ edges contains an odd cycle, then it also contains a triangle.

9.12 (Dirac, 1963) Let $T_{r-1}(n)$ denote the balanced complete $(r-1)$-partite graph with n vertices, and let $n \geq r + 1$. Prove that any graph G with n vertices and at least $|E(T_{r-1}(n))| + 1$ edges contains a K_{r+1} from which an edge has been omitted.

9.13 (Erdős, 1962a; Edwards, 1975) Prove that there exist a positive constant c and a natural number n_0 such that any graph G with $n \geq n_0$ vertices and more than $\lfloor n^2/4 \rfloor$ edges contains cn triangles sharing an edge.

9.14 (Erdős 1964a) Prove that for $t > 0$, if a graph G with n vertices and $\lfloor n^2/4 \rfloor - t$ edges contains a triangle, then it contains at least $\lfloor n/2 \rfloor - t - 1$ triangles. Show that this bound is the best possible.

9.15 (Győri et al., 1991) Prove that among all triangle-free graphs with n vertices, the balanced complete bipartite graph $T_2(n)$ contains the largest number of cycles of length 4.

9.16 (Erdős, 1967b) Prove that every graph G has a bipartite subgraph H with $|E(H)| \geq |E(G)|/2$. Furthermore, we can also require that the sizes of the vertex classes of H differ by at most 1.

9.17 Show that for every r there exists $n_0 = n_0(r)$ such that any $K_{r,r}$-free graph with $n \geq n_0$ vertices has fewer than $n^{2-1/r}$ edges.

9.18★ (Simonovits, 1966) Prove that any graph G with $n \geq n_0$ vertices and $|E(T_r(n))| + n$ edges contains $K_{1,\underbrace{3,3,...,3}_{r \text{ times}}}$, a complete $(r+1)$-partite sub-graph with one vertex in its first class and three vertices in every other class (cf. Turán's theorem).

9.19★ (Erdős and Moser, 1959, 1970) Let G be a graph with $n > r$ vertices with the property that any r vertices have a common neighbor. Prove that

$$|E(G)| \geq (r-1)(n-1) - \binom{r-1}{2} + \left\lceil \frac{n-r+1}{2} \right\rceil,$$

and that this bound can be attained.

9.20 (Ramsey, 1930) Show that for any positive integers $r \geq k$ and $c \geq 2$, there exists $n_0 = n_0(r, k, c)$ satisfying the following condition. If we partition the set of all k-tuples of an n_0-element set X into c classes, then we can always find an r-element subset $Y \subseteq X$, all of whose k-tuples belong to the same class.

9.21 (Ramsey, 1930)

   (i) Show that any graph with infinitely many vertices contains either an infinite complete subgraph or an infinite empty subgraph (i.e., infinitely many pairwise nonadjacent vertices).

   (ii) Use part (i) to establish Theorem 9.13 for $k = 2$.

9.22 Construct a bipartite graph G with infinite vertex classes V_1 and V_2 such that there are no infinite subsets $V_1' \subseteq V_1, V_2' \subseteq V_2$, for which $V_1' \times V_2' \subseteq E(G)$ or $(V_1' \times V_2') \cap E(G) = \varnothing$. Does there exist an uncountable bipartite graph G with this property?

9.23 Let $r \geq 2$ be fixed. Show that any sufficiently long sequence of real numbers contains a monotone subsequence of length r.

9.24 Generalize the proof of the upper bound in Theorem 9.14 to establish Theorem 9.13.

9.25 (Moon and Moser, 1962) For any graph G with n vertices, let $k_r(G)$ denote the number of complete subgraphs of G with r vertices.

   (i) Show that

$$\frac{k_{r+1}(G)}{k_r(G)} \geq \frac{1}{r^2 - 1}\left(r^2 \frac{k_r(G)}{k_{r-1}(G)} - n\right).$$

   (ii) Let $|E(G)| = (n^2/2)(1 - 1/x)$ for some real number $x \geq r - 1$. Prove that

$$k_r(G) \geq \frac{\binom{x}{r}}{x^r} n^r.$$

9.26 (Erdős, 1955b) Let $\mathcal{P}$ denote the family of all subsets of the plane whose diameter is at most 1. For any $r \geq 2$, let

$$d_r = \max_{\substack{P \in \mathcal{P} \\ |P| = r}} \min_{p \neq q \in P} |p - q|.$$

Show that for any set $\{p_1, \ldots, p_n\} \subseteq \mathcal{P}$, the number of point pairs $p_i p_j$ ($i < j$) satisfying $|p_i - p_j| \leq d_{r+1}$ is at least $n^2/(2r) - n/2$.

9.27 (Bateman and Erdős, 1951; Erdős, 1955b) Given a set of n points $\{p_1, \ldots, p_n\} \subseteq \mathbb{R}^2$ with maximum distance 1, show that the number of pairs $p_i p_j$ ($i < j$) whose distance is larger than $1/\sqrt{2}$ is at most $\lfloor n^2/3 \rfloor$.

9.28 (Sós, 1970)   Given a set P of n points in $\mathbb{R}^d$, $2 \le r < n$, let $c_r^*(P)$ denote the smallest number with the property that one can select r elements of P such that any other element of P is within distance $c_r^*(P)$ from at least one of them. Show that the number of (unordered) pairs $p, q \in P$ ($p \ne q$) satisfying $|p - q| \ge c_r^*(P)$ is at least

$$(r - 1)(n - 1) - \binom{r - 1}{2} + \left\lceil \frac{n - r + 1}{2} \right\rceil.$$

Moreover, this bound can be attained.

9.29 Let $p_1, \ldots, p_r \in \mathbb{R}^2$ be r points such that no three of them are collinear and the orientation of the triangle $p_i p_j p_k$ is clockwise for every $1 \le i < j < k \le r$. Show that $p_1, \ldots, p_r$ are in convex position.

10

Repeated Distances in Space

In 1946, in a paper published in the *American Mathematical Monthly*, Paul Erdős raised two general questions about the distribution of distances determined by a finite set of points in a metric space:

1. At most how many times can a given distance occur among a set of n points?

2. What is the minimum number of distinct distances determined by a set of n points?

Because of the many failed attempts to give reasonable bounds for these functions, even in the case of planar sets, one had to realize that the above questions are not merely "gems" in recreational mathematics. They have raised deep problems, some of which can be solved using graph-theoretic and combinatorial ideas. In fact, the discovery of many important combinatorial techniques and results was strongly motivated by their expected geometric consequences. In this chapter we demonstrate how the extremal graph-theoretic tools introduced in Chapter 9 can be applied to problem 1 mentioned above and to some closely related questions.

UNIT DISTANCES IN THE PLANE

Definition 10.1. Given a set $P = \{p_1, \ldots, p_n\}$ of n points in d-dimensional Euclidean space $\mathbb{R}^d$, let $f(P)$ denote the number of pairs $p_i p_j$, $i < j$, such that their distance, $|p_i - p_j|$, is equal to 1. Let

$$f_d(n) = \max_{\substack{P \subseteq \mathbb{R}^d \\ |P| = n}} f(P).$$

In other words, $f_d(n)$ is the maximum number of unit distances determined by n points in $\mathbb{R}^d$.

It is easy to show that $f_1(n) = n - 1$, and that the only extremal configurations (i.e., the only n-element point sets for which this maximum is attained) are the arithmetic progressions of difference 1 and length n (see Exercise 10.3).

For $d = 2$, Erdős (1946) proved the following result:

Theorem 10.2 (Erdős). *Let $f_2(n)$ denote the maximum number of times the unit distance can occur among n points in the plane. Then there exists a constant $c > 0$ such that*

$$f_2(n) \leq cn^{3/2}.$$

Proof. Given a planar point set $P = \{p_1, \ldots, p_n\}$, let us define a graph G_P by connecting two points of P with an edge if and only if their distance is 1. That is, let

$$V(G_P) = P$$
$$p_i\, p_j \in E(G_P) \iff |p_i - p_j| = 1.$$

G_P is usually called the *unit distance graph* associated with P (see Figure 10.1). Evidently, G_P does not contain any subgraph isomorphic to $K_{2,3}$, a complete bipartite graph with two vertices in one of its classes and three in the other. Hence, we can apply Corollary 9.7 to obtain

$$|E(G_P)| \leq cn^{2 - 1/2}.$$

This proves the upper bound.   □

To obtain a lower bound on $f_2(n)$, we first need some preparation.

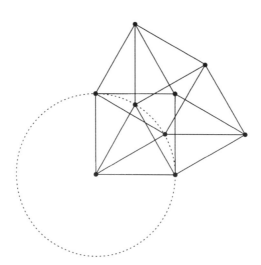

Figure 10.1. Unit distance graph.

Lemma 10.3. *Let $r(n)$ denote the number of different ways in which a natural number n can be represented as the sum of two squares. Then there exists a constant $c' > 0$ such that for infinitely many n,*

$$r(n) \geq n^{c'/\log\log n}.$$

Proof (Sketch). Assume, for simplicity, that $n = p_1 p_2 \cdots p_k$, where p_j is the jth smallest prime of the form $4m + 1$. By Theorem 1.12, p_j can be expressed as the sum of two squares, i.e.,

$$p_j = a_j^2 + b_j^2 = (a_j + b_j i)(a_j - b_j i),$$

where $i = \sqrt{-1}$.

For any subset $J \subseteq \{1, 2, \ldots, k\}$,

$$\prod_{j \in J}(a_j + b_j i)\prod_{j \notin J}(a_j - b_j i) = A_J + B_J i,$$

$$\prod_{j \in J}(a_j - b_j i)\prod_{j \notin J}(a_j + b_j i) = A_J - B_J i,$$

where A_J and B_J obviously satisfy

$$A_J^2 + B_J^2 = (A_J + B_J i)(A_J - B_J i) = \prod_{j=1}^{k} p_j = n.$$

By the unique factorization theorem for complex integers (see, e.g., Niven and Zuckerman, 1966), $A_J + B_J i$ is different for different choices of J, so we obtain

$$r(n) \geq 2^k.$$

It remains to give a lower bound on k in terms of n. It is well known that the primes of the form $4m + 1$ are "uniformly distributed" in the sequence of all primes. In particular,

$$c_1 k \log k < p_k < c_2 k \log k$$

for every $k > 2$, where c_1 and c_2 are suitable positive constants (see Hardy and Wright, 1960; LeVeque, 1956; or any other standard textbook on number theory). Thus,

$$n = \prod_{j=1}^{k} p_j < p_k^k < c_2(k \log k)^k,$$

which implies that $k > c' \log n / \log\log n$, and the result follows. $\square$

Now we can prove the desired lower bound on $f_2(n)$.

Theorem 10.4 (Erdős). *Let $f_2(n)$ denote the maximum number of times the unit distance can occur among n points in the plane. Then there exists a constant $c'' > 0$ such that for infinitely many values of n,*

$$f_2(n) > n^{1 + c''/\log\log n}.$$

Proof. Let N be an integer satisfying $r(N) \geq N^{c'/\log\log N}$ (see Lemma 10.3), and set $n = (2\lceil\sqrt{N}\rceil)^2$. Let P be a $2\lceil\sqrt{N}\rceil \times 2\lceil\sqrt{N}\rceil$ piece of square lattice of side length $1/\sqrt{N}$, i.e.,

$$P = \{ (i/\sqrt{N}, j/\sqrt{N}) \mid 1 \leq i, j \leq 2\lceil\sqrt{N}\rceil \}.$$

The distance between two points $(i/\sqrt{N}, j/\sqrt{N})$, $(i'/\sqrt{N}, j'/\sqrt{N}) \in P$ is 1 if and only if $(i - i')^2 + (j - j')^2 = N$. Thus, every point of P is at distance 1 from at least $r(N)$ elements of P. Therefore,

$$f(P) \geq \frac{nr(N)}{2} \geq n^{1 + c''/\log\log n},$$

as desired. It is easy to see that the same lower bound on $f_2(n)$ holds for every sufficiently large n. □

In the next chapter (see Corollary 11.11), we will return to the question of bounding $f_2(n)$ and improve the upper bound to $cn^{4/3}$. However, according to Erdős's conjecture, the true value of $f_2(n)$ is likely to be close to the lower bound. Erdős has repeatedly offered \$500 for a proof or disproof of the bound $f_2(n) \leq n^{1 + c/\log\log n}$.

We can slightly modify the above question by asking for the maximum number of pairs among n points in the plane which determine "nearly" the same distance. To avoid the trivial case, when all the $\binom{n}{2}$ distances are nearly 0, we assume that our points are "separated." The following result is due to Erdős, Makai, Pach, and Spencer (1991).

Theorem 10.5 (Erdős et al.). *Given $t > 0$ and a set of n points $p_1, \ldots, p_n$ in the plane with minimum distance at least 1, the number of pairs $p_i p_j$, $i < j$, whose distance is between t and $t + 1$, cannot exceed $\lfloor n^2/4 \rfloor$, provided that $n \geq n_0$. Furthermore, this bound can be attained for all n and for all sufficiently large $t > t_0(n)$.*

Proof. We define a graph G on the vertex set $\{p_1, \ldots, p_n\}$ by joining p_i and p_j with an edge if and only if $t \leq |p_i - p_j| \leq t + 1$. Suppose to the contrary that $|E(G)| \geq \lfloor n^2/4 \rfloor + 1$. Then, we can find an edge, say, $p_1 p_2 \in E(G)$ and $\lceil cn \rceil$ vertices $p_3, \ldots, p_{\lceil cn \rceil + 2}$ adjacent to both p_1 and p_2 (cf. Exercise 9.13). By the definition of our graph, all points $p_3, \ldots, p_{\lceil cn \rceil + 2}$ must lie in the intersection of two annuli R_1 and R_2, where R_j is centered at p_j ($j = 1, 2$), and its inner and

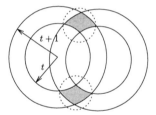

Figure 10.2. Illustration for the proof of Theorem 10.5.

outer radii are t and $t + 1$, respectively (see Figure 10.2). However, it is easy to see that $R_1 \cap R_2$ can be covered by two discs of radius $3/2$. The maximum number of points that can be chosen within a disc of radius of $3/2$ so that no two of them are closer than 1, is at most

$$\frac{(3/2 + 1/2)^2 \pi}{(1/2)^2 \pi} = 16.$$

Thus, we obtain that $cn \le 32$, contradicting our assumption that n is sufficiently large.

As to the lower bound, place n points on two vertical lines at distance t, each containing at least $\lfloor n/2 \rfloor$ points, and let the distance between two consecutive points on a line be 1. If t is chosen sufficiently large, then it is easy to see that the distance between any two points on different lines is between t and $t + 1$. $\square$

At this point we would like to mention two other problems which can be approached by extremal graph theory.

1. **The problem of incidences.**   Given a set $P = \{p_1, \ldots, p_m\}$ of m points and a set $L = \{\ell_1, \ldots, \ell_n\}$ of n lines in the plane, let $I(P, L)$ denote the number of incidences between them, that is, the number of pairs (i, j) such that $p_i \in \ell_j$. Determine

$$I(m, n) = \max_{\substack{|P| = m \\ |L| = n}} I(P, L).$$

2. **The problem of complexity of many cells.**   Given a set $L = \{\ell_1, \ldots, \ell_n\}$ of n lines in the plane, a connected component of $\mathbb{R}^2 - \bigcup_{i=1}^{n} \ell_i$ is called a *cell* of the arrangement. Each cell c is bounded by a number of segments and (possibly) half-lines, which are called the *edges* (or *sides*) of c. For a collection of cells $C = \{c_1, \ldots, c_m\}$, let $\mathcal{K}(C, L)$ denote the total number of edges of the cells in C. Determine

$$\mathcal{K}(m, n) = \max_{\substack{|C| = m \\ |L| = n}} \mathcal{K}(C, L).$$

Observe that

Lemma 10.6. *For any positive integers m and n,*

$$I(m,n) \leq \tfrac{1}{2}\mathcal{K}(m,2n) + m.$$

Proof. Let $P = \{p_1,\ldots,p_m\}$, $\mathcal{L} = \{\ell_1,\ldots,\ell_n\}$ be systems of points and lines, respectively. If a point of P is not an intersection point of two or more lines of $\mathcal{L}$, it contributes at most 1 to $I(P,\mathcal{L})$, so the total number of such incidences is at most m. Next, consider the points p_i lying on at least two members of $\mathcal{L}$. Let

$$\varepsilon = \min_{\substack{p \in P, \ell \in \mathcal{L} \\ p \notin \ell}} \mathrm{dist}(p,\ell),$$

i.e., the minimum distance between a point $p \in P$ and a line $\ell \in \mathcal{L}$ not passing through p. Translating each line of $\mathcal{L}$ by a distance of $\varepsilon/3$ in both directions, we obtain a set $\mathcal{L}'$ of $2n$ lines. Each p_i which belongs to at least two lines of $\mathcal{L}$ sits in a distinct cell c_i' determined by $\mathcal{L}'$, and c_i' has at least twice as many sides as the number of elements of $\mathcal{L}$ incident with p_i. Hence, the total number of sides of the distinct cells c_i' is at least $2(I(P,\mathcal{L}) - m)$. Equivalently, $\mathcal{K}(m,2n) \geq 2(I(m,n) - m)$. □

Theorem 10.7 (Canham, 1969). *Let $\mathcal{K}(m,n)$ denote the maximum number of edges in m distinct cells determined by an arrangement of n lines in the plane. Then*

$$\mathcal{K}(m,n) \leq c(m\sqrt{n} + n)$$

for some appropriate constant c.

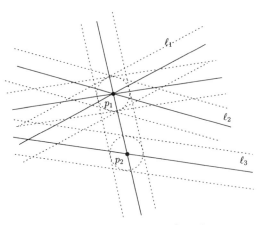

Figure 10.3. Transforming $\mathcal{L}$ to $\mathcal{L}'$.

Proof. Let $C = \{c_1, \ldots, c_m\}$ be a system of cells determined by an arrangement $\mathcal{L} = \{\ell_1, \ldots, \ell_n\}$ of n lines in $\mathbb{R}^2$. Construct a bipartite graph $G_{m,n}$ with vertex classes $V_1 = C$ and $V_2 = \mathcal{L}$. Let $c_i \ell_j \in E(G_{m,n})$ if and only if c_i contains a segment of ℓ_j on its boundary. Obviously, $\mathcal{K}(C, \mathcal{L}) = |E(G_{m,n})|$. It is easy to check that $K_{2,5} \not\subseteq G_{m,n}$ so that its first and second classes are contained in V_1 and V_2, respectively (see Exercise 10.5). Thus, we can apply Theorem 9.5 to complete the proof. $\qquad\square$

In Chapter 11 we combine this argument with some new ideas to determine the exact order of magnitude of $I(m, n)$ and $\mathcal{K}(m, n)$.

UNIT DISTANCES IN THE SPACE

Now we turn to the higher-dimensional versions of the unit distance problem.

Theorem 10.8 (Erdős, 1960). *Let $f_3(n)$ denote the maximum number of times the unit distance can occur among n points in $\mathbb{R}^3$. Then*

$$c_1 n^{4/3} \log\log n \le f_3(n) \le c_2 n^{5/3},$$

where c_1, c_2 are positive constants.

Proof. Consider a set P of at most n points forming a grid of size $\lfloor n^{1/3}\rfloor \times \lfloor n^{1/3}\rfloor \times \lfloor n^{1/3}\rfloor$,

$$P = \{(i, j, k) \mid 1 \le i, j, k \le \lfloor n^{1/3}\rfloor\}.$$

For any pair of points $p = (i, j, k)$, $p' = (i', j', k') \in P$,

$$|p - p'|^2 = (i - i')^2 + (j - j')^2 + (k - k')^2$$

is a positive integer that does not exceed $3n^{2/3}$. Hence, the number of distinct distances determined by P is at most $3n^{2/3}$. This implies that there is a distance that occurs at least

$$\binom{|P|}{2} \Big/ 3n^{2/3} \ge cn^{4/3}$$

times for some constant $c > 0$. Scaling our picture so that this distance becomes the unit distance, we get that $f_3(n) \ge cn^{4/3}$. A more delicate number-theoretic argument along the same lines gives the improvement by a factor of $\log\log n$.

To prove the upper bound, fix an n-element point set P in $\mathbb{R}^3$. Observe that the unit distance graph G obtained by connecting two points of P with an edge if and only if their distance is 1, does not contain $K_{3,3}$ as a subgraph. Thus, $f_3(n) \le c_3 n^{2-1/3}$ follows from Corollary 9.7. $\qquad\square$

It is very surprising that for $d \geq 4$, one can fairly easily describe the asymptotic behavior of $f_d(n)$.

Theorem 10.9 (Erdős, 1967a).   *Let $d \geq 4$. Then the maximum number of times the unit distance can occur among n points in $\mathbb{R}^d$ is*

$$f_d(n) = \frac{n^2}{2}\left(1 - \frac{1}{\lfloor d/2 \rfloor} + o(1)\right).$$

Proof.   The lower bound can be established by the following construction of Lenz (1955). Let $C_1, \ldots, C_{\lfloor d/2 \rfloor}$ be circles of radius $1/\sqrt{2}$ centered at the origin (of $\mathbb{R}^d$), and assume that the planes supporting these circles are mutually orthogonal. Let us pick n_i points on C_i, $n_i = \lfloor n/\lfloor d/2 \rfloor \rfloor$ or $n_i = \lceil n/\lfloor d/2 \rfloor \rceil$ so that $\sum_i n_i = n$ (see Figure 10.4). It is clear that any two points belonging to different C_i's are at distance 1 from each other. Hence, this point system determines at least as many unit distances as the number of edges in $T_{\lfloor d/2 \rfloor}(n)$, the balanced complete $\lfloor d/2 \rfloor$–partite graph with n vertices. This implies that

$$f_d(n) \geq |E(T_{\lfloor d/2 \rfloor}(n))| = \frac{n^2}{2}\left(1 - \frac{1}{\lfloor d/2 \rfloor} + o(1)\right)$$

(cf. Exercise 9.5).

We prove the upper bound by contradiction. Let $P = \{p_1, \ldots, p_n\} \subset \mathbb{R}^d$, and let G be the graph with these points as vertices, with p_i, p_j being joined by an edge if and only if they are at distance 1. Assume that n is sufficiently large and that

$$|E(G)| \geq \frac{n^2}{2}\left(1 - \frac{1}{\lfloor d/2 \rfloor} + \varepsilon\right)$$

$$C_2 : u^2 + v^2 = 1/2, \ x = y = 0$$

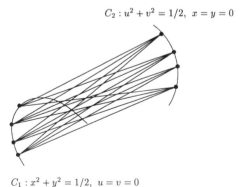

$$C_1 : x^2 + y^2 = 1/2, \ u = v = 0$$

Figure 10.4. Lenz's construction in four dimensions.

for some $\varepsilon > 0$. Then, by the Erdős–Stone theorem (Theorem 9.10), $G \supseteq K_{\underbrace{3,3,\ldots,3}_{\lfloor d/2 \rfloor + 1}}$; i.e., G contains a complete ($\lfloor d/2 \rfloor + 1$)–partite graph whose classes $V_1, V_2, \ldots, V_{\lfloor d/2 \rfloor + 1}$ are triples.

The points of V_i cannot be collinear, because they are at the same distance from any element of V_j ($j \neq i$), and no sphere contains three collinear points. Let Π_i denote the (affine) two-dimensional plane spanned by the elements of V_i ($1 \leq i \leq \lfloor d/2 \rfloor + 1$). Using the fact that for every $i \neq j$, all nine distances between the points of V_i and V_j are equal to 1, one can easily show that Π_i and Π_j are orthogonal to each other (see Exercise 10.6). However, this would force the existence of $\lfloor d/2 \rfloor + 1$ mutually orthogonal (two–dimensional) planes in $\mathbb{R}^d$, which is absurd. $\qquad \square$

UNIFORM HYPERGRAPHS

Before we turn to the next application of the graph-theoretic methods developed in Chapter 9, we extend some of our results to k-uniform hypergraphs.

Definition 10.10. A *k-uniform hypergraph* H consists of a finite vertex set $V(H)$ and a set $E(H)$ of k-tuples (i.e., k-element subsets) of $V(H)$. The elements of $E(H)$ are called the *hyperedges* (or, simply, *edges*) of H. Clearly, a graph can be regarded as a two-uniform hypergraph. We say that H' is a *subhypergraph* of H if $V(H') \subseteq V(H)$ and $E(H') \subseteq E(H)$.

The basic question of extremal graph theory can now be generalized, as follows. Given a (so-called *forbidden*) hypergraph H', what is the maximum number of hyperedges that a k-uniform hypergraph with n vertices can have without containing a subhypergraph isomorphic to H'? Unfortunately, even some of the simplest problems of this kind turned out to be hopelessly difficult (for a recent survey, see Füredi, 1991a). For example, let H' consist of all 3-tuples of a four-element set. Encouraged by the elegance of his graph theorem (Theorem 9.3), Turán conjectured that in this case the maximum number of edges of an H'-free three-uniform hypergraph with n vertices is roughly $\frac{5}{9}\binom{n}{3}$, and equality is attained for the following hypergraph H. Let us partition $V(H)$ into three parts V_0, V_1, V_2, as equally as possible, and let a 3-tuple belong to $E(H)$ if and only if it intersects all V_i's, or contains two vertices of some V_i and one vertex of $V_{i+1 \,(\mathrm{mod}\, 3)}$ (see Figure 10.5). Surprisingly, Brown (1983), Todorov (1983), and Kostochka (1982) exhibited a large number of H'-free hypergraphs with n vertices and exactly the same number of hyperedges as in the above construction. This suggests that if Turán's conjecture is true, its verification will not be easy. (Erdős has offered \$500 for a proof or disproof.)

Nevertheless, as was pointed out by Katona, Nemetz, and Simonovits (1964),

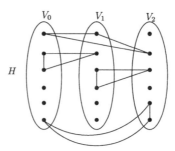

Figure 10.5. No four-element subset of $V(H)$ contains four 3-tuples from $E(H)$.

it is not hard to establish some nontrivial results related to Turán's problem for uniform hypergraphs.

The maximum number of pairwise nonadjacent vertices of a graph G is called the *independence number* of G, and is denoted by $\alpha(G)$. Turán's theorem (Theorem 9.3), applied to the complement of G, yields the fact that

$$\alpha(G) \geq \frac{|V(G)|}{2|E(G)|/|V(G)|+1} = \frac{|V(G)|^2}{2|E(G)|+|V(G)|} \tag{10.1}$$

holds for every graph. Note that $2|E(G)|/|V(G)|$ is the average degree of the vertices of G (see Exercise 9.6).

Given a k-uniform hypergraph H, a subset $A \subseteq V(H)$ is called *independent* if A contains no hyperedges. The *independence number* $\alpha(H)$ of H is defined as the maximum size of an independent subset of $V(H)$.

Refining the argument of Katona et al. (1964), Spencer (1972) extended (10.1) to k-uniform hypergraphs. [See also Spencer, 1987, 1990; Erdős and Spencer, 1974; and Thiele, 1993. The best known bounds are due to de Caen (1983a, 1983b, 1983c).]

Theorem 10.11 (Spencer, 1972).   *Let H be a k-uniform hypergraph with n vertices and $m \geq n/k$ edges, and let $\alpha(H)$ denote the independence number of H. Then*

$$\alpha(H) \geq \left(1 - \frac{1}{k}\right)\left\lfloor\left(\frac{1}{k}\frac{n^k}{m}\right)^{1/(k-1)}\right\rfloor.$$

Proof.   Every hyperedge of H is contained in $\binom{n-k}{r-k}$ r-element subsets of $V(H)$. Therefore, the average (expected) number of hyperedges contained in a randomly selected r-element subset of $V(H)$ is

$$\frac{m\binom{n-k}{r-k}}{\binom{n}{r}} \le \dot{m}\left(\frac{r}{n}\right)^k.$$

Let $0 < c < 1$ be a constant to be specified later. Set

$$r = \left\lfloor \left(c\frac{n^k}{m}\right)^{1/(k-1)} \right\rfloor.$$

Then one can choose an r-element subset $A \subseteq V(H)$ containing at most $m(r/n)^k \le cr$ hyperedges of H. For each hyperedge e contained in A, delete one of the vertices of e from A. We are now left with an independent subset $A' \subseteq A$ of size at least $(1 - c)r$. Thus, we obtain

$$\alpha(H) \ge |A'| \ge (1 - c)\left\lfloor \left(c\frac{n^k}{m}\right)^{1/(k-1)} \right\rfloor.$$

Choosing $c = 1/k$, the assertion follows. $\qquad\square$

We could terminate the above argument after the first step by choosing r to be the largest integer for which $m(r/n)^k < 1$. However, this would yield only a somewhat weaker bound

$$\alpha(H) \ge r = \left\lceil \left(\frac{n^k}{m}\right)^{1/k} \right\rceil - 1.$$

The trick applied in the second step of the proof is usually called the *deletion method*.

By using a slightly different averaging argument, one can also generalize Corollary 9.7 in the case $r = s$.

Let $K_r^{(k)}$ denote a k-uniform hypergraph, whose vertex set can be partitioned into k parts $V(K_r^{(k)}) = V_1 \cup V_2 \cup \cdots \cup V_k$, $|V_i| = r$ $(1 \le i \le k)$, and $E(K_r^{(k)})$ consists of all k-tuples containing exactly one point from each V_i. In particular, $K_r^{(2)} = K_{r,r}$, the complete bipartite graph with r vertices in each of its classes.

Theorem 10.12 (Erdős, 1964c). *Let k, $r \ge 2$ be fixed. Then there exists an integer $n_0 = n_0(k, r)$ such that every k-uniform hypergraph with $n \ge n_0$ vertices and at least $n^{k-1/r^{k-1}}$ hyperedges (k-tuples) contains a subhypergraph isomorphic to $K_r^{(k)}$.*

Proof. By induction on k. For $k = 2$ the statement is true (cf. Corollary 9.7 and Exercise 9.17). Assume that we have already established the theorem for some k, and we want to prove it for $k + 1$.

Let H be a $(k + 1)$-uniform hypergraph on the vertex set $V(H) = P$, $|P| = n$, and assume that the number of hyperedges satisfies $|E(H)| \ge n^{k+1-1/r^k}$.

Given any $y \in P$ (any k-tuple $X \subseteq P$), let $d(y)$ (resp. $D(X)$) denote the number of hyperedges in H containing y (resp. X). For any r-tuple $\{y_1, \ldots, y_r\} \subseteq P$, let

$$\varphi(\{y_1, \ldots, y_r\}) = |\{X \subseteq P \mid |X| = k, X \cup \{y_i\} \in E(H) \text{ for all } i\}|.$$

Just like in the proof of Theorem 9.5, we will make use of the fact that

$$f(z) = \begin{cases} \dfrac{z(z-1)\cdots(z-r+1)}{r!} & \text{if } z \geq r-1, \\ 0 & \text{if } z \leq r-1 \end{cases}$$

is a convex function.

Obviously, the average value of φ,

$$\frac{\displaystyle\sum_{\{y_1, \ldots, y_r\} \subseteq P} \varphi(\{y_1, \ldots, y_r\})}{\dbinom{n}{r}} = \frac{\displaystyle\sum_{X \subseteq P, |X| = k} \dbinom{D(X)}{r}}{\dbinom{n}{r}}$$

$$\geq \frac{\dbinom{n}{k}}{\dbinom{n}{r}} \cdot f\left(\sum_{X \subseteq P, |X| = k} D(X) \Big/ \dbinom{n}{k} \right)$$

$$= \frac{\dbinom{n}{k}}{\dbinom{n}{r}} \cdot f\left((k+1)|E(H)| \Big/ \dbinom{n}{k} \right).$$

$$\geq \frac{(n-k+1)^k}{k! n^r} \left(\frac{(k+1)n^{k+1-1/r^k}}{\dbinom{n}{k}} - r + 1 \right)^r$$

$$\geq n^{k-1/r^{k-1}},$$

provided that n is sufficiently large.

Thus, we can pick $y_1, \ldots, y_r$ such that

$$\varphi(\{y_1, \ldots, y_r\}) \geq n^{k-1/r^{k-1}}.$$

This in turn implies, by the induction hypothesis, that there is a $K_r^{(k)}$ in the k-uniform hypergraph H' whose edge set

$$E(H') = \{X \subseteq P \mid |X| = k \text{ and } X \cup \{y_i\} \in E(H) \text{ for all } i\}.$$

Hence, $H \supseteq K_r^{(k+1)}$.   □

We will also need the following straightforward generalization of Lemma 9.6 (see Exercise 10.7).

Lemma 10.13 (Erdős and Kleitman, 1968). *The vertex set of any k-uniform hypergraph H can be decomposed into k parts so that the number of those edges of H, which contain exactly one point from each part, is at least $(k!/k^k)|E(H)|$.*

NEARLY EQUAL DISTANCES IN THE PLANE

The aim of this section is to generalize Theorem 10.5. For any real numbers $d_1,\ldots,d_k$, and for any set of n points $\{p_1,\ldots,p_n\} \subseteq \mathbb{R}^2$ with minimum distance 1, we want to maximize the number of pairs $p_i p_j$ $(i<j)$, whose distance is nearly equal to some d_l $(1 \leq l \leq k)$.

Theorem 10.14 (Erdős, Makai, and Pach, 1993). *For any natural number k and any Δ, $\varepsilon > 0$, there exists an integer $n_0 = n_0(k, \Delta, \varepsilon)$ satisfying the following condition. For any set of $n \geq n_0$ points $P = \{p_1,\ldots,p_n\} \subseteq \mathbb{R}^2$ with minimum distance at least 1 and for any reals $d_1,\ldots,d_k \geq 1$, the number of pairs $p_i p_j$ $(i<j)$ such that*

$$|p_i - p_j| \in \bigcup_{l=1}^{k} [d_l, d_l + \Delta]$$

is at most

$$\frac{n^2}{2}\left(1 - \frac{1}{k+1} + \varepsilon\right).$$

To see that this bound is asymptotically tight, let

$$P = \left\{(in^2, j) \,\middle|\, 0 \leq i \leq k,\ 1 \leq j \leq \frac{n}{k+1}\right\}.$$

Observe that the distance between any two points with different x-coordinates is nearly ln^2 for some $1 \leq l \leq k$. Hence, there are at least $(n^2/2)(1 - 1/(k+1))$ pairs of points whose distances belong to the union of the intervals $[d_l, d_l + 1]$, $1 \leq l \leq k$.

The proof proceeds by induction on k.

For $k = 1$ the assertion is true (Theorem 10.5). So we can assume that $k \geq 2$, and that we have already proved the theorem for $k - 1$.

Fix a set $P = \{p_1,\ldots,p_n\} \subseteq \mathbb{R}^2$ with minimum distance at least 1, and suppose that there are some $d_1,\ldots,d_k \geq 1$ such that the number of pairs $p_i p_j$ $(i<j)$ satisfying

$$|p_i - p_j| \in \bigcup_{l=1}^{k} [d_l, d_l + \Delta]$$

is larger than $(n^2/2)(1 - 1/(k+1) + \varepsilon)$. We are going to show that this leads to a contradiction, provided that $\Delta > 1$, $k, \varepsilon > 0$, and n is sufficiently large in terms of Δ, k, and ε.

Claim A. If n is sufficiently large, then

$$\min_{1 \leq l \leq k} d_l \geq \frac{1}{10^3 k^2 \Delta} n.$$

Proof. Assume, for example, that d_k is smaller than the above bound. For any p_i, the number of points p_j with $|p_i - p_j| \in [d_k, d_k + \Delta]$ is at most $100\Delta(d_k + \Delta)$. Hence, the number of pairs $p_i p_j$ $(i < j)$ satisfying

$$|p_i - p_j| \in \bigcup_{l=1}^{k-1} [d_l, d_l + \Delta]$$

is larger than

$$\frac{n^2}{2}\left(1 - \frac{1}{k+1} + \varepsilon\right) - \frac{n}{2} 100\Delta(d_k + \Delta) > \frac{n^2}{2}\left(1 - \frac{1}{k} + \varepsilon\right),$$

which contradicts the induction hypothesis. $\square$

The combinatorial core of the proof lies in the following observation.

Claim B. There is a function $m(n)$ tending to ∞ such that one can choose disjoint $m(n)$-element subsets $P_i \subseteq P$ $(1 \leq i \leq k+2)$ with the following property. For any $1 \leq i < j \leq k + 2$, there exists $1 \leq l(i,j) \leq k$ such that

$$|p_i - p_j| \in [d_{l(i,j)}, d_{l(i,j)} + \Delta] \qquad \text{for all } p_i \in P_i, p_j \in P_j.$$

Proof. Let G denote the graph with vertex set P, in which two vertices are joined by an edge if and only if their distance belongs to $\bigcup_{l=1}^{k} [d_l, d_l + \Delta]$. By Corollary 9.18, we find that G contains at least $\varepsilon' n^{k+2}$ complete subgraphs with $k + 2$ vertices, for some $\varepsilon' > 0$ (depending on ε and k, but not on n). By Lemma 10.13, $V(G) = P$ can be partitioned into $k + 2$ parts $P_1^0, \ldots, P_{k+2}^0$ so that the number of $(k+2)$-tuples $(p_1, \ldots, p_{k+2})$, $p_i \in P_i^0$ $(1 \leq i \leq k+2)$, which induce complete subgraphs of G, is at least

$$\frac{(k+2)!}{(k+2)^{k+2}} \varepsilon' n^{k+2} = \varepsilon'' n^{k+2}.$$

For any such $(k + 2)$-tuple $(p_1, \ldots, p_{k+2})$, one can choose some integers $1 \le l(i,j) \le k$, $1 \le i < j \le k + 2$ such that $|p_i - p_j| \in [d_{l(i,j)}, d_{l(i,j)} + \Delta]$. The triangular array

$$
\begin{pmatrix}
l(1,2) & l(1,3) & \cdots & l(1, k + 2) \\
0 & l(2,3) & \cdots & l(2, k + 2) \\
0 & 0 & \cdots & l(3, k + 2) \\
\vdots & \vdots & \ddots & \vdots \\
0 & 0 & \cdots & l(k + 1, k + 2)
\end{pmatrix}
$$

is called the *type* of $(p_1, \ldots, p_{k+2})$. Since the number of different types is at most $k^{\binom{k+2}{2}}$, there exist at least $\varepsilon'' n^{k+2}/k^{\binom{k+2}{2}} = \varepsilon^* n^{k+2}$ complete subgraphs in G, whose vertex sets are of the same type. Now the claim follows from Theorem 10.12. □

In what follows we analyze the relative position of the sets P_i described in Claim B. Consider two sets P_1 and P_2, say, and assume that all distances between them belong to the interval $[d_1, d_1 + \Delta]$, i.e., $l(1,2) = 1$. For any p, $p' \in P_1$, all elements of P_2 must lie in the intersection of two annuli centered at p and p'. If $|p - p'| < 2d_1$, then the area of this intersection set is at most

$$
\frac{20d_1^2\Delta^2}{|p - p'|\sqrt{(2d_1)^2 - |p - p'|^2}},
$$

and we have

$$
m(n) = |P_2| \le \frac{50d_1^2\Delta^2}{|p - p'|\sqrt{(2d_1)^2 - |p - p'|^2}}.
$$

This yields that $|p - p'|/d_1$ is close to either 0 or 2. More exactly,

$$
|p - p'| \in \left[1, \frac{50\Delta^2}{m(n)}d_1\right] \cup \left[\left(2 - \frac{50\Delta^2}{m(n)}\right)d_1, 2d_1 + 2\Delta\right]
$$

holds for any $p, p' \in P_1$, provided that n is sufficiently large.

Now pick any point $q \in P_2$. P_1 must be entirely contained in the annulus around q, whose inner and outer radii are d_1 and $d_1 + \Delta$, respectively. Thus, if there exist two points $p, p' \in P_1$ with

$$
|p - p'| \ge \left(2 - \frac{50\Delta^2}{m(n)}\right)d_1,
$$

then all other points of P_1 must lie in the union of two discs of radius $(50\Delta^2/m(n))d_1$ centered at p and p'. In any case, there is a subset $Q_1 \subseteq P_1$

of size at least $m(n)/2$, whose diameter

$$\operatorname{diam} Q_1 \leq \frac{50\Delta^2}{m(n)} d_1 = o(1)d_1. \tag{10.2}$$

(As usual, $o(1)$ denotes a function of n, approaching 0 as n tends to infinity.)
Repeating this argument, we obtain the following.

Claim C. Let $l(i,j) = l(j,i)$, $i \neq j$, denote the same as defined in Claim B.
Then there exist pairwise disjoint subsets $Q_i \subseteq P_i$ ($1 \leq i \leq k + 2$), each of size
at least $m(n)/2$, such that the following conditions are satisfied.

(i) For any $1 \leq i \leq k + 2$,

$$\operatorname{diam} Q_i \leq o(1) \min_{j \neq i} d_{l(i,j)}.$$

(ii) There is a line ℓ such that the angle between ℓ and any line

$$p_i p_j \ (p_i \in Q_i, p_j \in Q_j, \ i \neq j)$$

is $o(1)$.

Proof. We only have to prove part (ii). Fix two subsets Q_i and Q_j ($i \neq j$).
By (i),

$$\max (\operatorname{diam} Q_i, \operatorname{diam} Q_j) \leq o(1)d_{l(i,j)},$$

so the angle between any two lines $p_i p_j$ and

$$p_i' p_j' \ (p_i, p_i' \in Q_i, \ p_j, p_j' \in Q_j)$$

is $o(1)$.

Let q_i and q_i' be two elements of Q_i, whose distance is maximum. Using the
fact that any two points of Q_i are at least 1 apart, we have

$$\frac{\sqrt{m(n)}}{10} \leq |q_i - q_i'|.$$

It is sufficient to show that for any $p_j \in Q_j$, the lines $q_i q_i'$ and $q_i p_j$ are almost
perpendicular. Indeed, by the law of cosines,

$$|\cos (\angle q_i' q_i p_j)| = \left| \frac{(|q_i - p_j| - |q_i' - p_j|)(|q_i - p_j| + |q_i' - p_j|) + |q_i - q_i'|^2}{2|q_i - p_j||q_i - q_i'|} \right|$$

$$\leq \frac{\Delta 2(d_{l(i,j)} + \Delta)}{2d_{l(i,j)}(\sqrt{m(n)}/10)} + \frac{|q_i - q_i'|}{2d_{l(i,j)}} \leq o(1). \qquad \square$$

Claim D. Let $3 \le h \le k+2$ be fixed, and assume that

$$\mathrm{diam}\,(Q_1 \cup \cdots \cup Q_h) = |p_1 - p_2| \qquad \text{for some } p_1 \in Q_1,\, p_2 \in Q_2.$$

Then, for any $1 \le i \ne j \le h$, $l(i,j) = l(1,2)$ if and only if $\{i,j\} = \{1,2\}$.

Proof. Suppose, in order to obtain a contradiction, that for some $2 \le i \ne j \le h$ there are two points $p_i' \in Q_i$, $p_j' \in Q_j$ with

$$|p_i' - p_j'| \in [d_{l(1,2)}, d_{l(1,2)} + \Delta].$$

By Claims A and C, all points of $Q_2 \cup \cdots \cup Q_h$ lie in a small sector (of angle $o(1)$) of the annulus around p_1, whose inner and outer radii are $n/(2 \times 10^3 k^2 \Delta)$ and $|p_1 - p_2|$, respectively (see Figure 10.6). Obviously, the diameter of this sector is $|u - v|$, where u (resp. v) is the intersection of one (the other) boundary ray with the inner (resp. outer) circle of the annulus. But then we have

$$
\begin{aligned}
|p_1 - p_2| - |p_i' - p_j'| &\ge |p_1 - p_2| - |u - v| \\
&= |p_1 - v| - |u - v| \\
&= \frac{2|p_1 - u||p_1 - v|\cos(\angle u p_1 v) - |p_1 - u|^2}{|p_1 - v| + |u - v|} \\
&\ge |p_1 - u|\cos(\angle u p_1 v) - \frac{|p_1 - u|}{2} \\
&\ge \frac{n}{10^3 k^2 \Delta}\left(\frac{1}{2} - o(1)\right) > \Delta
\end{aligned}
$$

if n is large enough, a contradiction. $\qquad\square$

Now we can easily complete the proof of Theorem 10.14. Assume without loss of generality that the diameter of $Q = Q_1 \cup \cdots \cup Q_{k+2}$ is attained

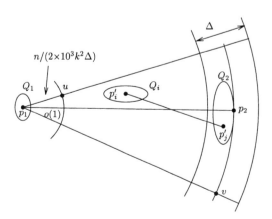

Figure 10.6. Illustration for Claim C ($j = 2$).

between a point of Q_1 and a point of Q_{j_1} for some $j_1 > 1$. By Claim D, no distance determined by the set $Q' = Q_2 \cup \cdots \cup Q_{k+2}$ belongs to the interval $[d_{l(1,j_1)}, d_{l(1,j_1)} + \Delta]$. Suppose that the diameter of Q' is attained between a point of Q_2 and a point of $Q_{j_2}, j_2 > 2$. Applying Claim D again, we obtain that none of the distances determined by the set $Q'' = Q_3 \cup \cdots \cup Q_{k+2}$ is in $[d_{l(2,j_2)}, d_{l(2,j_2)} + \Delta]$, where $l(2, j_2) \neq l(1, j_1)$. Proceeding like this, we can conclude that no distance determined by $Q_{k+1} \cup Q_{k+2}$ belongs to

$$\bigcup_{i=1}^{k} [d_{l(i,j_i)}, d_{l(i,j_i)} + \Delta],$$

where $\{l(i, j_i) | 1 \leq i \leq k\} = \{1, \ldots, k\}$. In other words, there exists no integer $l(k+1, k+2)$ satisfying the requirement in Claim B. This contradiction completes the proof of Theorem 10.14.

A closer look at the proof shows that Claim B holds with $m(n) = \lfloor b(\log n)^{1/(k+1)} \rfloor$ for a suitable constant $b > 0$ and that the whole argument goes through if $\Delta = \Delta(n)$ is a slowly growing function of n satisfying

$$\lim_{n \to \infty} \frac{\Delta(n)}{\sqrt{m(n)}} = 0.$$

[This condition is needed at (10.2).]

Notice in the above proof that we have not really used the fact that *all* distances between P_i and P_j ($i < j$) belong to the interval $[d_{l(i,j)}, d_{l(i,j)} + \Delta]$. It is sufficient to require that *many* distances have this property and there are much larger subsets $P_i \subseteq P$ ($1 \leq i \leq k + 2$) satisfying this weaker condition. In fact, the size of these sets can be chosen to be $m(n) = \lfloor b^* n \rfloor$, and Theorem 10.14 is true with any function

$$\Delta(n) \leq c \sqrt{n} \leq c^* \sqrt{m(n)}$$

for some $c, c^* > 0$. The proof of this stronger result, via Szemerédi's regularity lemma (Theorem 9.16) and Lemma 9.17, is left to the reader (Exercise 10.10).

DISTINCT DISTANCES DETERMINED BY SMALL SUBSETS OF A SET

Given a finite point set $P \subset \mathbb{R}^d$, let $\Delta(P)$ denote the number of distinct distances determined by P. The following problem was raised by Erdős and Purdy (1971): Determine the largest number $h_d(k) = h$ with the property that almost all k-element subsets of any n element set $P \subset \mathbb{R}^d$ determine at least h distinct distances, i.e.,

$$\min_{\substack{|P|\subset \mathbb{R}^d \\ |P|=n}} \frac{|\{P_k \subseteq P \mid |P_k| = k \text{ and } \Delta(P_k) \geq h\}|}{\binom{n}{k}} \longrightarrow 1,$$

for a fixed k, as n tends to infinity.

It is not difficult to see that $h_1(k) = h_2(k) = \binom{k}{2}$; that is, almost all k-tuples of a planar set are "in general position" (see Exercise 10.8).

Recall that $T_s(n)$ denotes the balanced complete s-partite graph with n vertices,

$$|E(T_s(n))| = \frac{1}{2}\left(1 - \frac{1}{s}\right)(n^2 - t^2) + \binom{t}{2},$$

where t is the remainder of n upon division by s (cf. Theorem 9.3 and Exercise 9.5). The vertices of $T_s(n)$ are divided into s almost equal classes whose sizes are $\lfloor n/s \rfloor$ or $\lceil n/s \rceil$.

Theorem 10.15 (Avis, Erdős, and Pach, 1991). *Let $n \to \infty$, $k, d \geq 3$ be fixed. Then the maximal number $h_d(k) = h$, with the property that almost all k-element subsets of any n-element set in $\mathbb{R}^d$ determine at least h distinct distances, can be expressed as*

$$h_d(k) = \binom{k}{2} - |E(T_s(k))| + \begin{cases} \lfloor k/2 \rfloor & \text{if } d = 3, \\ \lfloor k/s \rfloor + 1 & \text{if } d > 3 \text{ is odd}, \\ 1 & \text{if } d \text{ is even}, \end{cases}$$

where $s = \lceil d/2 \rceil$.

Proof. The upper bound on $h_d(k)$ can be established by a construction which coincides with Lenz's examples for even d.

If d is even, then, for any given large n, distribute n points as evenly as possible among $s = d/2$ mutually orthogonal circles with a common center and the same radii, so that each circle contains either $\lfloor n/s \rfloor$ or $\lceil n/s \rceil$ points. It is easy to see that a fixed positive proportion of all $\binom{n}{k}$ k-tuples of this set are also as evenly distributed among the s circles as possible. But every such k-tuple P_k determines

$$\Delta(P_k) \leq \binom{k}{2} - |E(T_s(k))| + 1$$

distinct distances.

If d is odd, then we distribute the n points as evenly as possible among $s - 1 = \lfloor d/2 \rfloor$ mutually orthogonal circles with the same center and radii, and the line orthogonal to these circles and passing through their common center

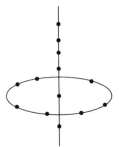

Figure 10.7. The lower bound of $h_3(k)$.

(see Figure 10.7). Now a positive proportion of all k-tuples P_k satisfy the following two conditions:

1. P_k is distributed as evenly as possible among the line and the $s-1$ circles.
2. The line contains $\lfloor k/s \rfloor$ points of P_k.

Obviously, for any such k-tuple P_k,

$$\Delta(P_k) \le \binom{k}{2} - |E(T_s(k))| + \lfloor k/s \rfloor + \begin{cases} 0 & \text{if } d = 3, \\ 1 & \text{if } d \ge 5. \end{cases}$$

We will establish the lower bound by contradiction. The proof is based on the same idea as in the preceding section. Let d, $k \ge 3$ be fixed. We can assume that $k \ge s$; otherwise, the statement is trivial. Suppose, in order to obtain a contradiction, that there exists a positive ε such that for infinitely many values of n, one can find an n-element point set $P \subset \mathbb{R}^d$ with the property that at least εn^k k-tuples of P determine at most h^* different distances, where h^* is smaller than the bound given in the theorem.

Let H denote the collection of these k-tuples. That is, H is k-uniform hypergraph, $|V(H)| = |P| = n$, $|E(H)| = \varepsilon n^k$. By Lemma 10.13, we can divide $V(H) = P$ into k disjoint parts $V_1, \ldots, V_k$ so that at least $(k!/k^k)|E(H)| \ge \varepsilon' n^k$ hyperedges intersect each V_i in exactly one point. Let H' denote the subhypergraph of H formed by these hyperedges.

Given two hyperedges $E = \{p_1, \ldots, p_k\}$, $E' = \{p'_1, \ldots, p'_k\} \in E(H')$, where $p_i, p'_i \in V_i$, we say that they have the same *type* if

$$|p_i - p_j| = |p_g - p_h| \Longleftrightarrow |p'_i - p'_j| = |p'_g - p'_h|$$

for all i, j, g, h. The number of different types is clearly at most $\binom{k}{2}^{\binom{k}{2}}$, hence there is a hypergraph $H^* \subseteq H'$ with

$$|E(H^*)| \geq \frac{\varepsilon'}{\binom{k}{2}^{\binom{k}{2}}} n^k = \varepsilon^* n^k,$$

such that any two hyperedges of H^* have the same type.

If n is sufficiently large, we can apply Theorem 10.12 to deduce that $H^* \supseteq K_3^{(k)}$. Let $T_1 \subseteq V_1, \ldots, T_k \subseteq V_k$ denote the vertex classes of $K_3^{(k)}$, $|T_i| = 3$ for all i. Obviously, $\Delta(\{p_1, \ldots, p_k\}) \leq h^*$ for any choice of $p_i \in T_i$. But in the next lemma we show that $\Delta(\{p_1, \ldots, p_k\}) > h^*$, a contradiction. Thus, the theorem is true. $\qquad \square$

Lemma 10.16. *Let $T_1, \ldots, T_k$ ($k \geq 2$) be triples of points in $\mathbb{R}^d$ such that $\{p_1, \ldots, p_k\}$ is of the same type for every choice of $p_i \in T_i$ ($i = 1, \ldots, k$). Then $\Delta(\{p_1, \ldots, p_k\})$ does not depend on the choice of $p_i \in T_i$. Letting $\Delta(T_1, \ldots, T_k) = \Delta(\{p_1, \ldots, p_k\})$, we have*

$$\Delta(T_1, \ldots, T_k) \geq \binom{k}{2} - |E(T_s(k))| + \begin{cases} \lfloor k/2 \rfloor & \text{if } d = 3, \\ \lfloor k/s \rfloor + 1 & \text{if } d > 3 \text{ is odd,} \\ 1 & \text{if } d \text{ is even,} \end{cases}$$

where $s = \lceil d/2 \rceil$.

Proof. We shall prove the lemma by induction on k. For $k \leq s - 1$, there is nothing to prove. Let $k \geq s$ and assume that the lemma is true for all $k' < k$.

Let us define an edge-colored graph G on the vertex set $V(G) = \{T_1, \ldots, T_k\}$ as follows. Choose a point $p_i \in T_i$, $1 \leq i \leq k$. Let $T_i T_j \in E(G)$ if and only if there exists $\{g, h\} \neq \{i, j\}$ such that

$$|p_i - p_j| = |p_g - p_h|. \tag{10.3}$$

(Note that in this case $T_g T_h$ is also an edge of G.) Furthermore, let two edges $T_i T_j, T_g T_h \in E(G)$ be colored with the same color if and only if (10.3) holds. According to our assumptions, the definition is independent of the choice of the points p_i representing the triples T_i. We have

$$\Delta(T_1, \ldots, T_k) = \binom{k}{2} - |E(G)| + (\text{number of edge colors in } G).$$

A vertex $T_i \in V(G)$ is called *linear* if its three points are on a line, and *circular* otherwise, i.e., when they determine a circle C_i.

Let $T_i T_j$ and $T_g T_h$ be two edges of G having the same color. It is easy to check (Exercise 10.6) that

(i) If i, j, g, h are all distinct, then T_i and T_j are circular and lie in orthogonal 2-planes.

(ii) If i, j, g are distinct and $h = j$, then T_i, T_g are circular, and the 2-plane or line spanned by T_j is orthogonal to the 2-planes spanned by T_i and T_g.

This implies that G satisfies the following important properties.

(1) $G \not\supseteq K_{s+1}$.
(2) If d is odd, then G does not have a subgraph isomorphic to K_s, all of whose vertices are circular.
(3) If T_i is linear, then either all edges of a given color are incident with T_i or none of them are.

Now we are in a position to prove the lemma.

If d is even, then by (1) we can use Turán's theorem (Theorem 9.3) to deduce that $|E(G)| \leq |E(T_s(k))|$. Hence,

$$\Delta(T_1, \ldots, T_k) = \binom{k}{2} - |E(G)| + (\text{number of edge colors in } G)$$

$$\geq \binom{k}{2} - |E(T_s(k))| + 1,$$

and the lemma follows.

Similarly, if $d \geq 5$ is odd and $G \not\supseteq K_s$, then

$$\Delta(T_1, \ldots, T_k) \geq \binom{k}{2} - |E(T_{s-1}(k))| + 1$$

$$\geq \binom{k}{2} - |E(T_s(k))| + \lfloor k/s \rfloor + 1,$$

and we are done (see Exercise 10.9).

Finally, suppose that d is odd and, say, $T_1, \ldots, T_s$ induce a complete subgraph in G. By (2), we can assume that T_1, say, is a linear vertex. Then, by (3), there is a color, say red, such that all red edges are incident with T_1. If $k = s$, then the lemma is true. If $k = s + 1$, then

$$\Delta(T_1, \ldots, T_{s+1}) \geq \Delta(T_2, \ldots, T_{s+1}) + 1,$$

and the lemma follows by the induction hypothesis.

Let $k \geq s + 2$. Since $G \not\supseteq K_{s+1}$, each of the $k - s \geq 2$ vertices of

$$G - \{T_1, \ldots, T_s\}$$

must be nonadjacent to at least one T_i, $1 \leq i \leq s$.

Let G' denote the edge-colored graph associated with the triple system $\{T_{s+1}, \ldots, T_k\}$, as described above. Denoting the complement of G by $\overline{G}$, we have

$$\binom{k}{2} - |E(G)| = |E(\overline{G})|$$

$$\geq |E(\overline{G'})| + k - s$$

$$= \binom{k-s}{2} - |E(G')| + k - s,$$

(number of edge colors in G) $\geq$ (number of edge colors in G') + 1.

Thus, using the inductive hypothesis for $\Delta(T_{s+1}, \ldots, T_k)$, we obtain

$$\Delta(T_1, \ldots, T_k) = \binom{k}{2} - |E(G)| + (\text{number of edge colors in } G)$$

$$\geq \binom{k-s}{2} - |E(G')| + k - s + (\text{number of edge colors in } G') + 1$$

$$= \Delta(T_{s+1}, \ldots, T_k) + k - s + 1$$

$$\geq \binom{k-s}{2} - |E(T_s(k-s))| + k - s + 1$$

$$+ \begin{cases} \lfloor (k-2)/2 \rfloor & \text{if } d = 3, \\ \lfloor (k-s)/s \rfloor + 1 & \text{if } d \geq 5 \text{ is odd,} \end{cases}$$

$$= \binom{k}{2} - |E(T_s(k))| + \begin{cases} \lfloor k/2 \rfloor & \text{if } d = 3, \\ \lfloor k/s \rfloor + 1 & \text{if } d \geq 5 \text{ is odd,} \end{cases}$$

as desired.

This completes the proof of the lemma, and hence the proof of the theorem.

$\square$

EXERCISES

10.1   (i) Show that one cannot choose more than $d + 1$ points in $\mathbb{R}^d$ such that any two of them are at distance 1.

    (ii) Let $d, k \geq 2$ be fixed. Can one choose arbitrarily many points in $\mathbb{R}^d$ such that they determine at most k distinct distances?

10.2   (Anning and Erdős, 1945; Erdős, 1945; Müller, 1953)

    (i) Let P be a set of infinitely many points in the plane such that all distinct distances determined by them are integers. Prove that all points of P are collinear.

(ii) Find a set of infinitely many points in the plane, not all on a line, such that all distances determined by them are rational numbers.

(iii) Show that for any $n \geq 3$, there exist n points in the plane, not all on a line, such that all distances determined by them are integers.

10.3 Prove that the unit distance cannot occur more than $n - 1$ times among n points on the line, and it occurs $n - 1$ times if and only if the points form an arithmetic progression of length n with difference 1.

10.4 (Erdős, Makai et al., 1991)   Given $t > 0$ and a set of n points $p_1, \ldots, p_n$ in $\mathbb{R}^d$ with minimum distance at least 1. Then the number of pairs $p_i p_j$, $i < j$, whose distance is between t and $t + 1$, cannot exceed $|E(T_d(n))|$, provided that n is sufficiently large. Furthermore, this bound can be attained for all n and $t > t_0(n)$. [$T_d(n)$ denotes the balanced complete d–partite graph on n vertices.]

10.5 Given a set of n lines $\ell_1, \ldots, \ell_n$ in the plane, the connected components of $\mathbb{R}^2 - \bigcup_{i=1}^{n} \ell_i$ are called cells. Prove that for any pair of cells c, c', there are at most four ℓ_i's such that both c and c' contain a portion of ℓ_i on their boundaries.

10.6 Show that if $\{p_1, p_2, p_3\}$, $\{q_1, q_2, q_3\}$ are two triples of points in $\mathbb{R}^d$ such that $|p_i - q_j| = 1$ for all $1 \leq i, j \leq 3$, then the p_i's induce a circle C and the q_j's induce a circle C' such that the affine two-planes containing C and C' are orthogonal.

10.7 (Erdős, 1964c)   Prove that the vertex set of any k-uniform hypergraph H can be decomposed into k parts so that the number of those k-tuples of H which contain exactly one point from each part is at least $(k!/k^k)|E(H)|$.

10.8 (i) Show that the number of isosceles triangles in a set of n points in the plane is $o(n^3)$.

(ii) Let k be fixed, $n \to \infty$. Show that almost all k-element subsets of any n-element point set in the plane are in general position, in the sense that all the $\binom{k}{2}$ distances determined by them are distinct.

10.9 Let $T_s(k)$ denote the balanced complete s partite graph with k vertices. Prove that

$$|E(T_s(k))| \geq |E(T_{s-1}(k))| + \lfloor k/s \rfloor$$

for any positive integers k, s.

10.10 (Erdős, Makai et al., 1993)   Prove that for any positive integer k and for any $\varepsilon > 0$, there exist $n_0 = n_0(k, \varepsilon)$, $c = c(k, \varepsilon) > 0$ with the following property. For any set of $n \geq n_0$ points in the plane and any $d_1, \ldots, d_k \geq 1$, the number of pairs of points whose distance is between d_l and $d_l+c\sqrt{n}$ for some l, is at most

$$\frac{n^2}{2}\left(1-\frac{1}{k+1}+\varepsilon\right).$$

10.11 Deduce the Erdős–Stone theorem (Theorem 9.10) from Corollary 9.18, Theorem 10.12, and Lemma 10.13.

11

Arrangement of Lines

A century ago, Sylvester (1893) posed the following problem in the *Educational Times:* "Prove that it is not possible to arrange any number of real points so that a right line through every two of them shall pass through a third, unless they all lie in the same right line." The problem remained unsolved until 1933, when it was raised independently by Erdős, and was solved shortly afterward by Gallai (see Erdős, 1943, 1982; Melchior, 1940; Coxeter, 1948, 1969; Dirac, 1951; Kelly and Moser, 1958; and Csima and Sawyer, 1993, 1995 for other solutions and generalizations). When more than 60 years ago Erdős and Gallai started asking questions that even Euclid would understand and appreciate, they initiated a revival of the most elementary concepts of geometry. Incidence relations between points and lines in the Euclidean and projective planes came to new life as objects of combinatorial study. Gallai's theorem has also contributed a great deal to the development of the theory of combinatorial designs, enriching combinatorics with a new algebraic flavor.

Notice that in the Fano plane **PG**(2,2), depicted in Figure 9.4, every line connecting two points passes through a third one. Hence, by Gallai's theorem, it is impossible to redraw this diagram by using only straight lines. Similarly, no finite projective plane **PG**$(2, p)$, $p > 2$, can be embedded in the Euclidean plane. [For the definition of **PG**(2,p), see, for example, the proof of Theorem 9.8.] There are many other striking differences between the combinatorial structures of the finite projective and the Euclidean plane. For instance, a finite projective plane of order p has $n = p^2 + p + 1$ points and n lines, each of them containing $p + 1 \geq \sqrt{n}$ points. On the other hand, it follows from a celebrated theorem of Szemerédi and Trotter (1983a) that given a set of n points in the Euclidean plane, the number of lines containing at least $\sqrt{n}$ points cannot exceed $c\sqrt{n}$, where c is a suitable constant. In fact, Szemerédi and Trotter (1983b) managed to prove a much stronger conjecture of Erdős, according to which the maximum number of incidences between m points and n lines in the plane is at most $c(m^{2/3}n^{2/3} + m + n)$, where c is some fixed constant. Moreover, this result is tight apart from the value of the constant.

In this chapter we shall present a fairly general and elegant approach to establish the above theorems and many related results, which has been developed by Clarkson, Edelsbrunner, Guibas, Sharir, and Welzl (1990), Clarkson (1987), and Clarkson and Shor (1989). Their idea is that by some simple probabilistic (or somewhat trickier constructive) arguments, one can obtain a cell decomposition of the plane satisfying some extremely pleasant technical conditions. Then one can apply the graph-theoretic tools discussed in previous chapters in each cell separately, to establish the desired results. The aim of this chapter is to demonstrate this method by presenting a simple proof of the Szemerédi–Trotter theorem (Corollary 11.8).

SUBDIVIDING AN ARRANGEMENT OF LINES

First we recall some definitions from previous chapters, and introduce a few new ones.

Let $L = \{\ell_1, \ell_2, \ldots, \ell_n\}$ be a set of n lines in the plane. The *arrangement* $\mathcal{A}(L)$, determined by them is defined as a planar cell complex whose *vertices* are the intersection points of the lines, whose *edges* (one-dimensional cells) are the maximal connected portions of lines containing no vertices, and whose *faces* (two-dimensional cells, or simply *cells*) are the connected components of $\mathbb{R}^2 - \bigcup_{i=1}^{n} \ell_i$ (see Figure 11.1). The number of edges bounding a face c in $\mathcal{A}(L)$ is often called the *complexity* of c.

Given a set $P = \{p_1, \ldots, p_m\}$ of m points, let $I(P, L)$ denote the number of pairs (i, j) such that $p_i \in \ell_j$. Let

$$I(m, n) = \max_{\substack{|P| = m \\ |L| = n}} I(P, L),$$

that is, the maximum number of incidences between m points and n lines in the plane. Similarly, if $C = \{c_1, \ldots, c_m\}$ is a collection of two-dimensional cells (faces) in $\mathcal{A}(L)$, then let $\mathcal{K}(C, L)$ denote the total number of edges bounding the members of C, and

$$\mathcal{K}(m, n) = \max_{\substack{|C| = m \\ |L| = n}} \mathcal{K}(C, L).$$

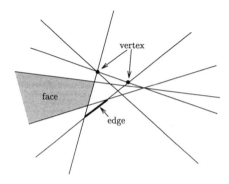

Figure 11.1. Arrangement of lines.

We have seen (Lemma 10.6) that

$$I(m, n) \leq \tfrac{1}{2}\mathcal{K}(m, 2n) + m.$$

Hence, to give a good upper bound on $I(m, n)$, it is sufficient to estimate $\mathcal{K}(m, n)$.

An arrangement of lines, $\mathcal{A}(L)$, is called *simple* if no two lines in L are parallel and no three lines are concurrent. A simple arrangement of n lines has $\binom{n}{2}$ vertices, n^2 edges, and $\binom{n}{2} + n + 1$ faces. Some of the edges and faces are unbounded. If we wish to bound $\mathcal{K}(m, n)$ from above, we can restrict our attention to simple arrangements. Therefore, in the rest of this chapter (unless stated otherwise) we assume that all arrangements $\mathcal{A}(L)$ are simple and no elements of L are horizontal or vertical. The next result is usually referred to in the literature as the *zone theorem* for lines.

Lemma 11.1 (Chazelle et al., 1985; Edelsbrunner, O'Rourke et al., 1986). *Given a family L of n lines in the plane and another line ℓ, the total number of edges of all faces in $\mathcal{A}(L)$ intersected by ℓ is at most $6n$.*

Proof. Assume without loss of generality that ℓ is horizontal and does not pass through any vertex of $\mathcal{A}(L)$.

First, we shall bound the total number of edges of the upper parts of all faces f intersected by ℓ, that is, those parts that lie above ℓ. The boundary of the upper part of such a face consists of two convex chains of edges (the *left* and *right* chains) and of a portion of ℓ. If the upper part of f is bounded, then the left and right chains meet at the topmost vertex of f. Otherwise, the last (topmost) edges of these chains are half-lines (see Figure 11.2). The edges belonging to the left (resp. right) chain, with the exception of the topmost edge, are called the *left* (resp. *right*) *edges* of f.

We claim that every line $\ell_i \in L$ contains at most one left edge. Suppose

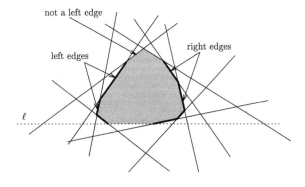

Figure 11.2. Left and right edges of a face.

to the contrary that ℓ_i has two portions e_1 and e_2 that are left edges of f_1 and f_2, respectively, where e_2 is above e_1. Then the line ℓ_j supporting the topmost edge of the left chain of f_1 would cross f_2, contradicting the fact that f_2 is a face in $\mathcal{A}(L)$. Hence, the total number of left (resp. right) edges of the upper parts of the faces intersected by ℓ is at most n. Taking into account the topmost edges of the chains, we obtain that the total number of edges of the upper parts of the faces intersected by ℓ is at most $4n$. The same argument can be repeated for the lower parts of these faces. Thus, the total number of edges in the faces intersected by ℓ is $8n - 2n = 6n$; the second term is due to the fact that the edges intersected by ℓ are counted twice.   □

Using a more careful counting argument, Bern et al. (1991) have shown that the bound in the above lemma can be improved to $5.5n$, and that this is tight up to an additive constant, as shown by Figure 11.3.

Next we show that given m points and n lines, the plane can be partitioned into a fixed number of trapezoids so that the average number of lines intersecting a trapezoid is small. This will be one of the crucial steps in the proof of Clarkson et al. (1990) for the Szemerédi–Trotter theorem (mentioned at the beginning of the chapter).

Theorem 11.2 (Clarkson et al., 1990).   *Given a set P of m points, a set L of n lines in general position, an integer $4 < r < n$, and a real number $0 < \alpha \leq 1$, the plane can be decomposed into $s \leq 3r^2$ (not necessarily bounded) trapezoids $\triangle_1, \triangle_2, \ldots, \triangle_s$ satisfying the following properties. If m_i and n_i denote the number of points in $P \cap \triangle_i$ and the number of lines in L intersecting the interior of $\triangle_i$, respectively, then*

(i) $\sum_{i=1}^{s} n_i \ \leq c_1 nr,$

(ii) $\sum_{i=1}^{s} m_i n_i^{\alpha} \leq c_2 m(n/r)^{\alpha},$

where c_1, c_2 are positive constants.

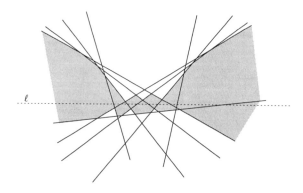

Figure 11.3. Lower bound construction for Lemma 11.1.

We present two different proofs of this theorem.

First proof (Clarkson et al., 1990).   Let $\mathcal{R}$ be any family of r lines in the plane in general position; i.e., no two lines of $\mathcal{R}$ are parallel, no three pass through the same point, and no two intersection points have the same x-coordinate. Subdivide the faces of the arrangement $\mathcal{A}(\mathcal{R})$ into trapezoids by drawing a vertical line in both directions through each vertex of $\mathcal{A}(\mathcal{R})$ until it hits another line of $\mathcal{R}$. The number of these trapezoids is

$$\binom{r}{2} + r + 1 + r(r-1) < 3r^2,$$

and some of them are degenerate; that is, some of the edges are single points or infinite rays. Denote this new (trapezoidal) planar map by $\mathcal{A}^*(\mathcal{R})$ (see Figure 11.4).

For each face $\triangle$ in $\mathcal{A}^*(\mathcal{R})$, let $r_\triangle \leq 4$ denote the size of the smallest subset $\mathcal{R}' \subseteq \mathcal{R}$ for which $\triangle$ is a face in $\mathcal{A}^*(\mathcal{R}')$. We shall say that $\triangle$ is *defined* by $r_\triangle$ lines.

From now on let $\mathcal{R}$ be a randomly selected r-element subset of L, where all the $\binom{n}{r}$ possible choices are equally likely. Let $\triangle_1, \triangle_2, \ldots, \triangle_s$ $(s < 3r^2)$ denote the faces of $\mathcal{A}^*(\mathcal{R})$, and let m_i and n_i $(1 \leq i \leq s)$ be defined as in the theorem (except that now all of them are random variables).

To prove (i), observe that every line $\ell \in L - \mathcal{R}$ intersects the interior of exactly $r + 1$ faces in $\mathcal{A}(\mathcal{R})$, and each face of $\mathcal{R}$ is subdivided into at most as many trapezoids as edges it has. Hence, we can apply Lemma 11.1 to conclude that every line $\ell \in L - \mathcal{R}$ intersects at most $6r$ trapezoids. This, in turn, implies that $\sum_{i=1}^{s} n_i \leq 6r(n-r)$.

To prove (ii), it is sufficient to show that the *expected value* of $\sum m_i n_i^\alpha$ over all possible choices of $\mathcal{R}$ is at most $c_2 m(n/r)^\alpha$. If q_j denotes the number of lines of L intersecting the interior of the (uniquely determined) trapezoid in $\mathcal{A}^*(\mathcal{R})$ which contains $p_j \in P$, then

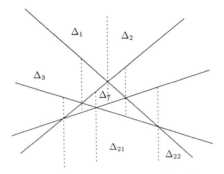

Figure 11.4. Subdividing an arrangement of lines into trapezoids.

$$\mathbf{E}\left[\sum_{i=1}^{s} m_i n_i^{\alpha}\right] = \mathbf{E}\left[\sum_{j=1}^{m} q_j^{\alpha}\right]$$

$$= \sum_{j=1}^{m} \mathbf{E}[q_j^{\alpha}]$$

$$\leq \sum_{j=1}^{m} (\mathbf{E}[q_j])^{\alpha}$$

(see Exercise 11.5). Thus, it suffices to show that $\mathbf{E}[q_j] \leq c_3 n/r$ for every j.

In the rest of the proof, let j be fixed and let $\triangle_{\mathcal{R}}$ denote the trapezoid in $\mathcal{A}^*(\mathcal{R})$ containing p_j.

Let $\mathcal{D}_L$ denote the family of all trapezoids containing p_j that can be defined by some subset $\mathcal{R}_\triangle \subset L$ with $|\mathcal{R}_\triangle| = r_\triangle \leq 4$. Given a trapezoid $\triangle \in \mathcal{D}_L$, let $n_\triangle$ denote the number of lines in L intersecting $\triangle$, and let $\mathrm{Pr}_\triangle$ be the probability that $\triangle_{\mathcal{R}} = \triangle$. Then

$$\mathbf{E}[q_j] = \sum_{\triangle \in \mathcal{D}_L} n_\triangle \mathrm{Pr}_\triangle.$$

Since any r-element subset $\mathcal{R} \subseteq L$ is chosen with the same probability, $\mathrm{Pr}_\triangle$ is equal to the number of subsets $\mathcal{R}$ for which $\triangle_{\mathcal{R}} = \triangle$, divided by $\binom{n}{r}$. On the other hand, evidently, $\triangle_{\mathcal{R}} = \triangle$ if and only if:

1. $\mathcal{R} \supseteq \mathcal{R}_\triangle$,
2. no line of $\mathcal{R}$ intersects the interior of $\triangle$.

Thus,

$$\mathrm{Pr}_\triangle = \binom{n - n_\triangle - r_\triangle}{r - r_\triangle} \Big/ \binom{n}{r}.$$

But

$$\binom{n - n_\triangle - r_\triangle}{r - r_\triangle} \leq \frac{n}{r - 4}\binom{n - n_\triangle - r_\triangle}{r - r_\triangle - 1};$$

therefore,

$$\mathbf{E}[q_j] \leq \frac{n}{r - 4} \sum_{\triangle \in \mathcal{D}_L} n_\triangle \binom{n - n_\triangle - r_\triangle}{r - r_\triangle - 1} \Big/ \binom{n}{r}.$$

However, the latter sum is bounded from above by 4 (see Exercise 11.6). Hence,

$$\mathbf{E}\left[\sum_{i=1}^{s} m_i n_i^{\alpha}\right] \le m \cdot \max_{1 \le j \le m} (\mathbf{E}[q_j])^{\alpha} \le m \left(\frac{4n}{r-4}\right)^{\alpha},$$

as desired.   □

To give a constructive proof of Theorem 11.2 (using no randomization), we have to analyze the finer structure of the system of edges of an arrangement of lines.

Definition 11.3. Let L be a set of n lines in the plane in general position, and let p be an arbitrary point. We say that $\ell \in L$ is above p if the open ray originating from p and pointing in the direction of the positive y-axis intersects ℓ. The *level* of p is the number of lines in L that lie above p, not including the line(s) passing through p itself. Clearly, the level of every point belonging to an (open) edge of the arrangement $\mathcal{A}(L)$ is the same. The closure of the union of all edges of $\mathcal{A}(L)$ whose points have level k is called the *k-level* of $\mathcal{A}(L)$ (see Figure 11.5). Thus, the k-level of $\mathcal{A}(L)$ is an x-monotone polygonal chain starting and ending with half-lines ($0 \le k < n$). We use the term *level* to mean a k-level for some $k \ge 0$. The number of edges in a level λ is denoted by $|\lambda|$, and is called the *size* of the level.

In the following two lemmata we describe two simple properties of the levels, which will facilitate a constructive proof of Theorem 11.2.

Lemma 11.4. *Let* $\Lambda_1, \Lambda_2, \ldots, \Lambda_u$ *be disjoint collections of levels in an arrangement of n lines. If each* Λ_i *contains at least* v *levels, then we can pick a level* $\lambda_i \in \Lambda_i$ $(1 \le i \le u)$ *such that*

$$\sum_{i=1}^{u} |\lambda_i| \le \frac{n^2}{v}.$$

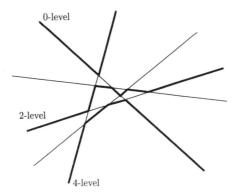

Figure 11.5. Levels in an arrangement of five lines.

Proof.   Let λ_i be the shortest level in Λ_i. Since no two levels share an edge,

$$\sum_{\lambda \in \Lambda_i} |\lambda| \geq v|\lambda_i|.$$

Since there are n^2 edges in an arrangement of n lines in general position,

$$n^2 \geq \sum_{i=1}^{u} \sum_{\lambda \in \Lambda_i} |\lambda| \geq v \sum_{i=1}^{u} |\lambda_i|. \qquad \square$$

In the next lemma (whose proof is left as an exercise) we show that if we "shortcut" a level, the resulting polygonal chain does not differ much from the original one.

Lemma 11.5 (Edelsbrunner and Welzl, 1986c). *Let $q_0, q_1, \ldots, q_t$ be the vertices of the k-level in $\mathcal{A}(L)$, in the order of their appearance, and let $d \geq 2$ be an integer. Let $\Pi(k, d)$ denote the polygonal path $q_0\, q_d\, q_{2d} \cdots q_{\lfloor t/d \rfloor d}\, q_t$ augmented with the initial and final rays of the k-level. Then*

   (i) *the interior of any edge of $\Pi(k, d)$ crosses at most $d - 2$ lines of L;*

  (ii) *$\Pi(k, d)$ lies between the $(k - \lfloor d/2 \rfloor)$ level and the $(k + \lfloor d/2 \rfloor + 1)$-level of $\mathcal{A}(L)$*

(see Exercise 11.7).

Second proof of Theorem 11.2.   We shall decompose the plane into $s \leq 3r^2$ trapezoids $\triangle_1, \triangle_2, \ldots, \triangle_s$ in three steps. Set $d = \lceil n/r \rceil$.

Step 1: Divide the levels of $\mathcal{A}(L)$ into groups of size d. That is, for every i ($1 \leq i \leq n/d$), let Λ_i denote the collection of all k-levels with $(i-1)d \leq k < id$. Assume that the k_i-level is the shortest member of Λ_i. By Lemma 11.4, the sum of the sizes of all k_i-levels ($1 \leq i \leq n/d$) is at most n^2/d.

Step 2: For every *even* i ($1 \leq i \leq n/d$), "shortcut" the k_i-level to obtain an unbounded polygonal path $\Pi(k_i, d)$, as described in Lemma 11.5. By Lemma 11.5(ii), the paths $\Pi(k_2, d), \Pi(k_4, d), \ldots$ are pairwise disjoint. Furthermore, the total number of vertices of $\Pi(k_i, d)$, for all even i, $1 \leq i \leq n/d$, is clearly at most $n^2/d^2 + n/(2d)$.

Step 3: Draw vertical segments (or rays) in both directions through every vertex of every $\Pi(k_i, d)$ until they hit $\Pi(k_{i-2}, d)$ or $\Pi(k_{i+2}, d)$, ($1 \leq i \leq n/d$, i is even). The resulting partition of the plane consists of at most

$$2\left(\frac{n^2}{d^2} + \frac{n}{2d}\right) + \frac{n}{2d} \leq 2r^2 + \frac{3}{2}r \leq 3r^2$$

(generalized) trapezoids $\triangle_1, \ldots, \triangle_s$.

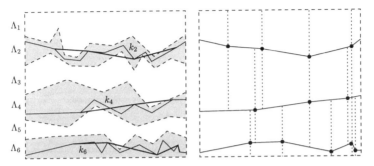

Figure 11.6. Constructing trapezoids with $r = 6, d = 3$; (a): steps 1, 2; (b) step 3.

By Lemma 11.5(i), the interior of any edge of $\Pi(k_i, d)$ intersects at most $d-2$ lines of $\mathcal{L}$. On the other hand, by the construction and by Lemma 11.5(ii), any vertical edge of a trapezoid can intersect at most

$$(3d - 1) + \left(\left\lfloor \frac{d}{2} \right\rfloor + 1\right) + 2 \leq \frac{7}{2}d + 2$$

lines. Thus, the interior of every trapezoid $\triangle_j$ is intersected by at most

$$2\left(\frac{7}{2}d + 2\right) + 2(d - 2) = 9d = 9\lceil n/r \rceil$$

lines of $\mathcal{L}$, and Theorem 11.2 follows. $\square$

As a matter of fact, we have obtained a stronger result.

Theorem 11.6 (Chazelle and Friedman, 1990; Matoušek, 1990).   *Given any set $\mathcal{L}$ of n lines in general position and any positive integer $r < n$, the plane can be decomposed into $s \leq 3r^2$ (not necessarily bounded) trapezoids $\triangle_1, \ldots, \triangle_s$ with the property that the interior of each trapezoid intersects at most cn/r lines of $\mathcal{L}$, where c is some suitable constant.*

COMPLEXITY OF A COLLECTION OF CELLS

Now we are in the position to prove the main theorem of this chapter.

Theorem 11.7 (Clarkson et al., 1990).   *Let $\mathcal{K}(m, n)$ denote the maximum total number of edges in m distinct faces of an arrangement of n lines in the plane. Then there exists a constant $c' > 0$ such that*

$$\mathcal{K}(m, n) \leq c'(m^{2/3}n^{2/3} + n).$$

Proof. Let $C = \{f^1, \ldots, f^m\}$ be a system of m distinct cells (faces) in a simple arrangement of n lines $\mathcal{A}(\mathcal{L})$ in the plane. For every $1 \le j \le m$, we pick a point $p^j \in f^j$ so that the vertical line through p^j does not contain any vertex of the arrangement $\mathcal{A}(\mathcal{L})$. Let $P = \{p^1, p^2, \ldots, p^m\}$.

Decompose the plane into $s \le 3r^2$ trapezoids $\triangle_1, \ldots, \triangle_s$ satisfying the properties in Theorem 11.2, where $r > 4$ is an integer to be specified later.

Let $\mathcal{L}_i \subseteq \mathcal{L}$ denote the collection of lines that intersect the interior of $\triangle_i$, and let $P_i = P \cap \triangle_i$. Set $n_i = |\mathcal{L}_i|$, $m_i = |P_i|$.

If $p^j \in \triangle_i$, then let f_i^j denote the face of $\mathcal{A}(\mathcal{L}_i)$ containing p^j. If f_i^j lies completely in the interior of $\triangle_i$, then $f_i^j = f^j$. Otherwise, the number of edges of f_i^j can be smaller than that of f^j, because f^j may have some edges lying outside $\triangle_i$. However, every edge of this type is composed of finitely many segments, such that each of them is an edge of some face of $\mathcal{A}(\mathcal{L}_k) \cap \triangle_k$ (for some $k \ne i$), incident with at least one side of $\triangle_k$. By Lemma 11.1, the total number of these edges is at most

$$\sum_{k=1}^{s} 4(6n_k) + \sum_{k=1}^{s} 2(n_k + 1),$$

where the second term is an upper estimate for the number of edges lying on some nonvertical side of a trapezoid $\triangle_k$. Theorem 11.2(i) yields that the above expression cannot exceed $c_1' rn$ for a suitable constant c_1'.

Thus, we have the following recurrence relation:

$$K(m, n) \le \sum_{i=1}^{s} K(m_i, n_i) + c_1' rn,$$

where $\sum_{i=1}^{s} m_i = m$ and conditions (i) and (ii) of Theorem 11.2 are also satisfied. We can apply Canham's theorem (Theorem 10.7) and Theorem 11.2 (with $\alpha = \frac{1}{2}$) to conclude that

$$K(m, n) \le c \sum_{i=1}^{s} (m_i \sqrt{n_i} + n_i) + c_1' rn$$

$$\le cc_2 m \sqrt{\frac{n}{r}} + (cc_1 + c_1')rn.$$

Choosing $r = \min \{\lceil m^{2/3}/n^{1/3} \rceil, n\}$, the result follows. $\square$

In view of Lemma 10.6, Theorem 11.7 immediately implies the Szemerédi–Trotter theorem mentioned at the beginning of this chapter.

Corollary 11.8 (Szemerédi and Trotter, 1983b). *Let $I(m, n)$ denote the maximum number of incidences between m points and n lines in the plane. Then there exists a constant $c > 0$ such that $I(m, n) \le c(m^{2/3}n^{2/3} + m + n)$.*

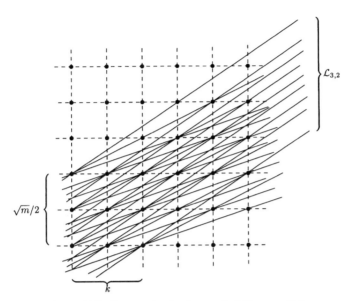

Figure 11.7. Construction for the proof of Theorem 11.9.

The tightness of this result is shown by the following construction of Erdős (see Edelsbrunner, 1987).

Theorem 11.9. *There exists a constant $c' > 0$ with the property that for any pair of positive integers m and n, one can find m points and n lines in the plane so that the number of incidences between them is at least $c'(m^{2/3}n^{2/3} + m + n)$.*

Proof. Trivial constructions show that $I(m, n) \geq \max\{m, n\}$. Thus, it suffices to show that $I(m, n) \geq c'm^{2/3}n^{2/3}$ for some $c' > 0$.

For the sake of simplicity, we shall ignore the "ceiling" and "floor" operators; that is, we assume that all numbers appearing in this construction are integers satisfying the necessary divisibility properties.

Let P be a $\sqrt{m} \times \sqrt{m}$ piece of the integer grid, i.e.,

$$P = \{(i,j) \mid 1 \leq i \leq \sqrt{m},\ 1 \leq j \leq \sqrt{m}\}.$$

Let $k = (cn/\sqrt{m})^{1/3}$, where c is a positive constant (to be specified later). For a fixed pair of relatively prime numbers (r, s) satisfying $1 \leq s < r \leq k$, let us define a collection $\mathcal{L}_{r,s}$ of $r\sqrt{m}/2$ lines, as follows:

$$\mathcal{L}_{r,s} = \left\{ y - j = \frac{s}{r}(x - i) \ \middle| \ 1 \leq j \leq \frac{\sqrt{m}}{2},\ 1 \leq i \leq r \right\}.$$

That is, $\mathcal{L}_{r,s}$ consists of all lines passing through (i,j) and $(i+r, j+s)$ for all $1 \le i \le r$, $1 \le j \le \sqrt{m}/2$. Let

$$\mathcal{L} = \bigcup_{\substack{\gcd(r,s)=1 \\ 1 \le s < r \le k}} \mathcal{L}_{r,s}.$$

The assumptions $\gcd(r,s)=1$ and $1 \le i \le r$ guarantee that all lines of $\mathcal{L}$ are distinct, i.e.,

$$|\mathcal{L}| = \left(\sum_{r=1}^{k} r\phi(r)\right)\frac{\sqrt{m}}{2} \le c^* k^3 \frac{\sqrt{m}}{2}$$

$$\le c^* \left(\frac{cn}{\sqrt{m}}\right)\frac{\sqrt{m}}{2} \le n$$

provided that $c \le 2/c^*$. Here ϕ is Euler's function, defined as

$$\phi(r) = |\{s \mid 1 \le s \le r \text{ and } \gcd(r,s)=1\}|,$$

and we have used the fact stated in Exercise 11.8(ii).

On the other hand, observe that the line $y - j = (s/r)(x-i)$ passes through one point in every rth column of the grid (starting from the ith column). Hence, the total number of incidences between the points of P and the lines of $\mathcal{L}$ is at least

$$\sum_{\substack{\gcd(r,s)=1 \\ 1 \le s < r \le k}} |\mathcal{L}_{r,s}|\frac{\sqrt{m}}{r} = \sum_{r=1}^{k} r\phi(r)\frac{\sqrt{m}}{2}\frac{\sqrt{m}}{r}$$

$$\ge \frac{m}{2}\tilde{c}k^2$$

$$= \frac{\tilde{c}c^{2/3}}{2}m^{2/3}n^{2/3},$$

as desired (see Exercise 11.8(i)).   □

It is not hard to modify the arguments presented in this chapter to obtain the following upper bounds for the number of incidences between m points and n circles.

Theorem 11.10 (Clarkson et al., 1990).   *There is a constant $c > 0$ such that:*

(i) *The maximum number of incidences between m points and n unit circles is at most $c(m^{2/3}n^{2/3} + m + n)$.*

(ii) *The maximum number of incidences between m points and n circles (not necessarily of the same radius) is at most $c(m^{3/5}n^{4/5} + m + n)$.*

Corollary 11.11 (Spencer, Szemerédi, and Trotter, 1984). *Let $f_2(n)$ be the maximum number of times the unit distance can occur among n points in the plane. Then*

$$f_2(n) \le cn^{4/3}$$

for a suitable constant $c > 0$.

Proof. Apply Theorem 11.10(i) with $m = n$ to any set P of n points in the plane, and to the family of all unit circles centered at points of P. Observe that any pair of points of P at distance 1 accounts for two incidences between the points and circles. □

This result is much better than the upper bound given in Theorem 10.2, but it is still far from the lower bound $n^{1+c/\log\log n}$, proved in Theorem 10.4, which is conjectured to be asymptotically tight.

EXERCISES

11.1 (Sylvester, 1893; Gallai, 1944) Let L be a set of n lines in the plane such that no two of them are parallel and not all of them pass through the same point. A point belonging to exactly two elements of L is called an *ordinary crossing*.

  (i) Show that the elements of L determine at least one ordinary crossing.

  (ii) Show that the number of intersection points of L is at least n.

11.2 Let L be a set of n lines in the plane such that no two of them are parallel and not all of them pass through the same point.

  (i) (Kelly and Moser, 1958) Show that the number of ordinary crossings is at least $3n/7$. (See Figure 11.8(i) for an arrangement of seven lines with three ordinary crossings.)

  (ii)⋆ (Csima and Sawyer, 1993, 1995) Show that for $n > 7$ the above bound can be improved to $6n/13$ (see Figure 11.8(ii)).

11.3 (Bern et al., 1991) Let $A(L)$ be an arrangement of n lines in the plane, and let C be a convex k-gon. Prove that the total number of sides of the polygons $C \cap f$ over all faces f of $A(L)$ intersecting the boundary of C, is at most $12n + k^2$.

11.4 Given an arrangement $A(L)$ of n lines, prove that

$$\sum_{f \in A(L)} |f|^2 \le cn^2,$$

where $|f|$ denotes the number of edges in the face f of $A(L)$, and $c > 0$ is a suitable constant.

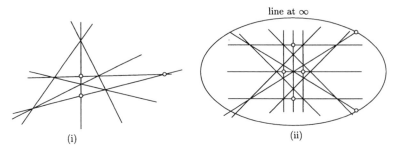

Figure 11.8. (i) An arrangement of seven lines with three ordinary crossings; (ii) an arrangement of thirteen lines with six ordinary crossings.

11.5 Let q be a random variable taking nonnegative values, and let $0 < \alpha < 1$. Show that the expected value of q^α,

$$\mathbf{E}[q^\alpha] \le (\mathbf{E}[q])^\alpha.$$

11.6 Prove that, using the notation in the first proof of Theorem 11.2,

$$\sum_{\triangle \in \mathcal{D}_L} n_\triangle \binom{n - n_\triangle - r_\triangle}{r - r_\triangle - 1} \bigg/ \binom{n}{r} \le 4.$$

11.7 (Edelsbrunner and Welzl, 1986c) Let $q_0, q_1, \ldots, q_t$ be the vertices of the k-level in $\mathcal{A}(L)$, and let $d \ge 2$ be a natural number. Let $\Pi(k, d)$ denote the infinite polygonal path obtained from the k-level by replacing the edges between $q_{(i-1)d}$ and q_{id} ($1 \le i \le \lfloor t/d \rfloor$), and the edge between $q_{\lfloor t/d \rfloor d}$ and q_t with straight-line segments. Show that

    (i) the interior of any edge of $\Pi(k, d)$ intersects at most $d - 2$ lines of L;

    (ii) $\Pi(k, d)$ lies between the $(k - \lfloor d/2 \rfloor)$-level and the $(k + \lfloor d/2 \rfloor + 1)$-level of $\mathcal{A}(L)$.

11.8 Let ϕ denote Euler's function, i.e.,

$$\phi(r) = |\{s \mid 1 \le s \le r \text{ and } \gcd(r, s) = 1\}|.$$

Prove that there are positive constants c^* and $\tilde{c}$ such that

    (i) $\displaystyle \tilde{c}k^2 \le \sum_{r=1}^{k} \phi(r) \le c^* k^2;$

    (ii) $\displaystyle \tilde{c}k^3 \le \sum_{r=1}^{k} r\phi(r) \le c^* k^3;$

    (iii) $\displaystyle \tilde{c} \log k \le \sum_{r=1}^{k} \frac{\phi(r)}{r^2} \le c^* \log k.$

11.9 Extend the first proof of Theorem 11.2 to nonsimple arrangements of lines, i.e., when some of the lines might be parallel and more than two lines can pass through the same point.

11.10 Let $L = \{\ell_1, \ldots, \ell_n\}$ be a set of n lines in $\mathbb{R}^2$, let $w_i \geq 0$ ($1 \leq i \leq n$) be a "weight" assigned to ℓ_i, and let $r \leq n$ be an integer. Show that one can partition the plane into $s \leq r^2$ trapezoids $\triangle_1, \ldots, \triangle_s$ so that for each $\triangle_i$, the total weight of lines intersecting its interior is at most $c \sum_{i=1}^{n} w_i/r$, where c is some suitable constant.

11.11 (Alon and Győri, 1986; Clarkson and Shor, 1989)   Let L be a set of n lines in the plane, and let $0 \leq k < n$ be an integer. Let $H_k(L)$ denote the number of vertices in the k-level of $\mathcal{A}(L)$, and put

$$H_{\leq k}(n) = \max_{|L|=n} \sum_{j=0}^{k} H_j(L).$$

Show that $H_{\leq k}(n) \leq cnk$, where c is a suitable constant.

11.12 (Aronov et al., 1993)   Let L be a collection of n lines in the plane in general position. For each triple of lines, consider the open circular disc inscribed in the triangle enclosed by them. Show that the number of such discs intersected by exactly k ($0 \leq k \leq n-3$) lines is $\binom{n-k-1}{2}$.

11.13 (Szemerédi and Trotter, 1983b)   Given a set of n points in the plane and an integer $k > 2$, prove that the number of distinct straight lines containing at least k points is at most $c(n^2/k^3 + n/k)$. Show that this result is tight, apart from the value of the constant c.

11.14⋆ (Szemerédi and Trotter, 1983b; Beck, 1983a)   Given a set P of n points in the plane, not all of them on a line, show that there is a point $p \in P$ that lies on at least cn distinct lines connecting p to other elements of P, where $c > 0$ is some suitable constant.

11.15 (Bárány et al., 1993)   Given an arrangement of n hyperplanes in $\mathbb{R}^d$, we call a cell determined by them *rich* if it has n facets (i.e., it contains a portion of each hyperplane).

  (i) Prove that any arrangement of $n \geq 5$ lines in the plane has at most one rich cell.

  (ii) Prove that, for $d \geq 3$, the number of rich cells is at most

$$\binom{n}{d-2} + Cn^{d-3},$$

where $C = C(d)$ is some constant depending on d. Show that this bound is asymptotically tight for large n.

12

Applications of the Bounds on Incidences

Erdős's questions about the distribution of distances (quoted in the introduction to Chapter 10) can be generalized in several directions. For instance, given an n-element point set P, we can try to find an upper bound on the number of k-tuples in P satisfying some special conditions ($k \geq 2$). In particular, for $k = 2$, we can ask for the maximum number of pairs $p_1, p_2 \in P$ such that the line $p_1 p_2$ bisects $P \subseteq \mathbb{R}^2$, $|P| = n$, into two (nearly) equal halves (see Lovász, 1971; Erdős, Lovász et al., 1973; Pach et al., 1992; Bárány et al., 1990; Aronov, Chazelle et al., 1991; Dey and Edelsbrunner, 1993; etc.). For $k = 3$, we may wish to bound the number of triangles $p_1 p_2 p_3$ ($p_1, p_2, p_3 \in P$) of the same area, of the same perimeter, or containing a given number of other elements of P in its interior. A number of such problems were raised by Erdős and Purdy (1971, 1975, 1976, 1989). In this chapter we settle some of these questions using the Szemerédi–Trotter theorem (Corollary 11.8). For many other applications, we need to utilize and, in certain cases, to generalize the powerful proof technique of Clarkson et al. (1990) presented in Chapter 11. In particular, we provide a partial answer to the second problem of Erdős mentioned at the beginning of Chapter 10: What is the minimal number of distinct distances determined by n points in space?

REPEATED ANGLES IN THE PLANE

The following theorem is an easy consequence of Corollary 11.8.

Theorem 12.1 (Erdős and Purdy, 1971). *There exists a constant $c > 0$ such that given any set P of n points in the plane, the number of triples spanning a triangle of unit area is at most $cn^{7/3}$.*

Proof. Fix a point $p \in S$. For every point $q \in P - \{p\}$, let ℓ_q and ℓ'_q denote the lines parallel to pq and at distance $2/|p - q|$ from it. Clearly, pqr defines a triangle of unit area if and only if $r \in \ell_q \cup \ell'_q$. Thus, the number of unit area triangles spanned by p and two other points of P is one half of the number of incidences between the $2(n-1)$ lines ℓ_q, ℓ'_q ($q \in P - \{p\}$) and the elements of $S - \{p\}$. However, by Corollary 11.8, this number is at most $cn^{4/3}$. Repeating the same argument for every $p \in S$, the result follows. $\square$

The (very similar) proof of the following result is left as an exercise (see Exercise 12.2).

Theorem 12.2. *There exists a constant $c > 0$ such that given any set P of n points in the plane, the number of triples spanning an isosceles triangle is at most $cn^{7/3}$.*

By the same argument we can show that the maximum number of triples that determine the same angle α is $cn^{7/3}$. However, this result can be improved, as follows.

Theorem 12.3 (Pach and Sharir, 1990). *Given any $0 < \alpha < \pi$, there exists a constant $c > 0$ such that the number of ordered triples of n points in the plane which determine the same angle α is at most $cn^2 \log n$. Furthermore, apart from the value of c, this bound is best possible for infinitely many α.*

Proof. We prove the upper bound only in the special case $\alpha = \pi/2$ so that we can use a rectilinear system of coordinates. The general case can be settled similarly.

Let P be a set of n points in the plane. First we shall bound the number of right-angle triangles spanned by three points of P and having two sides parallel to the axes. For any $p \in P$, let $h(p)$ and $v(p)$ denote the horizontal and vertical lines through p, respectively. Let $h_1, h_2, \ldots, h_k$ and $v_1, v_2, \ldots, v_l$ be the horizontal and vertical lines containing at least one element of P, and let

$$a_i = |h_i \cap P|, \quad b_j = |v_j \cap P|.$$

Clearly, there are $(|h(p) \cap P| - 1)(|v(p) \cap P| - 1)$ triangles in axis-parallel position, whose right angle is at p. Therefore, the total number of right-angle triangles of the desired form is at most

$$\sum_{h_i \cap v_j \in P} a_i b_j,$$

and we can ignore all lines h_i and v_j that contain just one point of P. We split this sum into three parts, as follows:

$$\underbrace{\sum_{\substack{h_i \cap v_j \in P \\ a_i > \sqrt{n}}} a_i b_j}_{\sigma_1} + \underbrace{\sum_{\substack{h_i \cap v_j \in P \\ a_i \leq \sqrt{n} < b_j}} a_i b_j}_{\sigma_2} + \underbrace{\sum_{\substack{h_i \cap v_j \in P \\ a_i, b_j \leq \sqrt{n}}} a_i b_j}_{\sigma_3}.$$

Then

$$\sigma_1 = \sum_{a_i > \sqrt{n}} a_i \left(\sum_{h_i \cap v_j \in P} b_j \right) \le n \sum_{a_i > \sqrt{n}} a_i,$$

$$\sigma_2 \le \sum_{b_j > \sqrt{n}} b_j \left(\sum_{h_i \cap v_j \in P} a_i \right) \le n \sum_{b_j > \sqrt{n}} b_j,$$

$$\sigma_3 \le \left(\sum_{\substack{h_i \cap v_j \in P \\ a_i \le \sqrt{n}}} a_i^2 \right)^{1/2} \left(\sum_{\substack{h_i \cap v_j \in P \\ b_j \le \sqrt{n}}} b_j^2 \right)^{1/2}$$

$$= \left(\sum_{a_i \le \sqrt{n}} a_i^3 \right)^{1/2} \left(\sum_{b_j \le \sqrt{n}} b_j^3 \right)^{1/2},$$

where the last inequality follows by the Cauchy–Schwarz inequality.

Let us turn our system of coordinates with angle φ around the origin so that there is at least one axis-parallel right-angle triangle in this position. Repeating the above analysis for all such $0 \le \varphi < \pi$, we obtain that the total number N of right-angle triangles spanned by P satisfies

$$N \le \sum_{\varphi} \left[\left(\sum_{a_i^\varphi \le \sqrt{n}} (a_i^\varphi)^3 \right)^{1/2} \left(\sum_{b_j^\varphi \le \sqrt{n}} (b_j^\varphi)^3 \right)^{1/2} + n \left(\sum_{a_i^\varphi > \sqrt{n}} a_i^\varphi + \sum_{b_j^\varphi > \sqrt{n}} b_j^\varphi \right) \right],$$

where a_i^φ (resp. b_j^φ) is the number of points of P on the ith (resp. jth) lines of orientation φ (resp. $\varphi + \pi/2$) that passes through at least two elements of P.

Let $\mathcal{L}_k, \mathcal{L}_{\ge k},$ and $\mathcal{L}_{\le k}$ denote the sets of all lines passing through k, at least k, and at most k (but at least two) elements of P, respectively.

Again using the Cauchy–Schwarz inequality,

$$N \le \left(\sum_{\varphi} \sum_{a_i^\varphi \le \sqrt{n}} (a_i^\varphi)^3 \right)^{1/2} \left(\sum_{\varphi} \sum_{b_j^\varphi \le \sqrt{n}} (b_j^\varphi)^3 \right)^{1/2} + n \left(\sum_{\varphi} \sum_{a_i^\varphi > \sqrt{n}} a_i^\varphi + \sum_{\varphi} \sum_{b_j^\varphi > \sqrt{n}} b_j^\varphi \right)$$

$$\le \sum_{\ell \in \mathcal{L}_{\le \sqrt{n}}} |\ell \cap P|^3 + 2n \sum_{\ell \in \mathcal{L}_{\ge \sqrt{n}}} |\ell \cap P|.$$

Let $c_k = |\mathcal{L}_k|, C_k = |\mathcal{L}_{\ge k}|$. Then $c_k = C_k - C_{k+1}$. Thus,

$$N \le \sum_{2 \le k \le \sqrt{n}} k^3 c_k + 2n \sum_{k \ge \sqrt{n}} k c_k$$

$$= \sum_{2 \le k \le \sqrt{n}} k^3 (C_k - C_{k+1}) + 2n \sum_{k \ge \sqrt{n}} k(C_k - C_{k+1})$$

$$\le \sum_{2 \le k \le \sqrt{n}} 3k^2 C_k + 2n \left(\lceil \sqrt{n} \rceil C_{\lceil \sqrt{n} \rceil} + \sum_{k > \sqrt{n}} C_k \right).$$

By Exercise 11.13 (which is an equivalent form of the Szemerédi–Trotter theorem, Corollary 11.8),

$$C_k \le c \cdot \max \left\{ \frac{n^2}{k^3}, \frac{n}{k} \right\}$$

for a suitable constant c. Hence,

$$N \le \sum_{2 \le k \le \sqrt{n}} 3k^2 \frac{cn^2}{k^3} + 2n \left(\lceil \sqrt{n} \rceil \frac{cn}{\lceil \sqrt{n} \rceil} + \sum_{k > \sqrt{n}} \frac{cn}{k} \right)$$

$$\le 3cn^2 \sum_{2 \le k \le \sqrt{n}} \frac{1}{k} + 2cn^2 + 2cn^2 \sum_{k > \sqrt{n}} \frac{1}{k}$$

$$\le 3cn^2 \log n,$$

as asserted.

Next we establish the lower bound. Let $0 < \alpha < \pi$ have the property that $\tan \alpha = a\sqrt{m}/b$ for some integers a, b, m, where $b > 0$ and $m > 0$ is not a square (but $m = 1$ is permitted). For any sufficiently large n, we shall construct an n-element point set P such that at least $c_\alpha n^2 \log n$ triples determine angle α ($c_\alpha > 0$).

Put $d_\alpha = 2 \max \{|a|m, b\}$, and assume without loss of generality that $n = (4k + 1)^2$ for some positive integer k. We define P to be the set of all "lattice points" of the form $(i, j\sqrt{m})$, where $-2k \le i, j \le 2k$.

To obtain a lower bound on the number of all ordered triples of P that determine the angle α, let p and q be relatively prime integers, $0 < p \le q \le k$, and consider the ray ρ from the origin $\mathbf{0} = (0, 0)$ to $(p, q\sqrt{m})$. Obviously, ρ contains at least

$$\left\lfloor \frac{k}{\max \{p, q\}} \right\rfloor = \left\lfloor \frac{k}{q} \right\rfloor$$

points of P distinct from the origin. Let ρ' denote a ray from the origin $\mathbf{0}$ to another point $(p', q'\sqrt{m})$, where p' and q' are relatively prime and $-k \le p', q' \le k$. Clearly, ρ' can be obtained from ρ by a counterclockwise rotation with α if and only if

$$\frac{\dfrac{q'\sqrt{m}}{p'} - \dfrac{q\sqrt{m}}{p}}{1 + \dfrac{q'\sqrt{m}}{p'}\dfrac{q\sqrt{m}}{p}} = \tan \alpha = \frac{a\sqrt{m}}{b},$$

i.e., if

$$\frac{p'}{q'} = \frac{pb - qam}{qb + pa}.$$

If this holds, then

1. there are at least

$$\left\lfloor \frac{k}{\max\{pb - qam, qb + pa\}} \right\rfloor \geq \left\lfloor \frac{k}{2\max\{p,q\}\max\{|a|m,b\}} \right\rfloor$$

$$= \left\lfloor \frac{k}{qd_\alpha} \right\rfloor$$

points of P on ρ';

2. the angle $x0x'$ is equal to α for all

$$x \in (\rho \cap P) - \{0\}, \quad x' \in (\rho' \cap P) - \{0\}.$$

Exactly the same argument can be repeated for the rays emanating from any "lattice point" $(i, j\sqrt{m})$, $-k \leq i, j \leq k$. Hence, the number of triples determining the angle α is at least

$$(2k + 1)^2 \sum_{q \leq k} \lfloor k/q \rfloor \left\lfloor \frac{k}{qd_\alpha} \right\rfloor \phi(q) \geq \frac{n^2}{100d_\alpha} \sum_{q \leq k/d_\alpha} \frac{\phi(q)}{q^2},$$

where $\phi(q)$ denotes the Euler function. Thus, we can use Exercise 11.8(iii) to conclude that at least $(\tilde{c}/200d_\alpha)n^2 \log n$ triples determine the angle α. $\qquad \Box$

SUBSETS WITH NO REPEATED DISTANCES

Erdős (1950) noticed that one can select infinitely many elements from any infinite set of points in the plane such that all mutual distances between them are distinct. Avis et al. (1991) established an analogous result for finite sets, which has been slightly improved by Thiele (1993).

Theorem 12.4 (Thiele).   *There exists a constant $c > 0$ with the property that any set of n points in the plane contains at least $cn^{2/9}$ elements such that all distances determined by them are distinct.*

Proof.   Let P be a fixed set of n points in the plane. A four-element subset $Q \subseteq P$ is said to be *regular* if all six distances determined by Q are distinct. Otherwise, Q is called *singular*.

There are two different types of singular quadruples. We say that Q is of type 1 if it contains three points forming an isosceles triangle, and Q is of type 2 if it consists of two disjoint pairs $\{q_1, q_2\}$ and $\{q_3, q_4\}$ with $|q_1 - q_2| = |q_3 - q_4|$. Note that some singular quadruples may have both types at the same time.

Let H be a four-uniform hypergraph on the vertex set $V(H) = P$, whose edges are the singular quadruples of P. That is, $E(H) = E_1 \cup E_2$, where E_i denotes the set of all singular quadruples of type i $(i = 1, 2)$. By Theorem 12.2,

$$|E_1| \le c_1 n^{10/3},$$

for a suitable constant $c_1 > 0$.

Let $\{d_1, \ldots, d_r\}$ be the set of all distinct distances determined by P, and let m_i denote the multiplicity of d_i, i.e., the number of pairs $\{p, q\} \subseteq P$ with $|p - q| = d_i$. Obviously,

$$\sum_{i=1}^{r} m_i = \binom{n}{2},$$

and Corollary 11.11 yields that $m_i \le c_2 n^{4/3}$ for every i. Thus,

$$
\begin{aligned}
|E_2| &\le \sum_{i=1}^{r} \binom{m_i}{2} \\
&\le \frac{\binom{n}{2}}{c_2 n^{4/3}} \binom{c_2 n^{4/3}}{2} \\
&\le c_3 n^{10/3}
\end{aligned}
$$

for some $c_3 > 0$.

Recall that the *independence number* $\alpha(H)$ of H is defined as the maximum size of a subset of $V(H)$ containing no element of $E(H)$. Theorem 10.11 implies that for $E(H) \ge n$,

$$
\begin{aligned}
\alpha(H) &\ge c_4 \left(\frac{n^4}{|E(H)|} \right)^{1/3} \ge c_4 \left(\frac{n^4}{|E_1| + |E_2|} \right)^{1/3} \\
&\ge c_4 \left(\frac{n^4}{(c_1 + c_3) n^{10/3}} \right)^{1/3} \ge c n^{2/9}.
\end{aligned}
$$

However, this means that P has a subset of size at least $c n^{2/9}$, all of whose quadruples are regular. $\square$

According to a recent result of Lefmann and Thiele (1995), the bound in Theorem 12.4 can be improved to $cn^{1/4}$. However, it seems quite likely that this estimate can be further improved to about $n^{1/3}$ (see Erdős and Guy, 1970).

The above argument can easily be adapted to obtain the following related result, whose proof is left to the reader (see Exercise 12.3).

Theorem 12.5 (Zhang, 1993).   *There exists a constant $c > 0$ such that for every positive integer n, one can select $m = \lfloor c(n/\log n)^{1/3} \rfloor$ points from the $\lfloor \sqrt{n} \rfloor \times \lfloor \sqrt{n} \rfloor$ grid*

$$S_n = \{(i,j) \mid 0 \le i,j < \lfloor \sqrt{n} \rfloor \text{ are integers}\},$$

with the property that all the $\binom{m}{2}$ lines determined by them have different slopes.

FAMILIES OF CURVES WITH BOUNDED DEGREES OF FREEDOM

Next we state a common generalization of the Szemerédi–Trotter theorem (Corollary 11.8), and the theorem of Clarkson et al. (Theorem 11.10), which has a number of interesting consequences. We shall only sketch the steps of the proof, because it follows by a fairly straightforward extension of the techniques developed in Chapter 11.

Let Γ be a class of curves in the xy-plane defined in terms of d real parameters. Suppose that there exists an integer s such that

(a) the dependence of the curves on x, y, and the parameters is algebraic of degree at most s;

(b) no two distinct curves of Γ intersect in more than s points;

(c) for any d points of the plane, there are at most s curves in Γ passing through all of them.

Then Γ is called a *regular class of curves with d degrees of freedom*. For example, the class of all circles (resp. unit circles) has three (resp. two) degrees of freedom.

Theorem 12.6 (Pach and Sharir).   *Let Γ be a regular class of curves with d degrees of freedom. Then there exists a constant $c = c_\Gamma$ such that the total number of incidences between m points and n curves in Γ is at most $c(m^{d/(2d-1)} n^{(2d-2)/(2d-1)} + m + n)$.*

Proof (Sketch).   Let $I_\Gamma(m,n)$ denote the maximum number of incidences between m points and n curves in Γ.

Step 1: By Theorem 9.5, we immediately obtain a weaker bound

$$I_\Gamma(m, n) \le c'(mn^{1-1/d} + n).$$

Step 2: Let P and $\Gamma' \subseteq \Gamma$ denote a fixed set of m points and a fixed collection of n curves, respectively. Choose a random r-element subset $\mathcal{R} \subseteq \Gamma'$. A point p is called a *critical point* of a curve if the curve either intersects itself or has a vertical tangent line at p. Subdivide the cells of the arrangement $\mathcal{A}(\mathcal{R})$ into trapezoid-like regions $\triangle_1, \ldots, \triangle_s$ ($s \le c''r^2$) by drawing a vertical line in both directions through every intersection point of a pair of curves and through every critical point of a curve in $\mathcal{R}$ until it hits another piece of some other element of $\mathcal{R}$. Let m_i and n_i denote the number of points in the interior (or on the left vertical side) of $\triangle_i$ and the number of curves in Γ' intersecting the interior of $\triangle_i$, respectively. As in the proof of Theorem 11.2, one can show that

$$\mathbf{E}\left[\sum_{i=1}^{s} n_i\right] < \tilde{c}nr,$$

$$\mathbf{E}\left[\sum_{i=1}^{s} m_i n_i^{1-1/d}\right] < \tilde{c}m(n/r)^{1-1/d}.$$

Hence, for some choice of $\mathcal{R}$,

$$\sum_{i=1}^{s} n_i \le 2\tilde{c}nr,$$

$$\sum_{i=1}^{s} m_i n_i^{1-1/d} \le 2\tilde{c}m(n/r)^{1-1/d}.$$

Step 3: Observe that the maximum number of incidences involving points sitting on the "upper" or "lower" boundary of some "trapezoid" is at most $\bar{c}nr + m$ (because every curve of $\Gamma' - \mathcal{R}$ has only a constant number of intersections with any element of $\mathcal{R}$). Thus,

$$I_\Gamma(m, n) \le \sum_{i=1}^{s} I_\Gamma(m_i, n_i) + \bar{c}nr + m$$

$$\le c' \sum_{i=1}^{s} (m_i n_i^{1-1/d} + n_i) + \bar{c}nr + m$$

$$\le 2c'\tilde{c}(m(n/r)^{1-1/d} + nr) + \bar{c}nr + m.$$

If $(c'm/c)^d \le n$, the theorem follows from the bound obtained in Step 1. Otherwise, choose $r = \min\left\{\lceil (m^d/n)^{1/(2d-1)}\rceil, n\right\} \ge 10$ to complete the proof. $\quad\square$

We mention only one immediate consequence of Theorem 12.6 (see also Exercise 12.4).

Corollary 12.7 (Pach and Sharir, 1990). *There is $c > 0$ such that any set of n points in the plane has at most $cn^{7/3}$ triples that determine a triangle of unit perimeter.*

REPEATED DISTANCES ON A SPHERE

Given a set P of n points on the unit sphere $S^2 \subset \mathbb{R}^3$ and a real number $0 < \alpha < 2$, let $h(P, \alpha)$ denote the number of pairs $p, q \in P$ at distance $|p - q| = \alpha$. Let

$$h(n, \alpha) = \max_{\substack{P \subset S^2 \\ |P| = n}} h(P, \alpha).$$

Corollary 12.8 (Erdős, Hickerson, and Pach, 1989; Clarkson et al., 1990). *There is a constant $c > 0$ such that*

$$h(n, \alpha) < cn^{4/3}.$$

Furthermore, this bound is asymptotically tight for $\alpha = \sqrt{2}$.

Proof. Let us draw a circle of radius α around every point of the sphere. After using stereographic projection, the images of these circles will form a regular class of curves (circles) in the plane with 2 degrees of freedom, and the first assertion follows from Theorem 12.6.

To show that $h(n, \sqrt{2}) > c'n^{4/3}$ for some $c' > 0$, recall Theorem 11.9. It implies that one can choose a set P of $\lfloor n/2 \rfloor$ points and a collection L of $\lceil n/2 \rceil$ lines in the plane $z = -1$ so that the number of incidences between them is at least $c'n^{4/3}$. For any point $p \in P$, let $\vec{u}_p$ be the unit vector pointing from the origin $\mathbf{0} = (0, 0, 0)$ in the direction of p. For any $\ell \in L$, let $\vec{v}_\ell$ be a unit vector from $\mathbf{0}$ perpendicular to the plane passing through ℓ and $\mathbf{0}$. Let u_p and v_ℓ denote the endpoints of $\vec{u}_p$ and $\vec{v}_\ell$, respectively. Evidently,

$$|u_p - v_\ell| = \sqrt{2} \Longleftrightarrow \vec{u}_p \perp \vec{v}_\ell \Longleftrightarrow p \in \ell.$$

Hence, $P^* = \{u_p \,|\, p \in P\} \cup \{v_\ell \,|\, \ell \in L\}$ is an n-element point set on the unit sphere around $\mathbf{0}$ such that $h(P^*, \sqrt{2}) \geq c'n^{4/3}$. $\square$

However, for $\alpha \neq \sqrt{2}$ we only have a much weaker lower bound on $h(n, \alpha)$.

The *iterated logarithm* of n is the height of the shortest exponential "tower" of 2's, whose value is at least n, that is, it is defined as the minimum integer k

such that

$$n \le 2^{2^{\cdot^{\cdot^{2}}}} \Big\} k \text{ times} .$$

Theorem 12.9 (Erdős, Hickerson, and Pach, 1989). *Let $h(n, \alpha)$ denote the maximum number of times the distance α can occur among n points on the unit sphere. Then there exists a constant $c > 0$ such that for any $\alpha > 0$ and any positive integer n,*

$$h(n, \alpha) > cn \log^* n,$$

where $\log^*$ denotes the iterated logarithm function.

Proof. Let S_0 be the equator of $\mathbb{S}^2$, and S_ε the 'strip' of width ε around S_0, defined as

$$S_\varepsilon = \{(x, y, z) \mid x^2 + y^2 + z^2 = 1 \ \text{ and } \ |z| \le \varepsilon\} \quad \text{for } \varepsilon \ge 0.$$

Let $0 < \alpha < 2$ be a fixed number. Choose a small constant $\varepsilon > 0$ such that

$$2\sqrt{1 - \varepsilon^2} > \alpha;$$

that is, the diameter of the two circles bounding the strip S_ε is bigger than α.

We recursively construct a set of n_k points with the property that each point is at distance α from at least k other points.

Basis: For $k = 1$, let $n_1 = 2$ and take two points on the equator at distance α from each other.

Hypothesis: Suppose that we have already constructed a suitable set $P = \{p_1, p_2, \dots, p_{n_k}\}$ of n_k points, for some k, such that each p_i is at distance α from at least k other points. Assume that all points are in a narrow strip of width $\varepsilon_k < \varepsilon$, that is, $P \subset S_{\varepsilon_k}$.

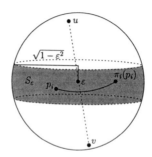

Figure 12.1. Strip S_ε.

Inductive Step: We construct a set P^* of n_{k+1} points in n_k steps, as described below. Let u and v be two antipodal points on the sphere such that u is at distance δ from the north pole $(0, 0, 1)$, for some sufficiently small $\delta > 0$.

Let $P^{(0)} = P$. Suppose that $P^{(0)}, P^{(1)}, \ldots, P^{(i-1)}$ have already been constructed for some $1 \leq i \leq n_k$, then we define $P^{(i)}$ as follows: Turn S^2 around the axis uv so as to bring $p_i \in P$ into a new position p_i' such that $|p_i - p_i'| = \alpha$. (If δ is sufficiently small, such a rotation is feasible.) More formally, let π_i be the isometry such that

$$\pi_i(u) = u, \quad \pi_i(v) = v, \quad \text{and} \quad |p_i - \pi_i(p_i)| = \alpha.$$

Set

$$P^{(i)} = \pi_i \left(P^{(0)} \cup P^{(1)} \cup \cdots \cup P^{(i-1)} \right).$$

Finally, let

$$P^* = P^{(0)} \cup P^{(1)} \cup \cdots \cup P^{(n_k)}.$$

If u, v (and δ) are chosen properly, the following conditions are satisfied:

(a) the above conditions are correct; i.e., all π_i exist;

(b) the sets $P^{(0)}, P^{(1)}, \ldots, P^{(n_k)}$ are pairwise disjoint;

(c) there exists an $\varepsilon_{k+1} < \varepsilon$ such that $P^* \subseteq S_{\varepsilon_{k+1}}$.

It follows from (b) that $|P^{(i)}| = 2^{i-1} |P^{(0)}|$ for $1 \leq i \leq n_k$, which implies that

$$|P^*| = n_{k+1} = n_k \cdot 2^{n_k}. \tag{12.1}$$

Now we show that for every $p_i \in P^*$, there are $k+1$ distinct points at distance α from p_i. Every $q \in P^*$ can be written as

$$q = \left(\prod_{i \in I} \pi_i \right) p_j,$$

for some appropriate subset $I \subset \{1, 2, \ldots, n_k\}$ and $p_j \in P$. By the inductive hypothesis, at least k points $p_{j_1}, \ldots, p_{j_k}$ of P are at distance α from p_j. Then the points

$$q_{j_t} = \left(\prod_{i \in I} \pi_i \right) p_{j_t}, \quad 1 \leq t \leq k,$$

are at distance α from q. Moreover, the point

$$
q' = \begin{cases} \left(\displaystyle\prod_{i \in I - \{j\}} \pi_i \right) p_j & \text{if } j \in I, \\[3mm] \left(\displaystyle\prod_{i \in I \cup \{j\}} \pi_i \right) p_j & \text{if } j \notin I \end{cases}
$$

is also at distance α from q, because π_j moves p_j by distance α. Furthermore, all q_{j_t} and q' are distinct. Thus, there are at least $k + 1$ points at distance α from q. Finally, condition (c) means that all points of P^* lie in the strip S_ε; i.e., the induction hypothesis is satisfied for $k + 1$.

It follows from the recursion formula (12.1) that $k \geq c \log^* n_k$, for some $c > 0$, and the construction can easily be extended for arbitrary values of n. $\square$

Now we are in the position to improve Theorem 10.9, i.e., to estimate the error term of $f_d(n)$, the maximum number of times the unit distance can occur among n points in $\mathbb{R}^d$, for $d \geq 4$.

Theorem 12.10 (Erdős and Pach, 1990).   *Let $d \geq 4$, and let $f_d(n)$ denote the maximum number of times the unit distance can occur among n points in $\mathbb{R}^d$. Then*

$$
f_d(n) = \frac{n^2}{2}\left(1 - \frac{1}{\lfloor d/2 \rfloor}\right) + \begin{cases} n - O(d) & \text{if } d \text{ is even,} \\ \Theta(n^{4/3}) & \text{if } d \text{ is odd.} \end{cases}
$$

[$O(f(n))$ denotes a function $g(n)$ satisfying $g(n) \leq cf(n)$ and $\Theta(f(n))$ denotes a function $h(n)$ satisfying $c_1 f(n) < h(n) < c_2 f(n)$, for every sufficiently large n, where c, c_1, c_2 are positive constants.]

Proof.   If $d \geq 4$ is even, then we can obviously modify Lenz's construction (described in the proof of Theorem 10.9) so that for all but at most two points picked on the circle C_i there are two other points at distance 1 from it. Thus,

$$
f_d(n) \geq |E(T_{d/2}(n))| + n - \frac{d}{2}
$$

$$
= \frac{1}{2}\left(1 - \frac{1}{d/2}\right)(n^2 - s^2) + \binom{s}{2} + n - \frac{d}{2}
$$

$$
= \frac{n^2}{2}\left(1 - \frac{1}{d/2}\right) + n + \frac{s^2}{d} - \frac{s}{2} - \frac{d}{2},
$$

where s is the remainder of n upon dividing by $d/2$. The above expression is minimized when $s = d/4$. Hence,

$$
f_d(n) \geq \frac{n^2}{2}\left(1 - \frac{1}{d/2}\right) + n - \frac{9}{16}d.
$$

If $d \geq 5$ is odd, then we can choose $\lfloor d/2 \rfloor - 1$ circles and a sphere of radius $1/\sqrt{2}$ such that all of them are centered at the origin and they induce mutually orthogonal subspaces. Let us pick $\lfloor n/\lfloor d/2 \rfloor \rfloor$ points on each circle and on the sphere. By the second assertion of Corollary 12.8, we can choose the points on the sphere so that they determine at least $c(\lfloor n/\lfloor d/2 \rfloor \rfloor)^{4/3} \geq c_d n^{4/3}$ unit distances. Hence, in this case

$$f_d(n) \geq |E(T_{\lfloor d/2 \rfloor}(n))| + c_d n^{4/3}$$

$$\geq \frac{n^2}{2}\left(1 - \frac{1}{\lfloor d/2 \rfloor}\right) + c_d n^{4/3} + O(d).$$

As for the upper bound, suppose first that d is even, n is large and let P be a set of n points in $\mathbb{R}^d$ having more than $|E(T_{\lfloor d/2 \rfloor}(n))| + n$ unit distance pairs. We can apply Exercise 9.18 to obtain that there are subsets $P_0, P_1, \ldots, P_{\lfloor d/2 \rfloor} \subset P$ with $|P_0| = 1$ and $|P_i| = 3$ ($1 \leq i \leq \lfloor d/2 \rfloor$) such that any two points belonging to different $P_i's$ are at distance 1. But then, by Exercise 10.6, the two-dimensional planes induced by $P_1, \ldots, P_{\lfloor d/2 \rfloor}$ must be mutually orthogonal, and the point P_0 must lie on a line orthogonal to all of them. This contradiction completes the proof of the upper bound if d is even. For odd values of d, the proof is somewhat more complicated, and the interested reader can consult Erdős and Pach (1990). $\qquad \square$

DISTINCT DISTANCES DETERMINED BY POINTS

In the last part of this chapter we apply the theorems of Clarkson et al. (1990) (Theorem 11.2 and 11.10) to obtain a partial answer to the second question raised by Erdős in his celebrated paper published in the *American Mathematical Monthly* in 1946 (and quoted at the beginning of Chapter 10): What is the minimum number, $g_d(n)$, of distinct distances determined by n points in $\mathbb{R}^d$? Obviously,

$$f_d(n) \cdot g_d(n) \geq \binom{n}{2}. \qquad (12.2)$$

In particular, Corollary 11.11 immediately yields that $g_2(n) \geq cn^{2/3}$ for some $c > 0$, which had been shown by Moser (1952) much before Corollary 11.11 was known. This bound was subsequently improved by Chung (1984) and Beck (1983a). In fact, Beck proved the stronger result that for any $\varepsilon > 0$, every set P of $n \geq n_0(\varepsilon)$ points in the plane contains an element realizing at least $n^{58/81-\varepsilon}$ distances.

Corollary 12.11 (Clarkson et al., 1990). *There exists $c > 0$ such that given any set P of n points in the plane, one can always pick $p \in P$ such that the number of distinct distances from p is at least $cn^{3/4}$.*

Proof. For any point $p \in P$, let $g_P(p)$ denote the number of distinct distances from p, i.e.,

$$g_P(p) = |\{|p - q| \mid p \neq q \in P\}|.$$

Put

$$G_P = \sum_{p \in P} g_P(p).$$

For every pair $p, q \in P$, let us draw circles of radius $|p - q|$ around p and q. Clearly, the total number of distinct circles is G_P. By Theorem 11.10(ii), the number of incidences between the circles and the elements of P is at most $c'((G_P)^{4/5}n^{3/5} + G_P + n)$. On the other hand, every element of P lies on exactly $n - 1$ circles. Therefore,

$$c'((G_P)^{4/5}n^{3/5} + G_P + n) \geq n(n - 1),$$

implying that

$$G_P \geq cn^{7/4},$$

for some constant $c > 0$. Hence, there is a $p \in P$ with $g_P(p) \geq cn^{3/4}$. $\qquad\square$

Recently, Chung, Szemerédi, and Trotter (1992) proved that

$$g_2(n) \geq n^{4/5} / \log^c n,$$

for some constant $c > 0$, provided that n is sufficiently large. However, it seems that their argument cannot easily be extended to show that there is always a single point from which there are at least that many distinct distances.

One of the most challenging unsolved problems in this field is to decide whether the following upper bound on $g_2(n)$ is asymptotically tight.

Theorem 12.12 (Erdős, 1946). *Let $g_2(n)$ denote the minimum number of distinct distances determined by n points in the plane. Then*

$$g_2(n) \leq \frac{cn}{\sqrt{\log n}},$$

and the points of a $\lceil\sqrt{n}\,\rceil \times \lceil\sqrt{n}\,\rceil$ piece of the integer lattice determine that many distinct distances.

For $d = 3$, Clarkson et al. (1990) have proved that $f_3(n) \leq n^{3/2}\beta(n)$, where $\beta(n) = 2^{c\alpha^2(n)}$ and $\alpha(n)$ is the (extremely slowly growing) functional inverse of Ackermann's function. Therefore, it follows from (12.2) that $g_3(n) \geq n^{1/2}/2^{c'\alpha^2(n)}$.

For larger values of d, we have the following result.

Theorem 12.13 (Erdős, 1946). *Let $g_d(n)$ denote the minimum number of distinct distances determined by n points in $\mathbb{R}^d$ ($d \geq 2$). There are suitable constants c_d, c'_d such that*

$$c_d n^{3/(3d-2)} \leq g_d(n) \leq c'_d n^{2/d}$$

holds for all n.

Proof. Assume, for simplicity, that $n^{1/d}$ is an integer, and let P be the set of vertices of an $\underbrace{n^{1/d} \times n^{1/d} \times \cdots \times n^{1/d}}_{d \text{ times}}$ piece of the integer lattice in $\mathbb{R}^d$. Since the distance between any two points in P is of the form $\sqrt{x_1^2 + x_2^2 + \cdots + x_d^2}$, where each $0 \leq x_i < n^{1/d}$ is an integer, there are at most $dn^{2/d}$ distinct distances. Hence, $g_d(n) \leq c'_d n^{2/d}$.

Next we prove the lower bound. Let $g'_d(n)$ denote the minimum number of distinct distances determined by n points on the d-dimensional sphere $\mathbb{S}^d \subseteq \mathbb{R}^{d+1}$ ($d \geq 2$). We will prove the following stronger result:

$$g_d(n) \geq c_d n^{3/(3d-2)}, \quad g'_d(n) \geq c''_d n^{3/(3d-2)}.$$

Observe that the proof of Corollary 12.11 also yields the fact that $g'_2(n) \geq c_2 n^{3/4}$; i.e., the assertion is true for $d = 2$.

Suppose that we have already verified the statement for $d - 1$ and want to prove it for d. To check the first inequality, observe that if our point set P ($|P| = n$) has an element p_i with

$$g_P(p_i) \leq n^{3/(3d-2)},$$

then there is a $(d-1)$-dimensional sphere centered at p_i, which contains at least

$$\frac{n-1}{n^{3/(3d-2)}} \geq \frac{1}{2} n^{1-3/(3d-2)}$$

elements. In this case, by the induction hypothesis, the number of distinct distances determined by P is at least

$$c''_{d-1} \left(\frac{1}{2} n^{1-3/(3d-2)} \right)^{3/(3(d-1)-2)} \geq \frac{c''_{d-1}}{2} n^{3/(3d-2)},$$

as required. The second inequality can be verified in exactly the same way. □

Starting the induction with the three-dimensional bound of Clarkson et al., one can further improve the lower bound in Theorem 12.13 to about $n^{1/(d-1)}$.

EXERCISES

12.1 (Erdős and Purdy, 1971)   Give a lower bound for the maximum number of triangles of unit area determined by a set of n points in the plane.

12.2 Show that for any set of n points in the plane, the number of triples that determine an isosceles triangle is $O(n^{7/3})$.

12.3 (Zhang, 1993)

(i) There exists a constant $c > 0$ such that for every positive integer n, one can select $m = \lfloor c(n/\log n)^{1/3} \rfloor$ points from the $\lfloor \sqrt{n} \rfloor \times \lfloor \sqrt{n} \rfloor$ grid

$$S_n = \{(i,j) \mid 0 \le i, j < \lfloor \sqrt{n} \rfloor \text{ are integers}\},$$

with the property that all the $\binom{m}{2}$ lines determined by them have different slopes.

(ii) (Erdős, Graham et al., 1992)   Show that it is impossible to select more than $c'n^{2/5}$ points from the grid S_n such that all lines determined by them have different slopes.

12.4 (Pach and Sharir, 1990)   Show that for any set of n points in the plane,

(i) the maximum number of triples that determine a triangle of unit perimeter is $O(n^{7/3})$;

(ii) the maximum number of triples that determine a triangle whose inscribed circle has radius 1 is $O(n^{7/3})$.

12.5 (Elekes and Erdős, 1994; Pach)   Two sets in $\mathbb{R}^d$ are *homothetic* if they are similar to each other and they are in parallel position.

(i) Prove that any n-element point set $P \subseteq \mathbb{R}^d$ has at most $n^{1+1/d}$ $(d+1)$-tuples that determine pairwise homothetic simplices.

(ii) Prove that this bound is asymptotically tight.

12.6 (Elekes and Erdős, 1994)   Show that there exists $c > 0$ such that for any triangle T and $n \ge 3$, there is an n-element point set in $\mathbb{R}^2$ containing at least cn^2 triples which induce a triangle similar to T.

12.7 (Elekes and Erdős, 1994)   Let α be a transcendental number, and let $T = \{0, 1, 2, \alpha\}$. Show that for any $\varepsilon > 0$ and sufficiently large n, there exists a set of n points on a line containing at least $n^{2-\varepsilon}$ similar copies of T.

12.8 (Erdős, Füredi et al., 1993; Alon)   Prove that for every $\varepsilon > 0$ there exists $C = C(\varepsilon) > 0$ with the following property. Let P be a set of Cn points in a disc of radius n such that the distance between any two points is at least 1. Then for any α ($0 \le \alpha < \pi$), we can find $p, q, r \in P$ such that $|{\llcorner} pqr - \alpha| \le \varepsilon$.

12.9   (i) (Croft, 1967; Conway et al., 1979)   Prove that the maximum number of times that a fixed angle α can be induced by triples of an n-element point set in $\mathbb{R}^3$ is $o(n^3)$.

   (ii) (Purdy, 1988)   Construct an n-element point set $P \subset \mathbb{R}^4$ such that at least cn^3 triples determine the angle $\pi/2$.

12.10   The distance between two disjoint compact sets C and D is defined as $\min_{c \in C, d \in D} |c - d|$. Let $\{C_1, \ldots, C_n\}$ be a collection of pairwise disjoint translates of a compact convex set in the plane. Show that there are at most $cn^{4/3}$ pairs C_i, C_j whose distance is 1.

12.11   (Chung, 1989)   Let $f_3(n)$ denote the maximum number of times that the unit distance can occur among n points in $\mathbb{R}^3$. Show, without using the result of Clarkson et al. (1990), quoted before Theorem 12.13, that

$$f_3(n) < cn^{11/7}$$

for a suitable constant c.

12.12   (Elekes)   Prove that for every $\varepsilon > 0$ there exists a constant $C = C(\varepsilon)$ with the following property. The number of pairwise similar $\lfloor \varepsilon n \rfloor$-element subsets of an n-element set $P \subset \mathbb{R}$ cannot exceed Cn.

13

More on Repeated Distances

In the previous chapters (particularly in Chapter 10) we established several results on the distribution of distances among n points in d-dimensional Euclidean space $\mathbb{R}^d$. We addressed the problem of estimating $f_d(n)$, the maximum number of times the same distance can occur among n points in $\mathbb{R}^d$ and developed some powerful techniques for studying these questions, but even in the plane ($d = 2$) we are very far from having satisfactory answers. However, if we impose some *special conditions* on the point sets (e.g., they are in convex position, or in general position, etc.), or if we are interested in the maximum number of times some *special distance* (e.g., the maximum distance or the minimum distance) can get repeated, then the analogous questions often become less difficult. The aim of this chapter is to present some results of this kind, but even in this area there are more unsolved problems than solved ones.

POINT SETS IN CONVEX POSITION

Given a set $P = \{p_1, \ldots, p_n\}$ of n points in the plane, we say that P is in *convex position* if it forms the vertex set of a convex polygon. Let $f(P)$, as usual, denote the number of pairs (p_i, p_j), $i < j$, such that the distance between p_i and p_j, $|p_i - p_j|$, is equal to 1. Further, let

$$f^{\mathrm{conv}}(n) = \max_P f(P),$$

where the maximum is taken over all sets of n points in convex position.

Erdős and Moser (1970) (see also Erdős, 1980) conjectured a long time ago that $f^{\mathrm{conv}}(n) \leq cn$ for some constant c. The question is still open. A nontrivial upper bound has been given by Füredi (1990), who established the following result.

Theorem 13.1 (Füredi). *Let $f^{\mathrm{conv}}(n)$ denote the maximum number of times the unit distance can occur among n points in convex position. Then there exists a constant $c > 0$ such that*

$$f^{\mathrm{conv}}(n) \leq cn \log n.$$

To prove this theorem, we need an auxiliary result on the maximum number of 1's in an m by n $(0, 1)$-matrix that does not contain a submatrix of some given type. It is clear that many problems in extremal graph theory can be reformulated in terms of "forbidden" submatrices. For instance, Theorem 9.5 can be interpreted as an upper bound on the number of 1's in an m by n $(0,1)$-matrix containing no r by s submatrix, all of whose entries are 1's. Many other results of this type were obtained by Anstee (1985); Anstee and Füredi (1986); Bienstock and Győri (1991); Frankl, Füredi, and Pach (1987); and Füredi and Hajnal (1992).

Here we shall use the following result.

Lemma 13.2 (Füredi; Bienstock and Győri).   *Let M be an m by n $(0,1)$-matrix that does not have*

$$M_0 = \begin{pmatrix} 1 & 1 & * \\ 1 & * & 1 \end{pmatrix}$$

*as a submatrix, where * stands for an unspecified value. Then the number of 1's in M cannot exceed $m + (m + n)\lfloor \log n \rfloor$.*

Proof.   To give an upper bound on the number of the 1's in M, we label some of the 1's, and bound the number of labeled and unlabeled 1's separately.

An entry $m_{ij} = 1$ is labeled (j, k), if there exist integers p, q $(j < p < q \le n)$ such that

$$m_{i,p} = m_{i,q} = 1,$$
$$p - j < 2^k \le q - j$$

$(1 \le i \le m,\ 1 \le j < n,\ 1 \le k \le \lfloor \log n \rfloor)$. Thus, the number of distinct labels is at most $n \lfloor \log n \rfloor$.

We claim that no two distinct nonzero entries can have the same label. Suppose, for the sake of contradiction, that two distinct entries $m_{i_1, j} = m_{i_2, j} = 1$ $(i_1 < i_2)$ get the same label (j, k); see Figure 13.1. But then we can choose integers p, q $(j < p < q \le n)$ so that

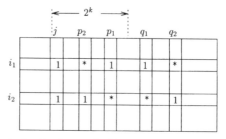

Figure 13.1. Illustration for the proof of Lemma 13.2; choose $p = p_1$ and $q = q_2$.

$$m_{i_1,p} = m_{i_2,q} = 1,$$
$$p - j < 2^k \leq q - j.$$

This implies that the rows i_1, i_2 and the columns j, p, q induce a submatrix of type M_0, which is a contradiction.

Next, we are going to estimate the number of unlabeled 1's in a single row of M. Let $m_{i,j_1} = m_{i,j_2} = \cdots = m_{i,j_t}$ ($j_1 < j_2 < \cdots < j_t$) be the nonzero entries in the ith row of M, having no label. Then, for every $1 \leq s < t$, we have

$$j_t - j_s > 2(j_t - j_{s+1}),$$

because otherwise

$$j_t - j_s \geq 2(j_{s+1} - j_s),$$

and m_{i,j_s} would have been labeled by $(j_s, \lfloor \log(j_t - j_s) \rfloor)$. Hence, if $t > 1$, then

$$j_t - j_1 \geq 2(j_t - j_2) + 1 \geq 2^2(j_t - j_3) + 2 + 1 \geq \cdots$$
$$\geq 2^{t-2}(j_t - j_{t-1}) + 2^{t-3} + 2^{t-4} + \cdots + 1$$
$$\geq 2^{t-1} - 1.$$

This yields $2^{t-1} \leq n$, i.e., the number t of unlabeled entries in the ith row cannot exceed $1 + \lfloor \log n \rfloor$.

Thus, the total number of 1's in M is at most

$$n\lfloor \log n \rfloor + m(1 + \lfloor \log n \rfloor) = m + (m + n)\lfloor \log n \rfloor. \qquad \square$$

Lemma 13.3. *Let C_1 and C_2 be two disjoint arcs of a convex polygon, containing n_1 and n_2 vertices, respectively. Then the number of pairs (p, q), where p is a vertex of C_1, q is a vertex of C_2, and their distance $|p - q| = 1$ is at most*

$$2n_1 + 2(n_1 + n_2)\lfloor \log n_2 \rfloor.$$

Proof. Let $p_1, \ldots, p_{n_1}$ and $q_1, \ldots, q_{n_2}$ be the vertices of C_1 and C_2, listed in counterclockwise and in clockwise order, respectively.

Define an n_1 by n_2 $(0,1)$-matrix $M = (m_{i,j})$ whose ith row and jth column are associated with p_i and q_j, respectively, and

$$m_{i,j} = \begin{cases} 1 & \text{if } |p_i - q_j| = 1, \\ 0 & \text{otherwise.} \end{cases}$$

Color the nonzero entries of M by two colors, as follows. The entry $m_{i,j} = 1$ is colored red if C has a supporting line ℓ at p_i such that the angle from ℓ to $p_i q_j$ in clockwise direction is acute. Otherwise, $m_{i,j}$ is colored blue.

Now $M = M_r + M_b$, where M_r (resp. M_b) is the $(0,1)$-matrix, whose 1's are the red (resp. blue) entries of M.

We claim that M_r does not contain

$$M_0 = \begin{pmatrix} 1 & 1 & * \\ 1 & * & 1 \end{pmatrix}$$

as a submatrix. Suppose to the contrary that there are indices $i_1 < i_2$ and $j_1 < j_2 < j_3$ such that

$$|p_{i_1} - q_{j_1}| = |p_{i_1} - q_{j_2}| = |p_{i_2} - q_{j_1}| = |p_{i_2} - q_{j_3}| = 1,$$

and all of these unit distances give rise to red entries. Consider the convex quadrilateral $p_{i_1} p_{i_2} q_{j_3} q_{j_1}$ (see Figure 13.2). The angles $\angle q_{j_1} p_{i_1} p_{i_2}$ and $\angle p_{i_2} q_{j_3} q_{j_1}$ are obviously acute, as they are symmetric angles of the isosceles triangles $\triangle q_{j_1} p_{i_1} p_{i_2}$ and $\triangle p_{i_2} q_{j_3} q_{j_1}$, respectively. The angle $\angle q_{j_3} q_{j_1} p_{i_1}$ is acute because by the convexity of C, it is smaller than $\angle q_{j_2} q_{j_1} p_{i_1}$, which is again a symmetric angle of an isosceles triangle. Finally, $\angle p_{i_1} p_{i_2} q_{j_3}$ is acute, because m_{i_2, j_3} was colored red. Thus, all four angles of the quadrilateral $p_{i_1} p_{i_2} q_{j_3} q_{j_1}$ are smaller than $\pi/2$, a contradiction.

Hence, by Lemma 13.2, M has at most $n_1 + (n_1 + n_2)\lfloor \log n_2 \rfloor$ red entries, and we can obtain the same upper bound for the number of blue entries.   □

Proof of Theorem 13.1.   Let C be a convex n-gon. Partition the plane into equal squares by drawing horizontal and vertical lines at distance $(1/\sqrt{2}) - \varepsilon$ from each other. Let C_i denote the part(s) of C lying inside the ith square. The distance between any two points in the same square is less than 1. On the other hand, if $\varepsilon > 0$ is chosen sufficiently small, then each point that can be at unit distance from some element of C_i must lie in one of the 24 squares closest to the ith square. Let $J(i)$ be the set of indices of these 24 squares. Denoting the number of vertices in C_i by n_i, it follows from Lemma 13.3 that the number of times that the unit distance can occur among the vertices of C is at most

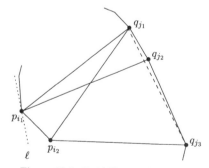

Figure 13.2. Forbidden configuration.

$$\frac{1}{2} \sum_{i} \sum_{j \in J(i)} (2n_i + 2(n_i + n_j)\lfloor \log n_j \rfloor)$$

$$< 24 \sum_{i} n_i(1 + \lfloor \log n \rfloor) + \sum_{i} \sum_{j \in J(i)} n_j \lfloor \log n_j \rfloor$$

$$\leq 24n(1 + 2\lfloor \log n \rfloor). \qquad \qquad \square$$

The best known lower bound for $f^{\mathrm{conv}}(n)$ is due to Edelsbrunner and Hajnal (1991).

Theorem 13.4. (Edelsbrunner and Hajnal). *Let $f^{\mathrm{conv}}(n)$ denote the maximum number of times the unit distance can occur among the vertices of a convex n-gon. Then*

$$f^{\mathrm{conv}}(n) \geq 2n - 7.$$

Proof. Let p, q, and r be the vertices of an equilateral triangle of side length 1, in clockwise order. Connect q and r with a circular arc $\overset{\frown}{qr}$ of radius 1 around p. Similarly let $\overset{\frown}{rp}$ and $\overset{\frown}{pq}$ be circular arcs around q and r, respectively. The region enclosed by these three arcs is called a *Reuleaux triangle*.

Let p_0, q_0, and r_0 denote the midpoints of the arcs $\overset{\frown}{qr}$, $\overset{\frown}{rp}$, and $\overset{\frown}{pq}$, respectively. Furthermore, let γ_p, γ_q, and γ_r denote the unit circles centered at p_0, q_0, and r_0, respectively (see Figure 13.3).

Pick a point p_1 on γ_p, which is close to p and comes after p in the clockwise direction. Then there is a point $q_1 \in \gamma_q$, close to q such that $|p_1 - q_1| = 1$. It is easy to see that q_1 lies after q in the clockwise direction, and $|q_1 - q| < |p_1 - p|$. Next we choose a point $r_1 \in \gamma_r$ satisfying $|q_1 - r_1| = 1$ and $|r_1 - r| < |q_1 - q|$. Proceeding in this manner, we can place further points $p_2, q_2, r_2, p_3, q_3, r_3, \ldots$ on the arcs γ_p, γ_q, and γ_r. Stop when the set $S = \{p_1, q_1, r_1, p_2, q_2, r_2, \ldots, \}$ has

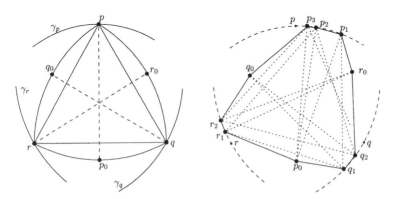

Figure 13.3. Construction of a convex n-gon with many unit distances.

exactly $n - 3$ points. If $|p_1 - p|$ was chosen sufficiently small, then the elements of $S \cup \{p_0, q_0, r_0\}$ are all distinct, and they are in convex position.

All the $n - 3$ points of S are at unit distance from p_0, q_0, or r_0. The unit distance occurs $n - 4$ times among the elements of S. Thus, $S \cup \{p_0, q_0, r_0\}$ determines at least $2n - 7$ unit distances. $\qquad\square$

Erdős has conjectured that there exists a natural number k with the following property. Every convex polygon has a vertex such that no k other vertices are equidistant from it. This conjecture, if true, would immediately imply that

$$f^{\mathrm{conv}}(n) < (k - 1)n.$$

Danzer (see Erdős, 1987) proved that if such a number k exists, it must be at least 4 (see Exercise 13.1). Later, Fishburn and Reeds (1993) exhibited a convex polygon with 20 vertices, each of whose vertices is at unit distance from three other vertices.

Let $g^{\mathrm{conv}}(n)$ denote the minimum number of distinct distances determined by the vertices of a convex n-gon. Altman (1963, 1972) showed that

$$g^{\mathrm{conv}}(n) = \left\lfloor \frac{n}{2} \right\rfloor.$$

This bound is attained for a regular n-gon.

Erdős conjectured that the following stronger statement is also true. Every convex n-gon has a vertex such that the number of distinct distances from this vertex is at least $\lfloor n/2 \rfloor$. The best known result in this direction is due to L. Moser (1952).

Theorem 13.5 (Moser). *Let C be the vertex set of a convex n-gon ($n \geq 3$). Then there exists $p \in C$ such that the number of distinct distances from p is at least $\lceil n/3 \rceil$.*

To establish this result, we need a simple lemma, whose proof is left as an exercise (Exercise 13.2).

Let D be a circular disc, and let pr be a chord of D. Then pr divides D into two parts, the smaller of which is called a *cap*. (If pr is a diameter of D, then both parts are considered caps.)

Lemma 13.6. *Let C' be a set of m points lying in a closed cap of a disc determined by a chord pr, and assume that $C' \cup \{p, r\}$ is in convex position. If $p \notin C'$, then all the m distances between p and the elements of C' are distinct.*

Proof of Theorem 13.5. Let D be the smallest disc containing all elements of C.

If only two points $p, r \in C$ lie on the boundary of D, then pr must be a diameter, and at least one of the two closed caps determined by pr contains at least $\lceil n/2 \rceil$ points of C different from p.

If D has more than two points on its boundary, and no two determine a diameter, then we can choose three of them, $p, q, r \in C$, so that no angle of $\triangle pqr$ is obtuse (for otherwise, D would not be minimum). Since the points of C are in convex position, no element of C is in the interior of $\triangle pqr$. Thus, at least one of the three caps determined by the sides of $\triangle pqr$ contains at least $\lceil n/3 \rceil$ points of C different from one of the endpoints of the corresponding side.

In any case, Theorem 13.5 follows from Lemma 13.6. $\qquad\square$

POINT SETS IN GENERAL POSITION

It seems very likely that Altman's above-cited result can be generalized to any set of points no three of which are collinear. That is, letting $g^{\text{noncoll}}(n)$ denote the minimum number of distinct distances determined by n points in the plane such that no three of them are collinear, it can be conjectured that

$$g^{\text{noncoll}}(n) = g^{\text{conv}}(n) = \left\lfloor \frac{n}{2} \right\rfloor.$$

The only known result in this direction is due to Szemerédi (see Erdős, 1987), and it can be viewed as a generalization of Theorem 13.5.

Theorem 13.7 (Szemerédi). *Let P be a set of n points in the plane, no three of which are collinear. Then there exists a $p \in P$ such that the number of distinct distances from p is at least $\lceil (n-1)/3 \rceil$.*

Proof. Assume that for every point $p \in P$, the number of distinct distances from p is at most k; i.e., every point $q \neq p$ is lying on one of the (at most) k concentric circles $C_1(p), \ldots, C_k(p)$ around p.

Let I denote the number of isosceles triangles spanned by triples of P, where an equilateral triangle is counted with multiplicity 3. Obviously,

$$I = \sum_{p \in P} \sum_{i=1}^{k} \binom{|C_i(p) \cap P|}{2},$$

which attains its minimum if the points belonging to $P - \{p\}$ are distributed among the circles $C_i(p)$ as uniformly as possible. In particular,

$$I \geq nk \frac{n-1}{k} \left(\frac{n-1}{k} - 1 \right) \Big/ 2.$$

On the other hand, every segment qr can be the base of at most two isosceles triangles determined by P, for otherwise the perpendicular bisector of qr would pass through at least three elements of P. Thus,

$$I \le 2 \binom{n}{2}.$$

Comparing these two inequalities, we get $k \ge \lceil (n-1)/3 \rceil$, as desired. $\square$

A set of points P in the plane is said to be in *general position* if no three of them are collinear and no four of them cocircular. Let $g^{\text{gen}}(n)$ denote the minimum number of distinct distances determined by a set of n points in the plane in general position.

Theorem 13.8 (Erdős, Füredi et al., 1993). *Let $g^{\text{gen}}(n)$ denote the minimum number of distinct distances determined by n points in the plane in general position. Then there exists a constant c such that*

$$g^{\text{gen}}(n) \le n2^{c\sqrt{\log n}}$$

for every n.

Proof. Assume, for the sake of simplicity, that $n = \lfloor 2^{d(d-2)}/d \rfloor$ for some natural number $d \ge 4$. Consider the set L of all lattice points $(x_1, \dots, x_d) \in \mathbb{R}^d$ with integer coordinates $0 \le x_i < 2^d$. The number of distinct distances determined by L is at most $d(2^d)^2$, because there are at most that many numbers of the form $\left(\sum_{i=1}^{d} (x_i - x_i')^2 \right)^{1/2}$, where $0 \le x_i, x_i' < 2^d$. In particular, there is a sphere around the origin that contains at least

$$\frac{|L|}{d(2^d)^2} = \frac{(2^d)^d}{d(2^d)^2} \ge \left\lfloor \frac{2^{d(d-2)}}{d} \right\rfloor = n$$

elements of L. Let P denote the set of these points.

Set $P + (-P) = \{p_1 - p_2 | p_1, p_2 \in P\}$, i.e., $P + (-P)$ is the set of all vectors determined by P (or the *difference region* of P). Observe that every element of $P + (-P)$ is a vector $(x_1, \dots, x_d) \in \mathbb{R}^d$ with integer coordinates $-2^d < x_i < 2^d$, hence

$$|P + (-P)| \le (2 \cdot 2^d)^d = 2^{d(d+1)} < n2^{8\sqrt{\log n}}.$$

Fix a two-dimensional plane Π in $\mathbb{R}^d$, and for any $p \in P$ let p' denote the orthogonal projection of p into Π. Evidently, we can choose Π so as to satisfy the following two conditions:

(i) $p_1' = p_2'$ if and only if $p_1 = p_2$ (for any $p_1, p_2 \in P$),
(ii) the point set $P' = \{p' | p \in P\} \subseteq \Pi$ is in general position.

Since that $p_1 - p_2 = p_3 - p_4$ implies that $|p_1' - p_2'| = |p_3' - p_4'|$, we have that the number of distinct distances determined by P' is at most

$$|P + (-P)| < n2^{8\sqrt{\log n}},$$

as required.

This argument easily extends to the general case when n can take any positive integer value. $\square$

It is an exciting open problem to decide whether

$$\lim_{n \to \infty} \frac{g^{\text{gen}}(n)}{n} = \infty.$$

However, a weaker version of this relation can easily be deduced from some deep results in the additive number theory.

Let $g^*(n)$ be the minimum number of distinct vectors determined by n points in the plane in general position. That is,

$$g^*(n) = \min_{P} |P + (-P)|,$$

where the minimum is taken over all n-element point sets in the plane in general position. Obviously, $g^*(n) \geq g^{\text{gen}}(n)$. In fact, the above argument shows that

$$g^*(n) \leq n2^{c\sqrt{\log n}}$$

for a suitable constant c; i.e., $g^*(n) \leq n^{1+\varepsilon}$ for every $\varepsilon > 0$, provided that n is sufficiently large.

Theorem 13.9 (Erdős, Füredi et al., 1993). *Let $g^*(n)$ be the minimum number of distinct vectors determined by n points in the plane, no three of which are on a line and no four on a circle. Then*

$$\lim_{n \to \infty} \frac{g^*(n)}{n} = \infty.$$

Proof. Suppose to the contrary that $g^*(n) < Cn$ for infinitely many values of n, where C is a constant.

Let P be an n-element point set in the plane in general position with

$$|P + (-P)| < Cn.$$

It can be shown (Exercise 13.3) that there exists another constant C' depending only on C such that

$$|P + P| = |\{p_1 + p_2 \mid p_1, p_2 \in P\}| < C'n.$$

We recall a celebrated result of Freiman (1966, 1973, 1987) (see Ruzsa, 1992, for a simpler proof).

Lemma 13.10 (Freiman). *For any integer C', there exists C'' with the property that any n-element set P in the plane with $|P+P| < C'n$ can be covered by the projection of a lattice of dimension C' and size $C''n$. That is,*

$$P \subseteq \{v_0 + m_1 v_1 + \cdots + m_{C'} v_{C'} \,|\, 1 \le m_i \le n_i\},$$

for suitable vectors $v_i \in \mathbb{R}^2$ and natural numbers n_i satisfying $\prod_{i=1}^{C'} n_i \le C''n$.

Without loss of generality assume that $n_1 \ge n^{1/C'}$. Obviously, we can fix some values $\overline{m}_2, \ldots, \overline{m}_{C'}$ so that

$$v_0 + m_1 v_1 + \overline{m}_2 v_2 + \cdots + \overline{m}_{C'} v_{C'} \in P$$

for at least

$$\frac{n}{n_2 n_3 \cdots n_{C'}} \ge \frac{n_1}{C''} \ge \frac{n^{1/C'}}{C''}$$

different integers m_1. However, the corresponding points of P are all on a line, contradicting our assumption that P is in general position.   □

MINIMUM AND MAXIMUM DISTANCES

As we have seen earlier, it is a very hard problem to determine the exact order of magnitude of $f_d(n)$, the maximum number of times the same distance can occur among n points in $\mathbb{R}^d$ ($n \to \infty$). We get a much simpler problem if we want to estimate $f_d^{\min}(n)$, the maximum number of occurrences of the *minimum* distance determined by an n-element point set in $\mathbb{R}^d$. That is, let

$$f_d^{\min}(n) = \max_{|P|=n} |\{pq \,|\, p,q \in P \quad \text{and} \quad |p-q| = 1\}|,$$

where the maximum is taken over all n-element point sets $P \subseteq \mathbb{R}^d$ satisfying

$$\min_{p \ne q \in P} |p-q| = 1.$$

(Each segment $pq = qp$ is counted at most once.)

Similarly, one can define $f_d^{\max}(n)$, the maximum number of occurrences of the *maximum* distance in an n-element point set in $\mathbb{R}^d$, i.e.,

$$f_d^{\max}(n) = \max_{|P|=n} |\{pq \,|\, p,q \in P \quad \text{and} \quad |p-q| = 1\}|,$$

where the maximum is taken over all n-element point sets $P \subseteq \mathbb{R}^d$ satisfying

$$\max_{p,q \in P} |p-q| = 1.$$

The problem of determining $f_d^{\min}(n)$ can be reformulated as follows. Given a packing $\mathcal{B} = \{B_1, B_2, \ldots, B_n\}$ of congruent balls in $\mathbb{R}^d$, we wish to maximize the number of *touching* (*kissing*) pairs (B_i, B_j).

Recall that the *Hadwiger (Newton) number* of a convex body $C \subseteq \mathbb{R}^d$ is defined as the largest number $H(C)$ (resp. $N(C)$) of nonoverlapping translates (resp. congruent copies) of C that can touch C (cf. Exercise 8.9). Let B^d denote, as usual, the d-dimensional unit ball. Obviously, $H(B^2) = 6$, but it is quite hard to determine the exact value of $H(B^3)$. This question caused a notorious dispute between Isaac Newton and David Gregory. Newton conjectured that $H(B^3) = 12$, while Gregory thought that the true value was 13. One hundred eighty years later, Hoppe (1874) finally proved that Newton was right. The following statement is obvious.

Proposition 13.11. *For every d and n,*

$$f_d^{\min}(n) \le \frac{nH(B^d)}{2}.$$

Evidently, for $d = 1$, $f_1^{\min}(n) = n - 1$. In the plane, by the above bound, $f_2^{\min}(n) \le 3n$, and in fact, it is not too difficult to show that $f_2^{\min}(n) \le 3n - c\sqrt{n}$ for a positive constant c. It is much more remarkable that in the plane one can actually determine the exact value of $f_2^{\min}(n)$. The following result due to Harborth (1974) settles a conjecture of Reutter (1972).

Theorem 13.12 (Harborth). *Let $f_2^{\min}(n)$ denote the maximum number of times the minimum distance can occur among n points in the plane. Then*

$$f_2^{\min}(n) = \lfloor 3n - \sqrt{12n - 3} \rfloor.$$

Proof. For simplicity, we write $f(n)$ for $f_2^{\min}(n)$. We have $f(0) = f(1) = 0$, $f(2) = 1$. In what follows, we can assume that $n \ge 3$. Consider a set P of n points with minimum distance 1, and connect two elements of P by a segment if and only if their distance is exactly 1. Thus, we obtain a graph G embedded in the plane. Assume that G has the largest possible number of edges; that is, $|E(G)| = f(n)$. It is easy to see that every vertex of G is adjacent to at least two other vertices. Moreover, G is *two-connected*; that is, G remains connected after the removal of any of its vertices.

The outer face of G is bounded by a simple closed polygon C. Let b and b_d denote the total number of vertices of this polygon and the number of those vertices that have degree d in G, respectively. Clearly, $b = b_2 + b_3 + b_4 + b_5$.

The internal angle of C at a vertex of degree d is at least $(d - 1)\pi/3$, and the sum of these angles is $(b - 2)\pi$. Hence,

$$b_2 + 2b_3 + 3b_4 + 4b_5 \le 3b - 6.$$

On the other hand, denoting by f_i ($i \ge 3$) the number of internal faces of G with i sides, we obtain from *Euler's polyhedral formula* (see, e.g., the proof of

Theorem 8.1) that

$$n - f(n) + f_3 + f_4 + \cdots = 1. \tag{13.1}$$

If we add up the number of sides of the internal faces of G, then every edge of C will be counted once and all other edges twice, i.e.,

$$b + 2(f(n) - b) = 3f_3 + 4f_4 + \cdots$$
$$\geq 3(f_3 + f_4 + \cdots) \tag{13.2}$$

Comparing (13.1) and (13.2), we get

$$n - b \geq f(n) - 2n + 3.$$

Assume that the theorem has already been established for all integers smaller than n. Delete from G the vertices of C and all edges incident to them. By the induction hypothesis, we obtain

$$f(n) - b - (b_3 + 2b_4 + 3b_5) \leq f(n - b),$$

$$\begin{aligned}
f(n) &\leq f(n - b) + b_2 + 2b_3 + 3b_4 + 4b_5 \\
&\leq f(n - b) + 3b - 6 \\
&\leq 3(n - b) - \sqrt{12(n - b) - 3} + 3b - 6 \\
&\leq 3n - 6 - \sqrt{12(f(n) - 2n + 3) - 3} \\
&\leq 3n - 6 - \sqrt{12(f(n) - 2n) + 33}.
\end{aligned}$$

Observe that for $f(n) = 3n \pm \sqrt{12n - 3}$, all of the above inequalities hold with equality. Thus, using the fact that $f(n) < 3n$, we have

$$f(n) \leq \lfloor 3n - \sqrt{12n - 3} \rfloor,$$

as desired.

The proof of the tightness of this bound is left to the reader (see Figure 13.4 and Exercise 13.8). □

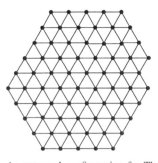

Figure 13.4. An extremal configuration for Theorem 13.12.

Brass (1992b) has shown that if P is an n-element point set in the plane, whose convex hull has k vertices, then the smallest distance cannot occur more than $3n - 2k + 4$ times between its points. This bound is tight for infinitely many values of n and k. Brass (1992a) has also proved that the second-smallest distance among n points in the plane can occur at most $(\frac{24}{7} + o(1))n$ times, as $n \to \infty$ (see Figure 13.5 for a lower bound of $\frac{24}{7} n$). His proof, which settles a conjecture of Vesztergombi (1987), uses linear programming. The same method can be applied to obtain nontrival upper bounds on the number of occurrences of the ith smallest distance for any fixed i (see Brass, 1992c).

Next we study the dual problem of bounding the number of occurrences of the largest distance.

Theorem 13.13 (Hopf and Pannwitz, 1934; Sutherland, 1935). *Let $f_2^{\max}(n)$ denote the maximum number of times the maximum distance can occur among n points in the plane. Then*

$$f_2^{\max}(n) = n$$

for every $n \geq 3$.

Proof. Let P be an n element point set in the plane and assume that

$$\max \{|p - q| \mid p, q \in P\} = 1.$$

Construct a graph G by connecting p and q by a line segment if and only if $|p - q| = 1$.

The theorem is obviously true for $n = 3$. Let $n > 3$, and assume that the assertion has already been proved for every integer smaller than n.

If G has a vertex p of degree at most 1, then we can apply the induction

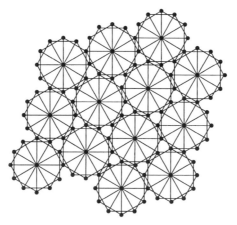

Figure 13.5. A lower-bound construction for the second smallest distance.

hypothesis to the set $P - \{p\}$ to obtain

$$|E(G)| \le f_2^{\max}(n - 1) + 1 = n.$$

Therefore, we can assume that every vertex of G is adjacent to at least two other vertices.

If all vertices of G have degree 2, then $|E(G)| = n$. Suppose that there is a point p connected to three other points $q, r, s \in P$, where $\angle qpr < \angle qps \le \pi/3$ (see Figure 13.6). Let t be a point of P different from p, which is also adjacent to r. Then either ps or pq does not intersect the segment rt. Assume without loss of generality that ps and rt are disjoint. However, this is impossible, because either pq intersects rt and $|s - t| > 1$, or it does not and $|p - t| > 1$.

Thus, $f_2^{\max}(n) \le n$, and the example on Figure 13.6 shows that this bound can be attained. $\square$

The maximum number of occurrences of the second largest distance determined by n points in the plane was investigated by Vesztergombi (1985, 1987). She proved that this number is at most $3n/2$ with equality if n is even.

The following three-dimensional analogue of Theorem 13.13 was conjectured by Vázsonyi and proved independently by Grünbaum (1956), Heppes (1956), and Straszewicz (1957).

Theorem 13.14 (Grünbaum, 1956; Heppes, 1956; Straszewicz, 1957). *Let $f_3^{\max}(n)$ denote the maximum number of times the maximum distance can occur among n points in $\mathbb{R}^3$. Then*

$$f_3^{\max}(n) = 2n - 2,$$

for every $n \ge 4$.

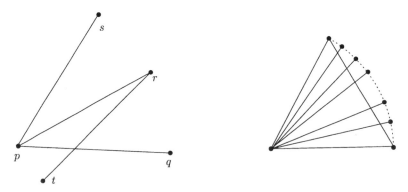

Figure 13.6. Illustration for the proof of Theorem 13.13.

Proof. The assertion is obviously true for $n = 4$. Assume that it has already been proved for every integer smaller than n. Fix a set P of n points in $\mathbb{R}^3$ whose diameter $\max_{p,q \in P} |p - q| = 1$. Connect a pair of elements of P by a line segment if their distance is 1.

If P has a point p connected to at most two other elements, then applying the induction hypothesis to $P - \{p\}$, we get that the number of unit distances determined by P is at most

$$f_3^{\max}(n - 1) + 2 = 2n - 2.$$

Therefore, we can assume that every element of P is adjacent to at least three points and is a vertex of the convex hull of P. Draw a unit ball $B(p)$ around each point $p \in P$, and set

$$C = \bigcap_{p \in P} B(p).$$

Obviously, C is a convex set (a *spherical polytope*) bounded by spherical "faces" and circular arcs ("edges") separating them. Let N, F, and E denote the number of vertices, faces, and edges of C, respectively. Notice that $F = n$, because each $B(p)$ contributes exactly one face to the boundary of C. On the other hand, $N \geq n$, where strict inequality holds if C has a vertex not belonging to P.

Every vertex x of C is incident with at least three edges of C. Furthermore, if $x \in P$, then the number of edges of C incident with x is equal to the number of points $q \in P$ with $|x - q| = 1$. Double-counting the edges of C, we obtain

$$2f_P + 3(N - n) \leq 2E,$$

where f_P denotes the number of unit distances determined by P.

By Euler's polyhedral formula,

$$N - E + F = 2.$$

Thus,

$$
\begin{aligned}
2f_P &\leq 2E - 3(N - n) \\
&= 2(N + F - 2) - 3(N - n) \\
&= 2(n + F - 2) - (N - n) \\
&\leq 2(n + n - 2) - 0 \\
&= 2(2n - 2),
\end{aligned}
$$

as required.

Figure 13.7 shows that $f_3^{\max}(n) \geq 2n - 2$. $\qquad\square$

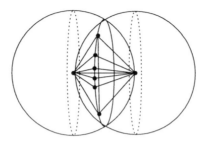

Figure 13.7. The largest distance can occur $2n - 2$ times ($n = 8$).

BORSUK'S PROBLEM

Recall that the diameter of a point set $P \subseteq \mathbb{R}^d$ is $\sup_{p, q \in P} |p - q|$. According to a famous conjecture of Borsuk (1933), any bounded set $P \subset \mathbb{R}^d$ can be partitioned into at most $d + 1$ subsets of smaller diameter.

As was pointed out by Heppes and Révész (1956), the results of the preceding section easily imply the following.

Corollary 13.15. *Let P be a finite set of points in $\mathbb{R}^d$, $d \leq 3$. Then P can be partitioned into at most $d + 1$ subsets of smaller diameter.*

Proof. Let $d = 3$. Proceed by induction on $|P| = n$. For $n \leq 4$, there is nothing to prove. Let $n > 4$, and assume that the assertion is true for every set of size at most $n - 1$.

Connect two points of P by an edge if and only if their distance is equal to the diameter of P. It follows from Theorem 13.14 that the resulting graph has a vertex p of degree at most 3. By the induction hypothesis, $P - \{p\} = P_1 \cup P_2 \cup P_3 \cup P_4$, where diam $P_i <$ diam P ($1 \leq i \leq 4$). Then there exists an index i such that p is not adjacent to any element of P_i. Hence, diam $(P_i \cup \{p\}) <$ diam P, and the result follows.

The case $d = 2$ can be settled in the same way using Theorem 13.13.   □

Borsuk's conjecture has been verified in many other special cases: for convex bodies with smooth boundaries (Hadwiger, 1945, 1946), for centrally symmetric sets (Riesling, 1971), for sets having the same symmetries as the regular simplex (Rogers, 1971), etc. See also Moser and Pach (1986) and Grünbaum (1963) for many additional references.

The aim of this section is to prove a surprising result of Kahn and Kalai (1993), which shows that Borsuk's conjecture is false (in a very strong sense) even for finite sets $P \subset \mathbb{R}^d$, provided that d is sufficiently large. We follow the presentation of Nilli (1993), who adapted an elegant argument of Alon, Babai, and Suzuki (1991).

The proof is based on a result of Frankl and Wilson (1981), which can be established by the so-called *linear algebra method* (see Babai and Frankl, 1988).

Theorem 13.16 (Frankl and Wilson). *Let $n = 4p$, where p is an odd prime. Let V denote the set of all vectors $v = (v_1, \ldots, v_n) \in \{+1, -1\}^n$, for which $v_1 = 1$ and the number of positive components is even. Then, for any subset $U \subseteq V$ containing no two orthogonal vectors,*

$$|U| \le \sum_{i=1}^{p-1} \binom{n}{i}.$$

Proof. Observe that by the evenness condition, the scalar product of any two vectors $u, v \in V$ satisfies $\langle u, v \rangle \equiv 0 \pmod 4$. Since the first component of every element of V is $+1$, we also have $\langle u, v \rangle \ne -4p = -n$. Thus, for any $u, v \in U$,

$$\langle u, v \rangle \equiv 0 \pmod p \Longleftrightarrow u = v. \tag{13.3}$$

Associate with every $u \in U$ a polynomial in n variables $(x_1, \ldots, x_n) = x$ over the finite field $GF(p)$, as follows:

$$f_u(x) = \prod_{i=1}^{p-1} (\langle u, x \rangle - i).$$

It follows from (13.3) that for any $v \in U$,

$$f_u(v) \equiv \begin{cases} (3p+1)(3p+2) \cdots (4p-1) \not\equiv 0 & \pmod p \quad \text{if } v = u, \\ 0 & \pmod p \quad \text{if } v \ne u. \end{cases} \tag{13.4}$$

Each $f_u(x)$ is a polynomial of degree at most $p - 1$; i.e., every term of $f_u(x)$ is of the form $a x_1^{k_1} x_2^{k_2} \cdots x_n^{k_n}$, where $a \in GF(p)$ and $k_1 + \cdots + k_n \le p - 1$. Replacing each such term by $a x_1^{\varepsilon_1} x_2^{\varepsilon_2} \cdots x_n^{\varepsilon_n}$, where

$$\varepsilon_i = \begin{cases} 1 & \text{if } k_i \text{ is odd}, \\ 0 & \text{if } k_i \text{ is even}, \end{cases}$$

we obtain a multilinear polynomial $\hat{f}_u(x)$. Notice that for any $\{+1, -1\}$-vector x, $\hat{f}_u(x) = f_u(x)$. Hence, (13.4) holds for f_u replaced by $\hat{f}_u$.

We claim that the polynomials $\hat{f}_u(x)$, $u \in U$, are linearly independent over $GF(p)$. Indeed, if

$$\sum_{u \in U} \lambda_u \hat{f}_u(x) = 0,$$

then by substituting $x = v$ we obtain that $\lambda_v \equiv 0 \pmod p$ for every $v \in U$. Therefore, $|U|$ cannot be larger than $\sum_{i=0}^{p-1} \binom{n}{i}$, the dimension of the space of multilinear polynomials of degree at most $p - 1$ in n variables. $\quad\Box$

Now we are ready to prove the main result of this section.

Theorem 13.17 (Kahn and Kalai).   *Let $b(d)$ denote the smallest integer such that every bounded set in $\mathbb{R}^d$ can be partitioned into at most $b(d)$ sets of smaller diameter. Then, $b(d) \geq (1.2)^{\sqrt{d}}$ for every sufficiently large d.*

Proof.   Let V denote the same set of $\{+1, -1\}$-vectors of n-components as in Theorem 13.16. For any $v = (v_1, \ldots, v_n) \in V$, let $p(v) = (p_{ij}(v) \mid 1 \leq i, j \leq n) \subset \mathbb{R}^{n^2}$, where $p_{ij}(v) = p_{ji}(v) = v_i v_j$. Set

$$P = \{p(v) \mid v \in V\}.$$

Notice that for any $u, v \in V$,

$$\langle p(u), p(v) \rangle = \langle u, v \rangle^2 \geq 0. \tag{13.5}$$

Since all vectors of P have the same length, (13.5) yields that

$$|p(u) - p(v)| = \operatorname{diam} P \iff \langle u, v \rangle = 0.$$

Thus, it follows from Theorem 13.16 that P cannot be divided into fewer than

$$|P| \Big/ \sum_{i=0}^{p-1} \binom{n}{i} = 2^{n-2} \Big/ \sum_{i=0}^{p-1} \binom{n}{i}$$

subsets of smaller diameter. Observe that all elements of P lie in a subspace of $\mathbb{R}^{n^2}$ whose dimension is $\binom{n}{2}$. Consequently, for any odd prime p and $d = \binom{n}{2} = \binom{4p}{2}$,

$$b(d) \geq 2^{n-2} \Big/ \sum_{i=0}^{p-1} \binom{n}{i} > (1.203)^{\sqrt{d}}.$$

In view of the prime number theorem, this implies that the result holds for any sufficiently large d. $\square$

Note that the smallest value of d, for which the above construction disproves Borsuk's conjecture in $\mathbb{R}^d$ is 946. The asymptotically best known upper bound on $b(d)$ is due to Schramm (1988), who showed that

$$b(d) \leq \left(\sqrt{\frac{3}{2}} + o(1) \right)^d.$$

[See Bourgain and Lindenstrauss (1991) for a different proof.]

EXERCISES

13.1   (Danzer, 1987)   Show that there exists a convex polygon in the plane, whose every vertex is equidistant from at least three other vertices.

13.2 Let C be a set of m points lying in a closed cap of a disc determined by a chord pr, and assume that $C \cup \{p, r\}$ is in convex position. If $p \notin C$, then all m distances between p and the elements of C are distinct.

13.3⋆ (Ruzsa, 1978)   Prove that for every $C > 0$ there exists C' with the property that if P is any set of n points in the plane such that $|P + (-P)| < Cn$, then $|P + P| < C'n$.

13.4 (L. Fejes Tóth, 1969, 1975)   Let C be a packing of unit discs in the plane, and let $C \in \mathcal{C}$. An element $C' \in \mathcal{C}$ is called a kth *neighbor* of C, if there is a chain of discs $C_0 = C, C_1, C_2, \ldots, C_k = C'$ in $\mathcal{C}$, such that C_i and C_{i-1} touch each other for all i $(1 \le i \le k)$. The kth *Hadwiger number*, H_k, is the largest number of kth neighbors of an element in any packing of unit discs. Prove that

$$\lim_{k \to \infty} \inf \frac{H_k}{k^2} \ge \frac{2\pi}{\sqrt{3}}.$$

13.5 (Erdős and Purdy, 1976).   Let $f^{\mathrm{noncoll}}(n)$ denote the maximum number of times the unit distance can occur among n points in the plane, no three of which are collinear. Prove that

$$f^{\mathrm{noncoll}}(n) > cn \log n$$

for a suitable constant $c > 0$.

13.6 (Erdős, Lovász, and Vesztergombi, 1989)   Let C be a set of n points in the plane in convex position, and let $d_1 > d_2 > \cdots$ denote the distinct distances determined by C. Let $G_k(C)$ be the graph on vertex set C obtained by joining two elements of C by an edge if their distance is at least d_k. Prove that

(i) $G_k(C)$ has a vertex of degree at most $3k - 1$;

(ii) $G_k(C)$ has at most $(3k - 1)n$ edges;

(iii) the *chromatic number* of $G_k(C)$ is at most $3k$ (i.e., one can color the points of C with $3k$ colors so that the distance between any two points of the same color is smaller than d_k).

13.7 Let $k(n) = k$ denote the smallest number such that in any set of n points in the plane one can increase the minimum distance by the deletion of at most k elements.

(i) (Pollack, 1985) Show that $k(n) \le \frac{3}{4}n$.

(ii) (Chung et al., 1987) Show that $k(n) \ge \frac{13}{19}n$ if n is divisible by 19.

13.8 (Reutter, 1972)   Let $f_2^{\min}(n)$ denote the maximum number of times the minimum distance can occur among n points in the plane. Prove that

$$f_2^{\min}(n) \geq \lfloor 3n - \sqrt{12n - 3} \rfloor.$$

13.9   (Avis, 1984; Edelsbrunner and Skiena, 1989)   Given a set P of n points in the plane, for any $p \in P$, let $\varphi_P(p)$ denote the number of *farthest neighbors* of p;

$$\varphi_P(p) = \left| \left\{ q \in P \mid |p - q| = \max_{r \in P} |p - r| \right\} \right|.$$

Let $\Phi(n) = \max_{|P| = n} \sum_{p \in P} \varphi_P(p)$. Prove that

  (i) $\Phi(n) \leq 3n - 2$;

  (ii) if P is in convex position, then

$$\sum_{p \in P} \varphi_P(p) \leq 2n,$$

and that this bound is the best possible in the worst case.

13.10* (Csizmadia, 1995; Avis et al., 1988)   Let $\Phi(n)$ be defined for $\mathbb{R}^3$ in the same way as in Exercise 13.9. Show that if n is sufficiently large, then

$$\Phi(n) = \frac{n^2}{4} + \frac{3n}{2} + \begin{cases} 3 & \text{if } n \equiv 0 \pmod{2}, \\ 9/4 & \text{if } n \equiv 1 \pmod{4}, \\ 13/4 & \text{if } n \equiv 3 \pmod{4}. \end{cases}$$

13.11   Let $f_d^{\max}(n)$ denote the maximum number of times the largest distance can occur among n points in $\mathbb{R}^d$. Prove that

$$f_d^{\max}(n) \geq \frac{n^2}{2} \left(1 - \frac{1}{\lfloor d/2 \rfloor} + o(1) \right),$$

provided that $d \geq 4$ and $n \to \infty$.

13.12   (Reddy and Skiena, 1995)   Let $d_1(n) > d_2(n) > \cdots$ denote the distinct distances determined by the n^2 points of an $n \times n$ piece of the integer grid, and let $s_i(n)$ denote the number of pairs of points whose distance is $d_i(n)$. Prove that $s_i(n)$ does not depend on n, provided that $n > i$.

13.13   (Lefmann and Thiele, 1995)   Let P be a set of n points in the plane, no three on a line. Let $d_1, d_2, \ldots, d_k$ denote the distinct distances determined by P, and let s_i be the multiplicity of d_i, that is, the number of point pairs whose distance is d_i $(1 \le i \le k)$. Prove that

$$\sum_{i=1}^{k} s_i^2 \le \frac{3}{4} n^2 (n-1).$$

13.14   Let P be a convex polytope in $\mathbb{R}^d$, $d \le 3$. Deduce from Corollary 13.15 that P can be partitioned into $d + 1$ subsets of smaller diameter.

13.15   (Berlekamp, 1969)   Let H be a hypergraph of n vertices such that $|E|$ is odd and $|E \cap F|$ is even for any pair of distinct hyperedges $E, F \in E(H)$. Prove that $|E(H)| \le n$.

13.16⋆   (Frankl and Wilson, 1981)   Let T be a set of nonnegative integers, and let H be a hypergraph of n vertices such that $|E \cap F| \in T$ for any pair of distinct hyperedges $E, F \in E(H)$. Prove that

$$|E(H)| \le \sum_{s=0}^{|T|} \binom{n}{s}.$$

13.17⋆   (Tóth, 1995)   Let $f(n)$ denote the maximum number of times the smallest distance can occur among n points in the plane, no three of which are collinear. Prove that

$$\left(2 + \tfrac{5}{16} - o(1)\right) n \le f(n) \le \left(2 + \tfrac{3}{7}\right) n.$$

14

Geometric Graphs

Most textbooks on graph theory and topology start with the discussion of the Königsberg bridge problem or Euler's polyhedral formula. Indeed, these two fields stem from the same root, and it is not surprising that the first genuine monograph on graph theory (König, 1936) had the subtitle *Kombinatorische Topologie der Streckenkomplexe*. However, in the past few decades the connection between graph theory and topology has somewhat faded away. In the most prolific new areas of graph theory (Ramsey theory, extremal graph theory, random graphs, etc.), graphs are regarded as abstract binary relations rather than one-dimensional simplicial complexes. More than 25 years ago, Turán (1970a) remarked that the part of graph theory "mainly relevant for applications belongs essentially to logic." Unfortunately, traditional graph theory is often incapable of providing satisfactory answers to questions arising in geometric applications (including a number of problems on repeated distances discussed in previous chapters). To address these questions, new techniques are required, combining combinatorial and geometric ideas. In this chapter we present some recent results of this kind, which indicate the emergence of a new trend in combinatorial geometry.

Definition 14.1. A *geometric graph* G is a graph drawn in the plane by (possibly crossing) straight-line segments; that is, it is defined as a pair $(V(G), E(G))$, where $V(G)$ is a set of points in the plane, no three of which are collinear, and $E(G)$ is a set of segments whose endpoints belong to $V(G)$. If $V(G)$ is the vertex set of a convex polygon, then G is called a *convex geometric graph*. $V(G)$ and $E(G)$ are the *vertex set* and the *edge set* of G, respectively.

We say that H is a *geometric subgraph* of G if $V(H) \subseteq V(G)$ and $E(H) \subseteq E(G)$.

FORBIDDEN GEOMETRIC SUBGRAPHS

Kupitz (1979), Erdős, and Perles initiated the investigation of the following general problem, similar to the fundamental question of extremal graph theory formulated in the introduction to Chapter 9. Given a class $\mathcal{H}$ of so-called *forbidden geometric subgraphs*, determine or estimate the maximum number $t(\mathcal{H}, n)$ (resp. $t_c(\mathcal{H}, n)$) of edges that a geometric graph (resp. a convex geometric graph) with n vertices can have without containing a subgraph belonging to $\mathcal{H}$.

For any $k \geq 1$, let $\mathcal{D}_k$ denote the class of all geometric graphs with $2k$ vertices, consisting of k pairwise disjoint edges. (Recall that, by definition, two disjoint edges cannot share even an endpoint.)

The proof of the following statement is almost identical to that of Theorem 13.13, so it is left to the reader (Exercise 14.1).

Theorem 14.2. *Let $t(\mathcal{D}_2, n)$ denote the maximum number of edges that a geometric graph with n vertices can have without containing two disjoint edges. Then*

$$t(\mathcal{D}_2, n) = n \qquad \text{for every } n \geq 3.$$

It is somewhat trickier to obtain a linear upper bound on $t(\mathcal{D}_3, n)$. The first such result was found by Alon and Erdős (1989), and it was subsequently improved by O'Donnell and Perles (1990) and Goddard et al. (1993).

Theorem 14.3 (Goddard et al.). *Let $t(\mathcal{D}_3, n)$ denote the maximum number of edges that a geometric graph with n vertices can have without containing three pairwise disjoint edges. Then*

$$t(\mathcal{D}_3, n) \leq 3n.$$

Proof. We say that an edge xy is to the *left* of edge xz if the ray $\overrightarrow{xz}$ can be obtained from the ray $\overrightarrow{xy}$ by a clockwise turn of less than π. A vertex x is called *pointed* if all edges incident with it lie in a half-plane whose boundary contains x.

Let G be a geometric graph with n vertices and at least $3n + 1$ edges. For each pointed vertex, delete from G the leftmost edge incident with it, and let $G_1 \subseteq G$ denote the resulting subgraph. Furthermore, for each vertex x, delete from G_1 all edges xy for which there are no two edges of G_1 incident with x to the right of xy.

Since for every vertex we have deleted at most three edges, the remaining graph has at least one edge $x_0 y_0$. Then there exist two edges $x_0 y_1, x_0 y_2 \in E(G_1)$ to the right of $x_0 y_0$, and two edges $y_0 x_1, y_0 x_2 \in E(G_1)$ to the right of $y_0 x_0 = x_0 y_0$ (in this order). We can also find an edge $x_2 y \in E(G)$ to the left of $x_2 y_0$, and an edge $y_2 x \in E(G)$ to the left of $y_2 x_0$ (see Figure 14.1).

We may assume, without loss of generality, that the intersection of the lines $y_0 x_2$ and $x_0 y_2$ is on the same side of $x_0 y_0$, as y_2 (or at infinity). Then $y_0 x_2$, $x_0 y_1$, and $y_2 x$ are three disjoint edges of G. (Observe that $y_0 x_2$ and $y_2 x$ must be disjoint because they lie on opposite sides of $x_0 y_2$.) □

As for the lower bound on $t(\mathcal{D}_3, n)$, it is not difficult to show that $t(\mathcal{D}_3, n) \geq \frac{5}{2} n - 4$ (Exercise 14.3).

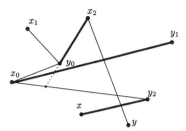

Figure 14.1. Illustration for the proof of Theorem 14.3.

Unfortunately, the generalization of the above argument for large (but fixed) values of k seems hopelessly complicated, and it is unlikely to yield a linear upper bound on $t(\mathcal{D}_k, n)$. In the next section we obtain such a bound by following a completely different approach.

Our problem becomes much simpler if we restrict our attention to convex geometric graphs G, i.e., if we assume that the vertices of G are in convex position.

Proposition 14.4 (Kupitz, 1982). *Let $t_c(\mathcal{D}_{k+1}, n)$ denote the maximum number of edges that a convex geometric graph with n vertices can have without containing $k + 1$ pairwise disjoint edges. Then for every k and $n \geq 2k + 1$,*

$$t_c(\mathcal{D}_{k+1}, n) = kn.$$

Proof. Let G be a convex geometric graph, whose vertices are denoted by $x_0, \ldots, x_{n-1}$, in cyclic order. Suppose, without loss of generality, that $\{x_0, \ldots, x_{n-1}\}$ is the vertex set of a regular n-gon. Partition the set of segments $x_i x_j$ ($i \neq j$) into n classes so that two segments belong to the same class if and only if they are parallel. Notice that if G has no $k + 1$ pairwise disjoint edges, then each class contains at most k elements of $E(G)$. Thus, $|E(G)| \leq kn$.

To see that this bound cannot be improved, consider the graph G whose edges are $x_i x_{i+\lfloor n/2 \rfloor + j}$ ($0 \leq i \leq n - 1$, $1 \leq j \leq k$), where the indices are taken modulo n (see Figure 14.2). $\qquad\square$

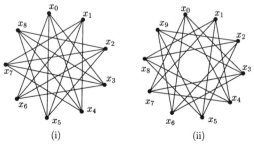

Figure 14.2. (i) $n = 9$, $k = 2$, (ii) $n = 10$, $k = 2$.

Next, we study the analogous question in the case when the forbidden configuration is a simple closed polygon with k edges. In fact, we can formulate our result in a slightly more general form.

Definition 14.5. A geometric graph is called *outerplanar* if it can be obtained by adding some noncrossing internal diagonals to a simple closed polygon in the plane. Two geometric graphs are said to be *isomorphic* if they are topologically equivalent, that is, if there exists a one-to-one continuous mapping $f: \mathbb{R}^2 \to \mathbb{R}^2$ that carries one into the other.

Theorem 14.6 (Pach; Perles). *Let $\mathcal{G}_k$ denote the class of all geometric graphs isomorphic to a fixed outerplanar graph G_k with k vertices ($k \geq 3$), and let $t(\mathcal{G}_k, n)$ denote the maximum number of edges that a geometric graph with n vertices can have without containing a geometric subgraph isomorphic to G_k. Then*

$$t(\mathcal{G}_k, n) = |E(T_{k-1}(n))|,$$

where $T_{k-1}(n)$ denotes the balanced complete $(k - 1)$-partite graph with n vertices (cf. the second section of Chapter 9).

The proof is based on the following lemma of Gritzmann et al. (1991) and Perles.

Lemma 14.7. *Let G_k be an outerplanar geometric graph with k vertices, and let P be any set of k points in the plane, no three of which are collinear. Then there exists a geometric graph G'_k isomorphic to G_k, with vertex set $V(G'_k) = P$.*

Proof. We can assume without loss of generality that G_k is triangulated; i.e., $|E(G_k)| = 2k - 3$. It will be more convenient to establish a stronger assertion.

Let x_1 and x_2 be any two consecutive vertices of the closed polygon bounding G_k, and let p_1 and p_2 be any two consecutive vertices of the convex hull of P. We are going to prove that there is a bijection $f: V(G_k) \to P$ such that $f(x_1) = p_1$, $f(x_2) = p_2$, and the open segments $(f(x), f(y))$, $xy \in E(G_k)$ are pairwise disjoint.

The proof is by induction on k. The assertion is trivial for $k = 3$. Suppose that $k > 3$, and let $x_1, x_2, \ldots, x_k$ denote the vertices of G_k, say, in counterclockwise order. Assume that the third vertex of the triangle containing x_1 and x_2 is x_i. We claim that there exists a point $p \in P - \{p_1, p_2\}$ satisfying the following two conditions:

   (i) the interior of the triangle $p_1 p_2 p$ contains no element of P;

  (ii) there is a line ℓ through p separating p_1 from p_2 such that $\ell \cap P = \{p\}$, and there are exactly $i - 2$ elements of P in the open half-plane bounded by ℓ which contains p_2.

To find such a point p, first choose a line $p_2 p'$ that separates exactly $i - 3$ elements of P from p_1. If there is no point of P in the interior of the triangle $p_1 p_2 p'$, then let $p = p'$. Otherwise, let p be the point in this region minimizing the angle $p_2 p_1 p$. By this choice, p satisfies (i), and there are at most $i - 2$ elements of P on the side of $p_1 p$ containing p_2. If we rotate this line about p so that its point of intersection with $p_1 p_2$ moves toward p_2, then it will reach a position ℓ satisfying (ii) before it would become parallel to $p_2 p'$.

Let H_1 and H_2 denote the closed half-planes bounded by ℓ, containing p_1 and p_2, respectively. Let G_1 and G_2 be the subgraphs of G induced by $\{x_i, x_{i+1}, \ldots, x_1\}$ and $\{x_2, x_3, \ldots, x_i\}$, respectively. By the induction hypothesis, we obtain two bijections $f_j : V(G_j) \to P \cap H_j$ ($j = 1, 2$), such that $f_1(x_1) = p_1$, $f_2(x_2) = p_2$, and $f_1(x_i) = f_2(x_i) = p$. Combining them into a single bijection $f : V(G_1) \cup V(G_2) \to P$, the assertion follows. $\square$

Proof of Theorem 14.6. Let G be a geometric graph with n vertices and more than $|E(T_{k-1}(n))|$ edges. By Turán's theorem (Theorem 9.3), G must contain a complete geometric subgraph of k points. That is, there exists a k-element subset $P \subseteq V(G)$ such that all the $\binom{k}{2}$ segments determined by P belong to $E(G)$. Lemma 14.7 implies that the complete geometric graph induced by the points of P contains a subgraph isomorphic to G_k.

On the other hand, it is easy to construct a convex geometric graph G with n vertices and $|E(T_{k-1}(n))|$ edges that does not contain any simple closed polygon of length k (see Exercise 14.5). By definition, every outerplanar graph G_k with k vertices contains such a polygon; hence $G_k \not\subseteq G$. $\square$

PARTIALLY ORDERED SETS

The edges of a geometric graph can be partially ordered in many natural ways, and these orderings store a lot of information about the topological structure of the original graph. In this section we exploit this information to generalize Theorems 14.2 and 14.3.

Definition 14.8. A *partially ordered set* is a pair $(X, \prec)$, where X is a set and $\prec$ is a reflexive, antisymmetric, and transitive binary relation on X. For any $x \ne y \in X$, we say that x and y are *comparable* in $(X, \prec)$, when either $x \prec y$ or $y \prec x$. If any two elements of a subset $C \subseteq X$ are comparable, then C is called a *chain*. If no two elements of C are comparable, then C is called an *antichain*.

We need the following basic result.

Theorem 14.9 (Dilworth, 1950). *Let $(X, \prec)$ be a finite partially ordered set.*

(i) *If the maximum length of a chain is k, then X can be partitioned into k antichains.*

(ii) *If the maximum length of an antichain is k, then X can be partitioned into k chains.*

Obviously, these statements cannot be improved, for a chain and an antichain can have at most one element in common (Figure 14.3).

Proof.   (i) For any $x \in X$, define the *rank* of x as the size of the longest chain whose maximal element is x. Clearly, $1 \leq \text{rank}(x) \leq k$, and the set of all elements of the same rank is an antichain.

(ii) The assertion is trivial if $|X| = 1$. Assume that it is valid for all partially ordered sets with fewer than n elements, and let $|X| = n \geq 2$. We can also assume that X cannot be partitioned into two nonempty subsets X_1 and X_2 such that any two elements belonging to different subsets are incomparable; otherwise, we can apply the induction hypothesis to $(X_1, \prec)$ and $(X_2, \prec)$ separately.

Let $x_{\min}$ and $x_{\max}$ be a minimal element and a maximal element of $(X, \prec)$, respectively, such that $x_{\min} \prec x_{\max}$. Deleting $x_{\min}$ and $x_{\max}$ from X, we get a partially ordered set of $n - 2$ elements, in which the maximum size of an antichain is at most k. If this set has no antichain of size k, then it can be partitioned into $k - 1$ chains, by the induction hypothesis. Adding the chain $\{x_{\min}, x_{\max}\}$, we obtain a partition we seek.

Suppose that $X - \{x_{\min}, x_{\max}\}$ contains an antichain A of size k, and let

$$A^+ = \{x \in X \,|\, \text{there exists } a \in A \text{ such that } a \prec x\},$$
$$A^- = \{x \in X \,|\, \text{there exists } a \in A \text{ such that } a \succ x\}.$$

Obviously, $A^+ \cup A^- = X$, $A^+ \cap A^- = A$, $x_{\min} \notin A^+$, and $x_{\max} \notin A^-$. Now both A^+ and A^- can be partitioned into k chains, each containing exactly one element of A. Gluing two such chains together if they contain the same element of A, we obtain a partition of X into k chains, as desired.   □

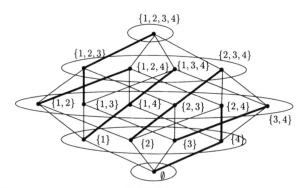

Figure 14.3. Partition of $(2^{\{1,2,3,4\}}, \subset)$ into chains and antichains.

Now we are in the position to prove the following result, which settles a conjecture of Avital and Hanani (1966), Kupitz (1979), and Erdős in the affirmative.

Theorem 14.10 (Pach and Törőcsik, 1993). *Let $t(\mathcal{D}_{k+1}, n)$ denote the maximum number of edges that a geometric graph with n vertices can have without containing $k + 1$ pairwise disjoint edges. Then for every $k, n \geq 1$,*

$$t(\mathcal{D}_{k+1}, n) \leq k^4 n.$$

Proof. Let G be a geometric graph with n vertices, containing no $k + 1$ pairwise disjoint edges. For any point v, let $x(v)$ and $y(v)$ denote the x- and y-coordinates of v, respectively. Assume without loss of generality that no two vertices of G have the same x-coordinates.

Let $uv, u'v' \in E(G)$, and suppose that $x(u) < x(v)$, $x(u') < x(v')$. We say that uv *precedes* $u'v'$ (in notation $uv \ll u'v'$) if $x(u) \leq x(u')$ and $x(v) \leq x(v')$. Furthermore, uv is said to lie *below* $u'v'$ if there is no vertical line ℓ that intersects both uv and $u'v'$ such that

$$y(\ell \cap uv) \geq y(\ell \cap u'v').$$

[Note that this latter relation is not necessarily antisymmetric or transitive on $E(G)$.]

Define four binary relations $\prec_i$ ($1 \leq i \leq 4$) on $E(G)$, as follows. Let uv and $u'v'$ be two disjoint edges of G such that uv lies below $u'v'$. Then

$$
\begin{aligned}
uv \prec_1 u'v' \quad &\text{if} \quad uv \ll u'v', \\
uv \prec_2 u'v' \quad &\text{if} \quad u'v' \ll uv, \\
uv \prec_3 u'v' \quad &\text{if} \quad [x(u), x(v)] \subset [x(u'), x(v')], \\
uv \prec_4 u'v' \quad &\text{if} \quad [x(u'), x(v')] \subseteq [x(u), x(v)]
\end{aligned}
$$

(see Figure 14.4). It follows readily from the definitions that

(a) $(E(G), \prec_i)$ is a partially ordered set ($1 \leq i \leq 4$);

(b) any pair of disjoint edges is comparable by at least one of the relations $\prec_i$ ($1 \leq i \leq 4$).

Observe that $(E(G), \prec_i)$ cannot contain a chain of length $k + 1$; otherwise, G would have $k+1$ pairwise disjoint edges ($1 \leq i \leq 4$). By Theorem 14.9(i), for any i, $E(G)$ can be partitioned into at most k classes, so that no two edges belonging to the same class are comparable by $\prec_i$. Overlaying these four partitions, we obtain a decomposition of $E(G)$ into at most k^4 classes E_j ($1 \leq j \leq k^4$) such that no two elements of E_j are comparable by any of the relations $\prec_i$. Hence, by (b), none of the classes E_j contains two disjoint edges. Applying Theorem

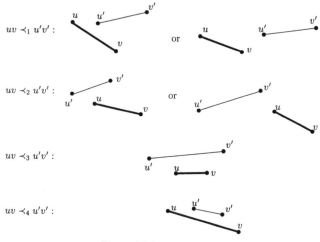

Figure 14.4. Relations $\prec_i$.

14.2 to the graphs $(V(G), E_j)$, we obtain

$$|E_j| \leq t(\mathcal{D}_2, n) \leq n \qquad (1 \leq j \leq k^4),$$

$$|E(G)| = \sum_{j=1}^{k^4} |E_j| \leq k^4 n. \qquad \qquad \Box$$

CROSSING EDGES

We say that two edges of a geometric graph *cross* each other if they have an interior point in common. For any $k \geq 2$, let C_k denote the family of all geometric graphs consisting of k pairwise crossing edges. Since a planar graph of n vertices can have at most $3n - 6$ edges, we have the following.

Proposition 14.11. *Let $t(C_2, n)$ denote the maximum number of edges that a geometric graph with n vertices can have without containing two crossing edges. Then for every $n \geq 3$,*

$$t(C_2, n) = 3n - 6.$$

Encouraged by the results of the preceding section, one is tempted to conjecture that for every $k \geq 3$, there exists a constant c_k such that $t(C_k, n) \leq c_k n$. This has been confirmed for $k = 3$ by Agarwal et al. (1995), but the problem is open for $k > 3$.

In order to give a nontrivial upper bound on $t(C_3, n)$, we need a quantitative result that guarantees that any geometric graph with a large number of edges contains many crossings.

Theorem 14.12 (Ajtai et al., 1982; Leighton, 1983). *Let $\kappa(n,m)$ denote the minimum number of crossing pairs of edges in a geometric graph with n vertices and m edges. If $m \geq 4n$, then*

$$\kappa(n,m) \geq \frac{1}{100} \cdot \frac{m^3}{n^2}.$$

Proof. For any fixed $n \geq 3$, it follows readily by induction on m that

$$\kappa(n,m) \geq m - 3n + 6. \tag{14.1}$$

Indeed, if $m \leq 3n - 6$, then there is a nothing to prove. Let G be a geometric graph with n vertices and $m > 3n - 6$ edges. Then it must contain at least one crossing pair of edges. Applying the induction hypothesis to the graph obtained from G by the deletion of any edge involved in a crossing, (14.1) follows.

We shall prove, by induction on n, that

$$\kappa(n,m) \geq \frac{3}{64} \cdot \binom{n}{4} \cdot \left(m \Big/ \binom{n}{2}\right)^3 \qquad \text{for every } m \geq 4n. \tag{14.2}$$

If $n \leq 8$, then G cannot have $4n$ edges. If $n = 9$, then G must be a complete graph with $m = \binom{9}{2} = 36$ edges, and (14.2) follows from (14.1). In fact, (14.1) also yields (14.2) for $n \geq 10$, $4n \leq m \leq 5n$.

Hence, we can assume that G is a geometric graph with $n \geq 10$ vertices and $m > 5n$ edges, and that (14.2) is valid for any geometric graph with fewer than n vertices. Let $\kappa(G)$ denote the number of crossing pairs of edges in G. Obviously,

$$\sum_{x \in V(G)} \kappa(G - x) = (n - 4)\kappa(G),$$

where $G - x$ denotes the graph obtained from G by the deletion of x and all edges incident to x. Since $G - x$ has at least $m - (n - 1) \geq 5n - (n - 1) > 4(n - 1)$ edges, we can use the induction hypothesis to get that

$$\kappa(G - x) \geq \frac{3}{64} \cdot \binom{n - 1}{4} \cdot \left(m_x \Big/ \binom{n - 1}{2}\right)^3,$$

where m_x is the number of edges in $G - x$. Using the fact that

$$\sum_{x \in V(G)} m_x = m(n - 2),$$

by Jensen's inequality,

$$\kappa(G) = \frac{1}{n-4} \sum_{x \in V(G)} \kappa(G-x)$$

$$\geq \frac{1}{n-4} \cdot \frac{3}{64} \cdot \frac{\binom{n-1}{4}}{\binom{n-1}{2}^3} \sum_{x \in V(G)} m_x^3$$

$$\geq \frac{1}{n-4} \cdot \frac{3}{64} \cdot \frac{\binom{n-1}{4}}{\binom{n-1}{2}^3} \cdot n\left(\frac{m(n-2)}{n}\right)^3$$

$$= \frac{3}{64}\binom{n}{4}\left(m\Big/\binom{n}{2}\right)^3,$$

as asserted. Observe that the right-hand side of (14.2) is greater than $\frac{1}{100}(m^3/n^2)$, whenever $n \geq 9$. □

The order of magnitude of the above lower bound on $\kappa(n,m)$ cannot be improved. To see this, distribute n points as evenly as possible on $n^2/(2m)$ disjoint circles in the plane, and join two points by a segment if they lie on the same circle.

Theorem 14.12 can be used to obtain nontrivial (i.e., subquadratic) upper bounds for $t(C_{k+1}, n)$ for every fixed $k \geq 2$.

Corollary 14.13. *Let $t(C_3, n)$ denote the maximum number of edges that a geometric graph with n vertices can have without containing three pairwise crossing edges. Then*

$$t(C_3, n) < 13n^{3/2}.$$

Proof. Let G be a geometric graph with n vertices and $m \geq 4n$ edges. Then, by Theorem 14.12, there exists an edge e crossing at least $m^2/(50n^2)$ other edges of G. If G has no three pairwise crossing edges, the edges that cross e form a planar graph of $n-2$ vertices. Hence,

$$m^2/(50n^2) \leq t(C_2, n-2) \leq 3(n-2) - 6,$$

and the result follows. □

Iterating this argument, we obtain that for every k there exists a constant c_k such that

$$t(C_k, n) \leq c_k n^{2 - 1/2^{k-2}}$$

(see Exercise 14.8).

In the next section we substantially improve these naive upper bounds by using a different approach.

As in Proposition 14.4, where the forbidden configuration consisted of pairwise disjoint edges, again we have an exact result for convex geometric graphs.

Theorem 14.14 (Capoyleas and Pach, 1992). *Let $t_c(C_{k+1}, n)$ denote the maximum number of edges that a convex geometric graph with n vertices can have without containing $k + 1$ pairwise crossing edges. Then for every k and $n \geq 2k + 1$,*

$$t_c(C_{k+1}, n) = 2kn - \binom{2k + 1}{2}.$$

Proof.   Let $k \geq 1$ be fixed. For $n = 2k + 1$ the statement is trivial, as

$$2k(2k + 1) - \binom{2k + 1}{2} = \binom{2k + 1}{2}.$$

Assume that we have already established the theorem for every number smaller than n, and consider a graph G consisting of the maximum number of line segments connecting the vertices of a convex n-gon such that no $k + 1$ cross each other.

We need the following easy observation.

Claim A: If two vertices are separated by fewer than k vertices along the boundary of our convex n-gon, they are connected by an edge of G.

Indeed, by adding an edge between two such vertices, we cannot create $k + 1$ pairwise crossing edges in G.

If there is no edge whose endpoints are separated by at least k vertices along (both arcs of) the boundary of the n-gon, then

$$|E(G)| = t_c(C_{k+1}, n) = kn \leq 2kn - \binom{2k + 1}{2}.$$

So we may assume that there is such an edge $uv \in E(G)$, and let

$$u, x_1, \ldots, x_{n_1}, v, y_1, \ldots, y_{n_2} \qquad (n_1, n_2 \geq k)$$

denote the vertices of G, listed in clockwise order.

Define a partial order "$\prec$" on the set of all edges of G that cross uv, as follows. Two edges $x_i y_j$ and $x_{i'} y_{j'}$ are comparable if and only if they cross each other, and

$$x_i y_j \prec x_{i'} y_{j'} \iff i < i' \quad \text{and} \quad j < j'.$$

This is clearly a transitive relation.

As in the proof of Theorem 14.9(i), define the *rank* of an edge $x_i y_j$ as the size of the longest chain whose maximal element is $x_i y_j$. The set of all elements of the same rank is an *antichain*; i.e., no two edges of the same rank cross each other. Furthermore, for every edge

$$1 \le \operatorname{rank}(x_i y_j) \le k - 1;$$

otherwise, G would have $k + 1$ pairwise crossing edges.

Next, we define a convex geometric graph G_1 with $n_2 + k + 1$ vertices,

$$u, x_1^*, \ldots, x_{k-1}^*, v, y_1, \ldots, y_{n_2}$$

(in clockwise order), as follows. Let G_1 be the same as G, when restricted to $\{u, v, y_1, \ldots, y_{n_2}\}$. Let $x_r^* y_j \in E(G_1)$ if and only if there exists $x_i y_j \in E(G)$ with $\operatorname{rank}(x_i y_j) = r$. Finally, let $\{u, x_1^*, \ldots, x_{k-1}^*, v\}$ induce a subgraph in G_1, whose only edge is uv.

Claim B: G_1 has no $k + 1$ pairwise crossing edges.

It will be more convenient to establish the stronger statement that if there are t pairwise crossing edges $x_{r_1}^* y_{j_1}, \ldots, x_{r_t}^* y_{j_t}$ in G_1 ($r_1 > \cdots > r_t$, $j_1 > \cdots > j_t$), then one can find t pairwise crossing edges in G such that all of them cross uv and have an endpoint belonging to $\{y_j \mid j_1 \ge j \ge j_t\}$.

Pick t edges $x_{i_1} y_{j_1}, \ldots, x_{i_t} y_{j_t} \in E(G)$ with

$$\operatorname{rank}(x_{i_1} y_{j_1}) = r_1, \ldots, \operatorname{rank}(x_{i_t} y_{j_t}) = r_t.$$

Let $x_{p_1} y_{q_1} = x_{i_1} y_{j_1}$, and suppose by induction that we have already found a sequence of pairwise crossing edges $x_{p_1} y_{q_1}, \ldots, x_{p_{t-1}} y_{q_{t-1}} \in E(G)$ such that

1. $\operatorname{rank}(x_{p_1} y_{q_1}) = r_1, \ldots, \operatorname{rank}(x_{p_{t-1}} y_{q_{t-1}}) = r_{t-1}$;
2. $q_1 = j_1$, $q_2 \ge j_2, \ldots, q_{t-1} \ge j_{t-1}$.

There exists an edge $x_p y_q$ of rank r_t, which crosses $x_{p_{t-1}} y_{q_{t-1}}$. If $q \ge j_t$, then choose this edge to be $x_{p_t} y_{q_t}$, and we are done. If $q < j_t$, then $p \ge i_t$, because $x_p y_q$ and $x_{i_t} y_{j_t}$ (two edges of the same rank) cannot cross. Now let $x_{p_t} y_{q_t} = x_{i_t} y_{j_t}$. This edge must cross $x_{p_{t-1}} y_{q_{t-1}}$ and hence all edges $x_{p_1} y_{q_1}, \ldots, x_{p_{t-1}} y_{q_{t-1}}$. This proves Claim B.

Claim C: $|E(G_1)| \le t_c(C_{k+1}, n_2 + k + 1) - k^2 + k$.

In view of Claim B, it is sufficient to show that one can add $k^2 - k$ edges to G_1 without creating $k + 1$ pairwise crossing edges. By Claim A, any edge connecting a pair of points separated by fewer than k vertices can be added to G_1. Observe that for every edge $x_i y_j \in E(G)$, $1 \le j \le k - 1$,

$$\operatorname{rank}(x_i y_j) \le j.$$

Thus,

$$x_r^* y_j \notin E(G_1) \qquad \text{whenever } r > j.$$

Furthermore, all edges within $\{u, x_1^*, \ldots, x_{k-1}^*, v\}$ except uv are missing from G_1. Hence, at least

$$\sum_{j=1}^{k-1}(k - 1 - j) + \binom{k+1}{2} - 1 = k^2 - k$$

edges can be added to G_1 without creating $k + 1$ pairwise crossing edges. This proves Claim C.

Similarly, we can define a convex geometric graph G_2 on the vertex set

$$\{u, x_1, \ldots, x_{n_1}, v, y_1^*, \ldots, y_{k-1}^*\}$$

by connecting x_i and y_r^* with an edge of G_2 if and only if there exists an edge $x_i y_j \in E(G)$ with rank $(x_i y_j) = r$, and letting $\{u, x_1, \ldots, x_{n_1}, v\}$ induce the same subgraph in G_2 as in G. Just as before,

$$|E(G_2)| \leq t_c(C_{k+1}, n_1 + k + 1) - k^2 + k.$$

Let $d_{G_i}(z)$ denote the degree of $z \in V(G_i)$ in G_i ($i = 1, 2$), and let e_r denote the number of edges in G whose rank is r ($1 \leq r \leq k - 1$). Since the edges of rank r form a *forest* (i.e., an acyclic subgraph) in G with $d_{G_1}(x_r^*) + d_{G_2}(y_r^*)$ nonisolated vertices, we have the following.

Claim D: $e_r \leq d_{G_1}(x_r^*) + d_{G_2}(y_r^*) - 1$ for every $1 \leq r \leq k - 1$.

Now we can easily complete the proof of Theorem 14.14. Since $uv \in E(G_1) \cap E(G_2)$, by Claim D,

$$|E(G)| = |E(G_1)| + |E(G_2)| - 1 - \sum_{r=1}^{k-1}(d_{G_1}(x_r^*) + d_{G_2}(y_r^*) - e_r)$$

$$\leq |E(G_1)| + |E(G_2)| - k.$$

Combining this with Claim C, we obtain

$$|E(G)| \leq t_c(C_{k+1}, n_1 + k + 1) + t_c(C_{k+1}, n_2 + k + 1) - 2k^2 + k.$$

According to our assumptions, $n_1 + n_2 + 2 = n$, $n_1 \geq k$, and $n_2 \geq k$. Therefore,

$$2k + 1 \leq n_i + k + 1 < n \qquad \text{for } i = 1, 2,$$

and we can apply the induction hypothesis to obtain

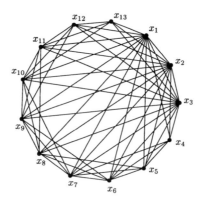

Figure 14.5. $n = 13$, $k = 3$.

$$\|E(G)\| = t_c(C_{k+1}, n)$$

$$\leq 2k(n_1 + k + 1) + 2k(n_2 + k + 1) - 2\binom{2k+1}{2} - 2k^2 + k$$

$$= 2k(n_1 + n_2 + 2) - 2k^2 - k$$

$$= 2kn - \binom{2k+1}{2},$$

as desired.

This bound is tight, as is shown by the following example. Let $x_1, \ldots, x_n$ be the vertices of a convex n-gon in clockwise order. Connect x_i and x_j $(i \neq j)$ by an edge if and only if they are separated by fewer than k vertices along the boundary of the polygon, or $i \leq k$ (see Figure 14.5).   ☐

CROSSING NUMBER AND BISECTION WIDTH

In Corollary 14.13 we proved that the number of edges of a geometric graph with n vertices, containing no three pairwise crossing edges, cannot exceed $13n^{3/2}$. The proof was based on a general lower bound for the minimum number of crossing pairs of edges in a geometric graph with n vertices and m edges (cf. Theorem 14.12). Our arguments suggest that if we want to improve Corollary 14.13, we have to analyze the structure of graphs that can be drawn with relatively few crossings. In the present section we follow this approach. Our method can be applied not only to geometric graphs, but to any graph drawn in the plane by Jordan arcs, no two of which meet more than once.

Definition 14.15. Let G be a simple graph of n vertices. A *drawing* of G is a representation of G in the plane such that every vertex corresponds to a point, and every edge is represented by a Jordan arc connecting the corresponding two points without passing through any vertex other than its endpoints. Two arcs are said to cross each other if they have an interior point in common. The *crossing number* cr (G) of G is defined as the minimum number of crossing pairs of arcs in a drawing of G.

It is easy to show that this minimum can always be attained for a drawing that satisfies the following conditions (see Exercise 14.12):

(i) No two arcs meet in more than one point (including their endpoints);
(ii) no three arcs share a common interior point.

For any partition of the vertex set of a graph G into two disjoint parts, V_1 and V_2, let $E(V_1, V_2)$ denote the set of edges with one endpoint in V_1 and the other endpoint in V_2. Define the *bisection width* of G as

$$b(G) = \min_{|V_1|, |V_2| \leq 2n/3} |E(V_1, V_2)|,$$

where the minimum is taken over all partitions $V(G) = V_1 \cup V_2$ such that $|V_1|, |V_2| \leq 2n/3$.

We need the following result, which is an easy consequence of a weighted version of the Lipton–Tarjan separator theorem for planar graphs (Theorem 8.3).

Lemma 14.16. *Let G be a graph of n vertices with degrees $d_1, \ldots, d_n$. Then*

$$b^2(G) \leq (1.58)^2 \left(16 \operatorname{cr}(G) + \sum_{i=1}^{n} d_i^2 \right),$$

where $b(G)$ and cr (G) denote the bisection width and the crossing number of G, respectively.

Proof. Consider a drawing of G with cr (G) crossing pairs of arcs satisfying conditions (i) and (ii). Introducing a new vertex at each crossing, we obtain a planar graph H with $N = n + \operatorname{cr}(G)$ vertices. Assign weight 0 to each new vertex and weights of $1/n$ to all other vertices. By Exercise 8.5, H can be separated into two parts H_1 and H_2 by deleting at most

$$1.58 \left(16 \operatorname{cr}(G) + \sum_{i=1}^{n} d_i^2 \right)^{1/2}$$

edges, such that each of the sets $V_1 = V(H_1) \cap V(G)$ and $V_2 = V(H_2) \cap V(G)$ has at most $2n/3$ elements. Hence,

$$b(G) \leq |E(V_1, V_2)| \leq 1.58 \left(16 \operatorname{cr}(G) + \sum_{i=1}^{n} d_i^2 \right)^{1/2}. \qquad \square$$

Now we are ready to prove the main result of this section.

Theorem 14.17 (Pach et al., 1994). *Let G be a graph with n vertices, and let k be a positive integer. If G has a drawing such that no two arcs meet in more than one point and there are no k + 1 pairwise crossing arcs, then*

$$|E(G)| \le 3n(10 \log n)^{2k-2}.$$

Proof. We prove the theorem by double induction on k and n. The assertion is true for $k = 1$ and for all n. It is also true for any $k > 1$ and $n \le 6 \cdot 10^{2k-2}$, because for these values the above upper bound exceeds $\binom{n}{2}$.

Assume that we have already proved the theorem for some k and all n. We now prove it for $k + 1$ and all n. Let $n \ge 6 \cdot 10^{2k}$, and suppose that the theorem holds for $k + 1$ and for all graphs having fewer than n vertices.

Let G be a graph of n vertices, which has a drawing such that no two arcs meet in more than one point and there are no $k + 2$ pairwise crossing edges. For the sake of simplicity, this drawing will also be denoted by $G = (V(G), E(G))$. For any arc $e \in E(G)$, let G_e denote the graph consisting of all arcs that cross e. Clearly, G_e has no $k + 1$ pairwise crossing arcs. Thus, by the induction hypothesis,

$$\text{cr}(G) \le \frac{1}{2} \sum_{e \in E(G)} |E(G_e)|$$

$$\le \frac{1}{2} \sum_{e \in E(G)} 3n(10 \log n)^{2k-2}$$

$$\le \frac{3}{2} |E(G)| \cdot n(10 \log n)^{2k-2}.$$

Since $\sum_{i=1}^{n} d_i^2 \le 2|E(G)| \cdot n$ holds for every graph G with degrees $d_1, \ldots, d_n$, Lemma 14.16 implies that

$$b(G) \le 1.58 \left(16 \, \text{cr}(G) + \sum_{i=1}^{n} d_i^2 \right)^{1/2}$$

$$\le 9 \sqrt{n|E(G)|} \cdot (10 \log n)^{k-1}.$$

Consider a partition of $V(G)$ into two parts V_1 and V_2, each containing at most $2n/3$ vertices, such that the number of edges connecting them is $b(G)$. Let G_1 and G_2 denote the subgraphs of G induced by V_1 and V_2, respectively. Since neither G_1 nor G_2 contains $k + 2$ pairwise crossing edges and each of them has fewer than n vertices, by the induction hypothesis,

$$|E(G)| = |E(G_1)| + |E(G_2)| + b(G)$$

$$\le 3n_1(10 \log n_1)^{2k} + 3n_2(10 \log n_2)^{2k} + b(G),$$

where $n_i = |V_i|$ ($i = 1, 2$). Combining the last two inequalities, we get

$$|E(G)| - 9\sqrt{n}(10\log n)^{k-1}\sqrt{|E(G)|}$$
$$\leq 3n_1\left(10\log \frac{2n}{3}\right)^{2k} + 3n_2\left(10\log \frac{2n}{3}\right)^{2k}$$
$$\leq 3n(10\log n)^{2k}\left(1 - \frac{k}{\log n}\right).$$

If the left-hand side of this inequality is negative, then $|E(G)| \leq 3n(10\log n)^{2k}$ and we are done. Otherwise,

$$f(x) = x - 9\sqrt{n}(10\log n)^{k-1}\sqrt{x}$$

is a monotone increasing function of x when $x \geq |E(G)|$. An easy calculation shows that

$$f(3n(10\log n)^{2k}) > 3n(10\log n)^{2k}\left(1 - \frac{k}{\log n}\right).$$

Hence,

$$f(|E(G)|) < f(3n(10\log n)^{2k}),$$

which in turn implies that

$$|E(G)| < 3n(10\log n)^{2k},$$

as desired.   □

Corollary 14.18. *Let* $t(C_{k+1}, n)$ *denote the maximum number of edges that a geometric graph with n vertices can have without containing* $k + 1$ *pairwise crossing edges. Then for every* $k, n \geq 1$,

$$t(C_{k+1}, n) \leq 3n(10\log n)^{2k-2}.$$

EXERCISES

14.1 (i) Prove Theorem 14.2 by using the idea of Theorem 13.13.

(ii) Deduce Theorem 13.13 from Theorem 14.2.

14.2 Let $G = (V(G), E(G))$ be a geometric graph containing no two disjoint edges. Show that there is a bijection f from $E(G)$ to a suitable subset $V' \subseteq V(G)$ such that $f(e) \in e$ for every $e \in E(G)$.

14.3 Show that for every $n \geq 4$, there exists a geometric graph with n vertices and at least $\frac{5}{2}n - 4$ edges that does not contain three pairwise disjoint edges.

14.4   For every k and $n \geq 2k + 1$, construct a convex geometric graph with n vertices and kn edges, which has k vertices of degree $n - 1$, and does not contain $k + 1$ pairwise disjoint edges (cf. Proposition 14.4).

14.5   For every n, $k \geq 3$, construct a geometric graph with n vertices and $|E(T_{k-1}(n))|$ edges, with the property that no k edges form a simple closed polygon (cf. Theorem 14.6).

14.6   (Sperner, 1928)   Given an n-element subset X, let 2^X denote the family of all subsets of X. Show that in the partially ordered set $(2^X, \subseteq)$

    (i) one can find $\binom{n}{\lfloor n/2 \rfloor}$ chains whose union covers 2^X,

    (ii) the maximum size of an antichain is $\binom{n}{\lfloor n/2 \rfloor}$.

14.7   (Larman et al., 1994)

    (i) Prove that any family of n closed convex sets in the plane has at least $n^{1/5}$ members that are either pairwise disjoint or pairwise intersecting.

    (ii) Construct a family of n line segments in general position in the plane, which does not contain more than $n^{\log_5 2}$ members that are either pairwise disjoint or pairwise intersecting.

14.8   Let $t(C_k, n)$ denote the maximum number of edges that a geometric graph with n vertices can have without containing k pairwise crossing edges. Deduce from Theorem 14.12 that for every $k \geq 3$, there exists a constant c_k such that

$$t(C_k, n) \leq c_k n^{2 - 1/2^{k-2}}.$$

14.9   (A. Bialostocki and P. Dierker)   Let G be a complete convex geometric graph with n vertices, whose $\binom{n}{2}$ edges are colored with two colors. Show that G has

    (i) a monochromatic spanning tree, whose edges do not cross each other;

    (ii) $\lfloor (n + 1)/3 \rfloor$ pairwise disjoint edges of the same color.

    (iii) (Károlyi et al., 1995)   Show that (i) and (ii) remain true for any geometric graph with n vertices.

14.10   (Pach, 1991; Pach and Törőcsik, 1993)   Show that for every k there exists a constant $c_k > 0$ such that

    (i) every convex geometric graph with n vertices and $m \geq (2k + 1)n$ edges contains at least $c_k m^{2k+1}/n^{2k}$ $(k + 1)$-tuples of pairwise crossing edges;

    (ii) every geometric graph with n vertices and $m \geq (k^4 + 1)n$ edges contains at least $c_k m^{2k+1}/n^{2k}$ $(k+1)$-tuples of pairwise disjoint edges.

14.11 (Ding et al., 1994; Pach and Törőcsik, 1993)   Show that for every $d \geq 2$ there exists a polynomial $p_d(n)$ satisfying the following condition. Let P be any set of points, and let $\mathcal{B}$ be any set of axis-parallel boxes in $\mathbb{R}^d$ such that every member of $\mathcal{B}$ contains at least one point of P. Then, for every positive n,

   (a) either there are $n + 1$ boxes in $\mathcal{B}$ so that no point of P belongs to more than one of them, or

   (b) one can choose at most $p_d(n)$ points in P so that any member of $\mathcal{B}$ contains at least one of them.

14.12 Let G be a geometric graph with crossing number $\mathrm{cr}(G)$. Show that there is a drawing of G by Jordan arcs such that no two arcs meet in more than one point (including their endpoints), and there are exactly $\mathrm{cr}(G)$ pairs of crossing arcs.

14.13⋆ (Tutte, 1970)   Let G be a graph that has a drawing in the plane by Jordan arcs such that

   (a) no two arcs touch each other,

   (b) any two arcs that do not have an endpoint in common intersect each other in an even number of points.

   Show that G is planar.

14.14 J. Conway calls a graph drawing a *thrackle* if

   (a) no two arcs touch each other,

   (b) any pair of arcs meet in exactly one point (including their endpoints).

   Show that

   (i) a cycle of length k can be drawn as a thrackle if and only if $k \neq 4$;

   (ii) (Woodall, 1971) if a graph G can be drawn as a thrackle, then G does not contain two vertex-disjoint odd cycles;

   (iii) (Lovász et al., 1995) if G is a bipartite graph that can be drawn as a thrackle, then G is planar.

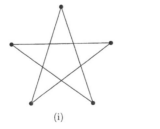 

(i)   (ii)

Figure 14.6. Cycles of length 5 and 10 drawn as thrackles.

15

Epsilon Nets and Transversals of Hypergraphs

For historical reasons, a finite set-system is often called a hypergraph. More precisely, a *hypergraph H* consists of a finite set $V(H)$ of *vertices* (points) and a family $E(H)$ of subsets of $V(H)$. The elements of $E(H)$ are usually called *hyperedges* (or, in short, *edges*). If the hyperedges of H are r-element sets, then H is said to be an *r-uniform hypergraph*. Using this terminology, a graph is a two-uniform hypergraph. In Chapter 10 we have extended some graph-theoretic results to r-uniform hypergraphs (cf. Theorems 10.11 and 10.12).

The concept of hypergraphs is a very general one, so it is not surprising that hypergraph theory has a large scale of applications in various fields of mathematics, including geometry. Given a hypergraph H, a subset $T \subseteq V(H)$ is called a *transversal* of H if $T \cap E$ is nonempty for every edge $E \in E(H)$. Many extremal problems from combinatorics and geometry can be reformulated as questions of the following type: What is the size of a smallest transversal in a given hypergraph H? This problem, in general, is known to be computationally intractable (cf. Garey and Johnson, 1979). However, under certain specific conditions on H, one can guarantee the existence of a relatively small transversal. The present chapter focuses on results of this kind. In particular, we shall see how a powerful probabilistic idea of Vapnik and Chervonenkis can be applied to obtain a number of interesting geometric and algorithmic results.

TRANSVERSALS AND FRACTIONAL TRANSVERSALS

Let H be a hypergraph with vertex set $V(H)$ and edge set $E(H)$. Let $\tau(H)$ denote the size of a smallest transversal of H, that is, the smallest number τ such that one can choose τ vertices with the property that any edge of H contains at least one of them. $\tau(H)$ is usually called the *transversal number* (or the *vertex-cover number*) of H.

The *packing number* (or *matching number*) of a hypergraph H is defined as the largest number $\nu = \nu(H)$ such that H has ν pairwise disjoint hyperedges. Obviously, $\nu(H) \leq \tau(H)$ for any hypergraph H. Typically, $\tau(H)$ is strictly larger than $\nu(H)$. In fact, $\tau(H)$ cannot even be bounded by any function of $\nu(H)$ (see Exercise 15.3).

Let $\mathbb{R}^+$ denote the set of all nonnegative real numbers. Let us call a function $t: V(H) \to \mathbb{R}^+$ a *fractional transversal* of H if

$$\sum_{x \in E} t(x) \geq 1 \qquad \text{for every hyperedge } E \in E(H). \tag{15.1}$$

The minimum of $\sum_{x \in V(H)} t(x)$ over all fractional transversals of H is called the *fractional transversal number* of H, and is denoted by $\tau^*(H)$. One can associate with each transversal T of H a function $t_T: V(H) \to \mathbb{R}^+$ defined as

$$t_T(x) = \begin{cases} 1 & \text{if } x \in T, \\ 0 & \text{if } x \notin T. \end{cases}$$

Since this function satisfies (15.1) and $\sum_{x \in V(G)} t_T(x) = |T|$, we have that $\tau^*(H) \leq \tau(H)$.

Similarly, a *fractional packing* of H is a nonnegative function $p: E(H) \to \mathbb{R}^+$ such that

$$\sum_{x \in E} p(E) \leq 1 \qquad \text{for every vertex } x \in V(H).$$

The maximum of $\sum_{E \in E(H)} p(E)$ over all fractional packings of H is called the *fractional packing number* of H, and is denoted by $\nu^*(H)$. As before, we have $\nu^*(H) \geq \nu(H)$.

It is easy to deduce directly from the definition that $\nu^*(H) \leq \tau^*(H)$ (see Exercise 15.1). In fact, these two numbers are always equal to each other. Moreover, the following is true.

Theorem 15.1. *For every hypergraph H,*

$$\nu(H) \leq \nu^*(H) = \tau^*(H) \leq \tau(H),$$

and the value of $\nu^*(H) = \tau^*(H)$ can be determined by linear programming.

Proof. Let x_i ($1 \leq i \leq n$) and E_j ($1 \leq j \leq m$) be the vertices and the edges of H, respectively. Let $A = (a_{ij})$ be the *incidence matrix* of H, i.e.,

$$a_{ij} = \begin{cases} 1 & \text{if } x_i \in E_j, \\ 0 & \text{if } x_i \notin E_j. \end{cases}$$

Let A^T denote the transpose of A, and let $\mathbf{1}_n$ denote the matrix consisting of one column of length n, all of whose entries are 1's. Given a function $t: V(H) \to \mathbb{R}$ (and $p: E(H) \to \mathbb{R}$), let $\underline{t}$ (resp. $\underline{p}$) denote a matrix consisting of one column whose ith entry is $t(x_i)$ (resp. $p(\overline{E}_i)$).

Observe that t is a fractional transversal of H if and only if

$$A^T \underline{t} \geq \mathbf{1}_m \quad \text{and} \quad \underline{t} \geq \mathbf{0}.$$

Similarly, p is a fractional packing of H if and only if

$$A\underline{p} \leq \mathbf{1}_n \quad \text{and} \quad \underline{p} \geq \mathbf{0}.$$

Thus,

$$\tau^*(H) = \min \{\mathbf{1}_n^T \underline{t} \, | \, A^T \underline{t} \geq \mathbf{1}_m, \underline{t} \geq \mathbf{0}\},$$
$$\nu^*(H) = \max \{\mathbf{1}_m^T \underline{p} \, | \, A\underline{p} \leq \mathbf{1}_n, \underline{p} \geq \mathbf{0}\}.$$

These two linear programming problems are dual to each other, so it follows immediately from the duality theorem of linear programming that their solutions, $\tau^*(H)$ and $\nu^*(H)$, are equal (see, e.g., Papadimitriou and Steiglitz, 1982; Chvátal, 1983; and Grötschel et al., 1987).   □

In general, $\tau^*(H)$ can be much smaller than $\tau(H)$ (see Exercise 15.3). The following theorem of Lovász (1975) shows that this is not the case when every point of H belongs to relatively few hyperedges.

Theorem 15.2 (Lovász).   *Let H be a hypergraph whose every vertex is contained in at most D edges. Then*

$$\tau^*(H) \leq \tau(H) \leq (\ln D + 1)\tau^*(H).$$

Proof.   We have to prove only the second inequality. Let $t: V(H) \to \mathbb{R}^+$ be a fractional transversal of H with $\sum_{x \in V(H)} t(x) = \tau^*(H)$.

We are going to select a set of vertices $x_1, x_2, \ldots$ by a *greedy* algorithm. Let x_1 be any vertex of H whose degree (i.e., the number of edges containing it) is maximal. Let D_1 denote the degree of x_1 in H. Set $H_1 = H - x_1$, that is, the hypergraph obtained from H by deleting the vertex x_1 and all edges containing x_1. If $x_1, \ldots, x_i \in V(H)$ have already been selected, then let $H_i = H - x_1 - x_2 - \cdots - x_i$. If H_i has no edges, we stop. Otherwise, let x_{i+1} be a vertex of H_i whose degree D_{i+1} is maximal, and so on. Clearly,

$$|E(H_i)| - |E(H_{i+1})| = D_{i+1}. \tag{15.2}$$

By the properties of t,

$$|E(H_i)| = \sum_{E \in E(H_i)} 1 \le \sum_{E \in E(H_i)} \sum_{x \in E} t(x)$$

$$= \sum_{x \in V(H_i)} t(x) \sum_{\substack{E \in E(H_i) \\ E \ni x}} 1$$

$$\le \sum_{x \in V(H_i)} t(x) D_{i+1}$$

$$\le D_{i+1} \tau^*(H).$$

Assume now that our procedure terminates in s steps; i.e., H_s is empty. Then, of course, $\tau(H) \le s$. Put $H_0 = H$. By (15.2), we have

$$s = \sum_{i=0}^{s-1} 1 = \sum_{i=0}^{s-1} \frac{|E(H_i)| - |E(H_{i+1})|}{D_{i+1}}$$

$$= \frac{|E(H)|}{D_1} + \sum_{i=1}^{s-1} |E(H_i)| \left(\frac{1}{D_{i+1}} - \frac{1}{D_i} \right).$$

Hence, using the inequality $|E(H_i)| \le D_{i+1} \tau^*(H)$ $(0 \le i \le s)$ and the fact $D_s \ge 1$, we obtain

$$s \le \tau^*(H) + \sum_{i=1}^{s-1} D_{i+1} \tau^*(H) \left(\frac{1}{D_{i+1}} - \frac{1}{D_i} \right)$$

$$= \tau^*(H) \left(1 + \sum_{i=1}^{s-1} \frac{D_i - D_{i+1}}{D_i} \right)$$

$$\le \tau^*(H) \left(1 + \sum_{i=1}^{s-1} \sum_{k=D_{i+1}+1}^{D_i} \frac{1}{k} \right)$$

$$= \tau^*(H) \left(1 + \sum_{k=D_s+1}^{D_1} \frac{1}{k} \right)$$

$$\le \tau^*(H)(1 + \ln D_1).$$

Thus,

$$\tau(H) \le s \le \tau^*(H)(1 + \ln D_1),$$

as desired.   □

VAPNIK–CHERVONENKIS DIMENSION

Suppose that for a public opinion poll we want to select a small number of individuals representing all major sections of the society. First, we have to choose certain categories of people and then decide which of these groups are considered "important." According to our democratic principles, we shall measure the "importance" of a group by its size (in the percentage of the population). Then the important groups will define a hypergraph H with the property that $|E| \geq \varepsilon|V(H)|$ for every edge $E \in E(H)$, where ε is some fixed constant $(0 < \varepsilon < 1)$. The smallest number of people representing all important groups is $\tau(H)$.

Clearly, the function $t(x) = 1/(\varepsilon|V(H)|)$, for all $x \in V(H)$, is a fractional transversal of H with $\sum_{x \in V(H)} t(x) = 1/\varepsilon$. Hence, $\tau^*(H) \leq 1/\varepsilon$, and Theorem 15.2 implies that

$$\tau(H) \leq \frac{1}{\varepsilon}(\ln D + 1), \tag{15.3}$$

where D is the maximum degree of the vertices of H. This bound is extremely poor if D is large.

In their seminal paper, Vapnik and Chervonenkis (1971) pointed out that if H satisfies certain natural conditions, the above upper bound can be replaced by a function depending only on ε. To specify these conditions, we need some preparation.

Definition 15.3. Let $H = (V(H), E(H))$ denote a hypergraph. A subset $A \subseteq V(H)$ is called *shattered* if for every $B \subseteq A$ there exists an $E \in E(H)$ such that $E \cap A = B$. The *Vapnik–Chervonenkis dimension* (or *VC dimension*) of H is the cardinality of the largest shattered subset of $V(H)$. It will be denoted by VC-dim (H).

The following theorem was proved independently by Shelah (1972), Sauer (1972), and Vapnik and Chervonenkis (1971).

Theorem 15.4. *Let H be a hypergraph with n vertices and VC-dimension d. Then*

$$|E(H)| \leq \binom{n}{0} + \binom{n}{1} + \cdots + \binom{n}{d},$$

and this bound cannot be improved.

First Proof. The assertion is trivial if $d = 0$ or $n \leq d$. Assume that we have already proved it for every hypergraph $\overline{H}$ with VC-dim $(\overline{H}) < d$, and for every hypergraph $\overline{H}$ with VC-dim $(\overline{H}) = d$ and $|V(\overline{H})| < n$.

Given a hypergraph H with n vertices and VC-dimension d, let us define two other hypergraphs, H_1 and H_2, as follows. Let $V(H_1) = V(H_2) = V(H) - \{x\}$ for some fixed $x \in V(H)$, and set

$$E(H_1) = \{E - \{x\} \mid E \in E(H)\},$$
$$E(H_2) = \{E \in E(H) \mid x \notin E \quad \text{and} \quad E \cup \{x\} \in E(H)\}.$$

Obviously, VC-dim $(H_1) \le d$ and VC-dim $(H_2) \le d - 1$.

On the other hand, by the induction hypothesis,

$$|E(H)| = |E(H_1)| + |E(H_2)|$$

$$\le \sum_{i=0}^{d} \binom{n-1}{i} + \sum_{i=0}^{d-1} \binom{n-1}{i}$$

$$= \sum_{i=0}^{d} \binom{n}{i}.$$

The tightness of this bound follows from the fact that if $E(H) = \{U \subseteq V \mid |U| \le d\}$, then VC-dim $(H) = d$. ☐

We also include a slightly more complicated proof due to Frankl and Pach (1983), because it is a good illustration of the so-called *linear algebra method* (see, e.g., Babai and Frankl, 1988).

Second Proof.   Let $E(H) = \{E_i \mid 1 \le i \le m\}$, and let X_j, $1 \le j \le \sum_{i=0}^{d} \binom{n}{i}$, be a list of all subsets of $V(H)$ of size at most d. Define an $m \times \sum_{i=0}^{d} \binom{n}{i}$ matrix $A = (a_{ij})$ by

$$a_{ij} = \begin{cases} 1 & \text{if } E_i \supseteq X_j, \\ 0 & \text{if } E_i \not\supseteq X_j. \end{cases}$$

Suppose, for contradiction, that $m > \sum_{i=0}^{d} \binom{n}{i}$. Then the rows of A are linearly dependent over the reals; thus there exists a nonzero function $f: E(H) \to \mathbb{R}$ such that

$$\sum_{E_i \supseteq X_j} f(E_i) = 0 \qquad \text{for every } X_j.$$

Let $A \subseteq V(H)$ be a *minimal* subset for which

$$\sum_{E_i \supseteq A} f(E_i) = \alpha \ne 0.$$

(Sets A with nonzero sums certainly exist, for we get a nonzero sum for any maximal element A of the family $\{A \in E(H) \mid f(A) \ne 0\}$.) Obviously, $|A| \ge d+1$. Given any $B \subseteq A$, let

$$F(B) = \sum_{E_i \cap A = B} f(E_i).$$

Thus, $F(A) = \alpha$, and setting $B = A - \{a\}$ for any fixed $a \in A$,

$$F(B) = \sum_{E_i \supseteq B} f(E_i) - \sum_{E_i \supseteq A} f(E_i)$$
$$= 0 - \alpha = -\alpha$$

In general, if B is any $(|A| - k)$-element subset of A ($0 \le k \le |A|$), then

$$F(B) = (-1)^k \alpha \ne 0.$$

This yields, in particular, that there exists at least one hyperedge E_i with $E_i \cap A = B$. Thus, A is shattered, contradicting our assumption that VC-dim $(H) = d$. $\square$

Vapnik and Chervonenkis (1971) discovered an ingenious probabilistic (counting) argument based on the above result, which leads to a substantial improvement of the bound (15.3). They showed (in a somewhat different setting) that there exists a function $f(d, \varepsilon)$ such that the transversal number of every hypergraph H of VC-dimension d, all of whose edges have at least $\varepsilon |V(H)|$ elements, is at most $f(d, \varepsilon)$ (see Exercise 15.6). The ideas of Vapnik and Chervonenkis have been adapted by Haussler and Welzl (1987) and Blumer et al. (1989) to obtain various upper bounds on $f(d, \varepsilon)$. These results were sharpened and generalized by Komlós, Pach, and Woeginger (1992), as follows. Given a finite set V, a function $\mu: V \to \mathbb{R}^+$ is called a *probability measure* if

$$\sum_{x \in V} \mu(x) = 1.$$

The measure of any subset $X \subseteq V$ is defined by $\mu(X) = \sum_{x \in X} \mu(x)$.

Theorem 15.5 (Komlós et al.).   *Let H be a hypergraph of VC-dimension d, let $\varepsilon > 0$, and let μ be a probability measure on $V(H)$ such that $\mu(E) \ge \varepsilon$ for every $E \in E(H)$. Then $\tau(H) \le t(d, \varepsilon)$, where $t(d, \varepsilon)$ denotes the smallest positive integer t satisfying*

$$2 \sum_{i=0}^{d} \binom{T}{i} \left(1 - \frac{t}{T}\right)^{(T-t)\varepsilon - 1} < 1$$

for some integer $T > t$. Consequently, for any $\varepsilon \le \frac{1}{2}$, we have

$$\tau(H) \le \frac{d}{\varepsilon} \left(\ln \frac{1}{\varepsilon} + 2 \ln \ln \frac{1}{\varepsilon} + 6 \right)$$

(cf. Exercise 15.9).

Proof. Let us select with possible repetition t random points of $V(H)$, where the selections are done with respect to the probability measure μ. We get a *random sample*

$$x \in [V(H)]^t = \underbrace{V(H) \times \cdots \times V(H)}_{t\text{times}}.$$

We say that x is a *transversal* of H if every edge $E \in E(H)$ contains at least one point of x. Let $I(E, x)$ denote the number of components of x that belong to E, counting with multiplicity. Then

$$\Pr[x \text{ is not a transversal of } H] = \Pr[\exists E \in E(H) : I(E, x) = 0].$$

Having picked the string x of length t, let us choose randomly another $T - t$ elements from $V(H)$. Let $y \in [V(H)]^{T-t}$ denote this new string, and let $z = xy \in [V(H)]^T$ stand for the full sequence. Furthermore, let $\langle z \rangle = \langle xy \rangle$ denote the *multiset* of all elements occurring in z (i.e., they are counted with multiplicities but their order is irrelevant).

For any $E \in E(H)$, $I(E, y)$ is a random variable having binomial distribution. Let m_E be the *median* of $I(E, y)$,

$$\Pr[I(E, y) > m_E] \le \tfrac{1}{2} \le \Pr[I(E, y) \ge m_E].$$

The following inequality is an immediate consequence of the independence of x and y.

$$\Pr[\exists E \in E(H) : I(E, x) = 0]$$

$$\le \frac{\Pr[\exists E \in E(H) : I(E, x) = 0 \text{ and } I(E, y) \ge m_E]}{\min_{E \in E(H)} \Pr[I(E, y) \ge m_E]}.$$

$$\le 2 \Pr[\exists E \in E(H) : I(E, x)$$

$$= 0 \text{ and } I(E, y) \ge m_E].$$

For a fixed $E \in E(H)$, the conditional probability for given $\langle z \rangle = \langle xy \rangle$

$$\Pr[\,I(E,x) = 0 \text{ and } I(E,y) \geq m_E \,|\, \langle z \rangle]$$

$$= \chi\,[I(E,z) \geq m_E]\,\frac{\dbinom{T-t}{I(E,z)}}{\dbinom{T}{I(E,z)}}$$

$$\leq \chi\,[I(E,z) \geq m_E]\left(1 - \frac{t}{T}\right)^{I(E,z)}$$

$$\leq \chi\,[I(E,z) \geq m_E]\left(1 - \frac{t}{T}\right)^{m_E}.$$

(Here $\chi\,[A]$ is the characteristic function of A, that is, $\chi\,[A] = 1$ if A is true, and 0 otherwise.)

By Theorem 15.4, a fixed multiset $\langle z \rangle$ has at most $\displaystyle\sum_{i=0}^{d}\binom{T}{i}$ different intersections with the edges of H. Thus,

$$\Pr[\exists\, E \in E(H):I(E,x) = 0 \text{ and } I(E,y) \geq m_E \,|\, \langle z \rangle]$$

$$\leq \left(\sum_{i=0}^{d}\binom{T}{i}\right)\left(1 - \frac{t}{T}\right)^{m},$$

where $m = \min_{E \in E(H)} m_E$. Using the known fact that the median of a binomial distribution is within 1 of the mean,

$$m \geq (T - t)\min_{E \in E(H)}\mu(E) - 1 \geq (T - t)\varepsilon - 1.$$

Hence, we obtain

$$\Pr[\exists\, E \in E(H):I(E,x) = 0] \leq 2\left(\sum_{i=0}^{d}\binom{T}{i}\right)\left(1 - \frac{t}{T}\right)^{(T-t)\varepsilon - 1}.$$

If the last expression is less than 1, then x is a transversal of H, with positive probability. This proves the first statement of the theorem. Choosing

$$t = \left\lfloor \frac{d}{\varepsilon}\left(\ln\frac{1}{\varepsilon} + 2\ln\ln\frac{1}{\varepsilon} + 6\right)\right\rfloor,$$
$$T = \left\lfloor \frac{\varepsilon}{d}t^2\right\rfloor,$$

we get after some calculations that

$$2\sum_{i=0}^{d}\binom{T}{i}\left(1 - \frac{t}{T}\right)^{(T-t)\varepsilon - 1} < 1,$$

provided that $\varepsilon \leq \frac{1}{2}$. $\square$

The above theorem is valid for any probability measure μ defined on the vertex set of H. In particular, one can choose μ to be constant; that is, $\mu(x) = 1/|V(H)|$ for every $x \in V(H)$. We can deduce another interesting result from Theorem 15.5 by applying it to the measure $\mu'(x) = t(x)/\tau^*(H)$, where $t: V(H) \rightarrow \mathbb{R}^+$ is a fractional transversal of H with $\sum_{x \in V(H)} t(x) = \tau^*(H)$. Observe that in this case

$$\mu'(E) = \sum_{x \in E} \mu'(x)$$

$$= \sum_{x \in E} \frac{t(x)}{\tau^*(H)} \geq \frac{1}{\tau^*(H)}$$

holds for every $E \in E(H)$. Thus, choosing $\varepsilon = 1/\tau^*(H)$ in Theorem 15.5, we obtain the following.

Corollary 15.6 (Komlós et al.).   *Let H be any hypergraph of VC-dimension d.*

(i) *If every edge of H has at least $\varepsilon|V(H)|$ elements for some $\varepsilon \leq \frac{1}{2}$, then*

$$\tau(H) \leq \frac{d}{\varepsilon}\left(\ln \frac{1}{\varepsilon} + 2\ln\ln \frac{1}{\varepsilon} + 6\right).$$

(ii) *If $\tau^*(H) \geq 2$, then*

$$\tau(H) \leq d\tau^*(H)\,(\ln \tau^*(H) + 2\ln\ln \tau^*(H) + 6).$$

Next we show that for $d \geq 2$, the bound given in Theorem 15.5 is close to being optimal.

Theorem 15.7 (Komlós et al.).   *Given any natural number $d \geq 2$ and any real $\gamma < 2/(d+2)$, there exists a constant $\varepsilon_{d,\gamma} > 0$ with the following property.*
For any $\varepsilon \leq \varepsilon_{d,\gamma}$, one can construct a hypergraph H of VC-dimension d, all of whose edges have at least $\varepsilon|V(H)|$ points, and

$$\tau(H) \geq (d - 2 + \gamma)\,\frac{1}{\varepsilon}\ln \frac{1}{\varepsilon}.$$

Proof.   Again, we use the probabilistic method. Let γ' be a fixed constant, $\gamma < \gamma' < 2/(d+2)$. Given a sufficiently small ε, let $n = (K/\varepsilon)\ln(1/\varepsilon)$, where K is a constant depending only on d, γ, and γ' (but not on ε), which will be specified later. Furthermore, let

$$r = \varepsilon n, \qquad p = \frac{\varepsilon^{1-d-\gamma'}}{\binom{n}{r}}, \qquad t = (d-2+\gamma)\frac{1}{\varepsilon}\ln\frac{1}{\varepsilon}.$$

We assume that n, r, t are integers, disregarding all roundoff errors.

Let V be a fixed n-element set. Construct a hypergraph H on the vertex set V by randomly selecting some r-element subsets of V, where each r-tuple is chosen independently with probability p. We are going to show that with high probability

(i) VC-dim $(H) \le d$, and

(ii) $\tau(H) > t$.

$\Pr[\text{VC-dim}(H) > d]$

$$\le \binom{n}{d+1} \Pr[\text{a fixed } (d+1)\text{-element subset } A \subseteq V \text{ is shattered by } H]$$

$$= \binom{n}{d+1} \prod_{B \subseteq A} \Pr[\exists E \in E(H) : E \cap A = B]$$

$$= \binom{n}{d+1} \prod_{B \subseteq A} \left(1 - (1-p)^{\binom{n-|A|}{r-|B|}}\right)$$

$$= \binom{n}{d+1} \prod_{j=0}^{d+1} \left(1 - (1-p)^{\binom{n-d-1}{r-j}}\right)^{\binom{d+1}{j}}$$

$$= \binom{n}{d+1} \prod_{i=0}^{d+1} \left(1 - (1-p)^{\binom{n-d-1}{r-d-1+i}}\right)^{\binom{d+1}{d+1-i}}$$

$$= \binom{n}{d+1} \prod_{i=0}^{d+1} \left(1 - (1-p)^{\binom{n-d-1}{r-d-1+i}}\right)^{\binom{d+1}{i}}$$

$$\le \binom{n}{d+1} \prod_{i=0}^{1} \left(p\binom{n-d-1}{r-d-1+i}\right)^{\binom{d+1}{i}}$$

$$= \binom{n}{d+1} p\binom{n-d-1}{r-d-1}\left(p\binom{n-d-1}{r-d}\right)^{d+1}$$

$$\le n^{d+1} p \binom{n}{r}\left(\frac{r}{n}\right)^{d+1}\left(p\binom{n}{r}\left(\frac{r}{n}\right)^{d}\right)^{d+1}$$

$$= \left(K \ln \frac{1}{\varepsilon}\right)^{d+1} \varepsilon^{2-(d+2)\gamma'},$$

which tends to 0 as $\varepsilon \to 0$. This proves (i).

Next we show that (ii) also holds with high probability.

$$\Pr[\tau(H) \le t] = \binom{n}{t} (1 - p)^{\binom{n-t}{r}}$$

$$\le \binom{n}{t} \exp\left[-p\binom{n-t}{r}\right]$$

$$\le \left(\frac{en}{t}\right)^t \exp\left[-p\binom{n}{r}\left(1 - \frac{r}{n-t+1}\right)^t\right].$$

From this, using the inequality $1 - ax > e^{-bx}$ for $b > a$, $0 < x < 1/a - 1/b$, we obtain the upper bound

$$\left(\frac{en}{t}\right)^t \exp\left[-p\binom{n}{r} e^{-K\varepsilon t/(K-d)}\right]$$

$$= \left(\frac{eK}{d-2+\gamma}\right)^t \exp\left[-\varepsilon^{1-d-\gamma'+K(d-2+\gamma)/(K-d)}\right],$$

which tends to 0 if

$$1 - d - \gamma' + K(d - 2 + \gamma)/(K - d) < -1,$$

i.e., if K is sufficiently large.   □

The condition $d \ge 2$ in Theorem 15.7 is not merely a technical assumption. In fact, it is not hard to characterize all finite hypergraphs H with VC-dimension 1, and one can check that $\tau(H) \le \lceil 1/\varepsilon \rceil - 1$, provided that every edge of H has at least $\varepsilon |V(H)|$ points for some $0 < \varepsilon < 1$ (see Exercise 15.8).

The following simple assertion will help us in deciding whether a given hypergraph has low VC-dimension.

Lemma 15.8. *Let H be a hypergraph of VC-dimension d, and let $\varphi(E_1, \ldots, E_k)$ be a set-theoretic formula of k variables (using $\cup, \cap, -$). If every edge E' of a hypergraph H' can be expressed as*

$$E' = \varphi(E_1, \ldots, E_k) \qquad \text{for suitable } E_i \in E(H),$$

then

$$\text{VC-dim}(H') \le 2dk \log(2dk).$$

Proof. Let A be a d'-element subset of $V(H') = V(H)$, which is shattered in H'. By Theorem 15.4,

$$|\{E \cap A \,|\, E \in E(H)\}| \le \sum_{i=0}^{d}\binom{d'}{i}.$$

Using the assumption on H', this yields

$$2^{d'} = |\{E' \cap A \,|\, E' \in E(H')\}| \le \left(\sum_{i=0}^{d} \binom{d'}{i} \right)^k.$$

Comparing the two sides of this inequality, we obtain that $d' \le 2dk \log (2dk)$, as required. □

Ding, Seymour, and Winkler (1994) have introduced another parameter of a hypergraph, closely related to its VC-dimension. They defined $\lambda(H)$ as the largest integer l such that one can choose l edges $E_1, E_2, \ldots, E_l \in E(H)$ with the property that for any $1 \le i < j \le l$, there is a vertex $x_{ij} \in E_i \cap E_j$ that does not belong to any other E_g ($g \ne i, j$). It is easy to see that VC-dim $(H) < \binom{\lambda(H) + 1}{2}$ for every hypergraph H. Combining Corollary 15.6 with Ramsey's theorem (Theorem 9.13), one can establish the following result.

Theorem 15.9 (Ding et al., 1994).   *For any hypergraph H,*

$$\tau(H) \le 6\lambda^2(H) \, (\lambda(H) + \nu(H)) \left(\frac{\lambda(H) + \nu(H)}{\lambda(H)} \right)^2.$$

At the beginning of this chapter we pointed out that in general it is impossible to bound τ from above by any function of ν. Gyárfás and Lehel (1983, 1985) initiated the investigation of certain classes of hypergraphs for which such functions exist. Theorem 15.9 provides a sufficient condition for a family of hypergraphs to have this property. It implies that if there exists a constant K such that $\lambda(H) \le K$ for all members of a family, then τ can be bounded from above by a polynomial of ν. For various geometric consequences of this fact, see Pach (1995).

RANGE SPACES AND ε-NETS

Haussler and Welzl (1987) were the first to recognize the relevance of the above machinery to geometric problems, and in fact they formulated and proved the first version of Theorem 15.5, too. It seems to capture the essence of the so-called *random* (or *probabilistic*) method in a large variety of geometric applications. This ready-to-use kit will save us a lot of time (and space) in situations where otherwise we would go through lengthy but routine calculations. However, the main significance of these ideas is that they shed some light on the general transversal problem. The transversal number is a *global* parameter of a set system. The results in the preceding section show that in any measure space of total measure 1, any system of large measurable sets admits a relatively small transversal, provided that its *local* behavior is nice (i.e., its VC-dimension is bounded).

In recent years, the above concepts and methods have found many interesting applications in discrete and computational geometry, learning theory, etc. (see Mulmuley, 1994; Anthony and Biggs, 1992). However, most of the literature in these fields is written in a different terminology. This language is so widely used that it is worth learning the basic definitions.

Definition 15.10.   A *range space* is a (finite or infinite) hypergraph, i.e., a pair $\Sigma = (X, \mathcal{R})$, where X is an underlying set and $\mathcal{R}$ is a family of subsets of X. (That is, $V(\Sigma) = X$ and $E(\Sigma) = \mathcal{R}$.) The elements of $\mathcal{R}$ (i.e., the hyperedges of Σ) are called *ranges*.

Given any $Y \subseteq X$, we define the *subspace* of Σ induced by Y as

$$\Sigma_Y = (Y, \{R \cap Y | R \in \mathcal{R}\}).$$

If $\Sigma = (X, \mathcal{R})$ is a finite range space and $\varepsilon > 0$, then let

$$\Sigma^\varepsilon = (X, \{R \in \mathcal{R} \,|\, |R| \geq \varepsilon|X|\}).$$

A subset $T \subseteq X$ is called an ε-*net* for Σ if it intersects every element of $\mathcal{R}^\varepsilon = \{R \in \mathcal{R} \,|\, |R| \geq \varepsilon|X|\}$, that is, every "large" range of Σ.

The *Vapnik–Chervonenkis dimension* (VC-dimension) of a range space is defined in exactly the same way as for hypergraphs. That is,

VC-dim $(\Sigma) = \max \{|A| \,|\, A \subseteq X$ and $\forall B \subseteq A \,\exists R \in \mathcal{R}$ such that $R \cap A = B\}$.

Of course, VC-dim $(\Sigma_Y) \leq$ VC-dim (Σ) for any subspace Σ_Y of Σ.

Let $f(d, \varepsilon)$ denote the maximum size of the smallest ε-net of Σ over all finite range spaces Σ of VC-dimension at most d.

Theorems 15.5 to 15.7 can now be summarized as follows.

$$d - 2 + \frac{2}{d+2} \leq \lim_{\varepsilon \to 0} \inf \frac{f(d, \varepsilon)}{\frac{1}{\varepsilon} \ln \frac{1}{\varepsilon}} \leq \lim_{\varepsilon \to 0} \sup \frac{f(d, \varepsilon)}{\frac{1}{\varepsilon} \ln \frac{1}{\varepsilon}} \leq d.$$

Next we give some important examples of range spaces whose VC-dimensions are finite (see also Exercise 15.11).

Example 15.11

(a) $X = \mathbb{R}^d$, $\mathcal{R} = \{$all half-spaces in $\mathbb{R}^d\}$;

(b) $X = \mathbb{R}^d$, $\mathcal{R} = \{$all open (unit) balls in $\mathbb{R}^d\}$;

(c) $X = \{$all hyperplanes in $\mathbb{R}^d\}$. Given any open segment s in $\mathbb{R}^d$, let R_s be the collection of all hyperplanes intersecting s. Set $\mathcal{R} = \{R_s | $ for all open segments $s\}$.

It is easy to check that the VC-dimension of each of the above range spaces is finite. We demonstrate this for (c). Assume, in order to obtain a contradiction, that for any large n we can find an n-element subset $A_n \subset X$ which is shattered, i.e.,

$$|\{R_s \cap A_n \,|\, R_s \in \mathcal{R}\}| = 2^n.$$

The hyperplanes belonging to A_n define a cell complex consisting of at most $O(n^d)$ cells. Observe that if $s = p_1 p_2$ and $t = q_1 q_2$ are two segments such that p_i, q_i belong to the same cell ($i = 1, 2$), then $R_s \cap A_n = R_t \cap A_n$. Since there are $O(n^{2d})$ possibilities for selecting two cells for the endpoints of s, we obtain

$$|\{R_s \cap A_n \,|\, R_s \in \mathcal{R}\}| = O(n^{2d}),$$

a contradiction.

Lemma 15.8 allows us to construct from (a), (b), (c) a number of additional examples of range spaces of finite VC-dimension. It implies that all range spaces whose ranges can be obtained from the ranges of a space Σ of finite dimension by using $\cap$, $\cup$, $-$ a bounded number of times, have finite VC-dimensions. In particular, this is true for the following range spaces.

(d) $X = \mathbb{R}^d$, $\mathcal{R} = \{$all open polyhedra in $\mathbb{R}^d$ with at most k vertices$\}$;

(e) $X = \mathbb{R}^d$, $\mathcal{R} = \{$all annuli in $\mathbb{R}^d\}$;

(f) $X = \{$all hyperplanes in $\mathbb{R}^d\}$. Given any k-dimensional simplex $S_k \subset \mathbb{R}^d$, let R_{S_k} be the collection of all hyperplanes intersecting the interior of S_k. Set $\mathcal{R} = \{R_{S_k} |$ for all simplices $S_k\}$.

A typical example of a geometric range space whose VC-dimension is not finite is the following.

(g) $X = \mathbb{R}^2$, $\mathcal{R} = \{$all open convex sets in $\mathbb{R}^2\}$

(see Exercise 15.18(i)).

SPANNING TREES OF LOW STABBING NUMBER

We are going to illustrate the power of the concepts and results described above by proving a beautiful theorem of Welzl (1988) (see also Chazelle and Welzl, 1989; and Welzl, 1992), and by indicating its algorithmic consequences.

Let S be a set of n points in the plane. A tree T whose vertices are the points of S and whose edges are straight-line segments is called a *spanning tree* of S. The *stabbing number* of a spanning tree T, denoted by $\sigma(T)$, is defined as the maximum number of segments in T that can be intersected by a single line.

Theorem 15.12 (Welzl).   *For any set S of n points in $\mathbb{R}^2$ in general position, there is a spanning tree T whose stabbing number is at most $c\sqrt{n}\log n$, where c is an appropriate constant.*

The proof is based on the following selection lemma.

Lemma 15.13.   *Given a set S of $n \geq 2$ points in $\mathbb{R}^2$ in general position, and a multiset L of m lines avoiding these points, one can always find a pair $x, y \in S$ such that the segment connecting them intersects at most $c'(m/\sqrt{n})\log n$ lines of L, where c' is a suitable constant.*

Proof.   One can assume without loss of generality that L is a *set* (i.e., it contains no repeated elements); otherwise, one can slightly change the position of some lines. Let us consider the subspace of the range space in Example 15.11(c) restricted to L $(d = 2)$. That is, let

$$\Sigma_L = (L, \{R_s \cap L \,|\, \text{for all open segments } s\}),$$

where $R_s \cap L$ is the set of all lines of L intersected by a given segment s.

As we have pointed out before, Σ_L has finite VC-dimension D. Hence, by Corollary 15.6(i), there exists a constant $c_D \geq 1$ such that Σ_L has an ε-net of size at most $c_D(1/\varepsilon)\log(1/\varepsilon)$ for any $0 < \varepsilon < 1$.

Setting $\varepsilon = c_D(\log n/\sqrt{n})$, we find that there exists a subset $L' \subseteq L$ with

$$|L'| \leq c_D \frac{1}{\varepsilon} \log \frac{1}{\varepsilon} < \frac{\sqrt{n}}{2},$$

and every open segment crossing at least $\varepsilon m = c_D(m/\sqrt{n})\log n$ elements of L intersects at least one line belonging to L'.

The lines of L' divide the plane into fewer than n cells. By the pigeonhole principle, at least one of them contains two points $x, y \in S$. The segment connecting them avoids every line of L'; therefore, it cannot intersect more than $c_D(m/\sqrt{n})\log n$ elements of L.   □

Proof of Theorem 15.12.   Fix a set $S = S_0$ of $n = n_0$ points in the plane in general position and a set L_0 consisting of $m_0 = \binom{n}{2} + 1$ lines avoiding these points such that they represent every possible partition of S into two parts by a line (see Exercise 15.10(iii)).

Applying Lemma 15.13 to the pair (S_0, L_0), we can find two points x_0, $y_0 \in S_0$ with the property that the segment connecting them intersects at most $c'(m_0/\sqrt{n_0})\log n_0$ elements of L_0. Let $S_1 = S_0 - \{x_0\}$, $n_1 = |S_1| = n-1$, and L_1 be the multiset of lines obtained from L_0 by duplicating each line intersected by the segment $x_0\,y_0$. Set $m_1 = |L_1|$. By Lemma 15.13, now there exists a pair of points $x_1, y_1 \in S_1$ such that the segment $x_1\,y_1$ intersects at most $c'(m_1/\sqrt{n_1})\log n_1$ lines of L_1.

Let $S_2 = S_1 - \{x_1\}$, $n_2 = |S_2| = n - 2$, and let L_2 be the multiset of lines obtained from L_1 by duplicating all lines intersected by the segment $x_1 y_1$. Set $m_2 = |L_2|$.

Proceeding in this way, for every $i \leq n - 1$, we can recursively define an $n_i = (n - i)$-element set $S_i \subseteq S$ and an m_i-element multiset of lines L_i. It follows from the construction that

$$m_{i+1} \leq m_i\left(1 + c' \frac{\log n_i}{\sqrt{n_i}}\right) \qquad \text{for every } 0 \leq i \leq n - 2.$$

Hence,

$$m_{n-1} \leq m_0 \prod_{i=0}^{n-2}\left(1 + c' \frac{\log n_i}{\sqrt{n_i}}\right)$$

$$\leq \left(\binom{n}{2} + 1\right) \exp\left(\sum_{i=0}^{n-2} c' \frac{\log n_i}{\sqrt{n_i}}\right)$$

$$\leq \left(\binom{n}{2} + 1\right) \exp\left(c' \log n \sum_{i=0}^{n-2} \frac{1}{\sqrt{n-i}}\right)$$

$$\leq 2^{c\sqrt{n}\log n},$$

for a suitable constant $c > 0$.

Observe that the segments $x_0 y_0, \ldots, x_{n-2} y_{n-2}$ form a spanning tree T of S. Assume that a line $\ell \in L_0$ intersects t edges of T. Then L_{n-1} contains 2^t copies of ℓ, and it follows from the last inequality that $t \leq c\sqrt{n}\log n$.

In view of the fact that the elements of L_0 exhaust all possible ways in which S can be bisected by a straight line, we can conclude that no line can cross more than $c\sqrt{n}\log n$ edges of T. $\square$

The statement of Theorem 15.12 can be strengthened slightly as follows.

Corollary 15.14. *For any set S of n points in the plane in general position, there is a spanning path P whose stabbing number is at most $C\sqrt{n}\log n$, where C is a suitable constant. Moreover, we can assume that P is a simple (non-self-intersecting) polygon.*

Proof. Let T be a spanning tree of S such that $\sigma(T) \leq c\sqrt{n}\log n$. Let $x_1, x_2, \ldots, x_{2n-2} = x_1$ be a sequence of points of S (with repetitions) such that

(i) $\{x_i, x_{i+1}\}$ is an edge of T, for $i < 2n - 2$,

(ii) for every edge e of T, $e = \{x_i, x_{i+1}\}$ for exactly two different values of i.

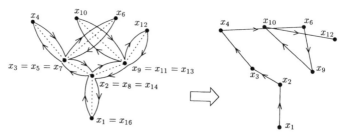

Figure 15.1. Converting a spanning tree to a spanning path.

Erase from this sequence every element x_j such that $x_j = x_i$ for some $i < j$ (see Figure 15.1). The resulting subsequence $x_{i_1} = x_1, x_{i_2}, \ldots, x_{i_n}$ contains every point of S exactly once, and it is easy to check that it induces a path P whose stabbing number is at most $2\sigma(T)$.

To prove the second statement, observe that if two edges of P cross each other, they are the diagonals of a convex quadrilateral Q. Replacing these two edges of P by two opposite sides of Q, we can obtain another spanning path P' with $\sigma(P') \leq \sigma(P)$. Since the total length of the edges of P' is strictly shorter than that of P, this algorithm will terminate in a finite number of steps. □

The bounds obtained in Theorem 15.12 and Corollary 15.14 can be improved by a factor of $\log n$ using a variant of Theorem 11.6 (see Exercise 15.13).

RANGE SEARCHING

The last result has some important algorithmic consequences. For example, it can be used to design an efficient algorithm for the *half-plane range searching* problem. In what follows, we briefly describe this problem and outline a solution.

Let S be a fixed set of n points in $\mathbb{R}^2$ in general position. We wish to build a linear-size *data structure* that allows us to answer efficiently questions (*queries*) of the following type: Given an open half-plane h, what is the number of points of S lying in h?

Let $P = x_1 x_2 \cdots x_n$ be a spanning path of S with stabbing number $\sigma(P) = O(\sqrt{n} \log n)$. Define a rooted binary tree T_P, as follows. Let the root r of T_P be associated with the entire path $x_1 \cdots x_n$. Assume that we have already defined a vertex v of T_p, and it was associated with the subpath $P(v) = x_i \cdots x_j$. If $j = i$, then v is a leaf of T_P. Otherwise, let v have two children v_1 and v_2 associated with the subpaths

$$P(v_1) = x_i \cdots x_{\lfloor \frac{i+j}{2} \rfloor}, P(v_2) = x_{\lfloor \frac{i+j}{2} \rfloor + 1} \cdots x_j.$$

Thus, T_P has exactly n leaves.

After some processing, we can store at each vertex $v \in T_P$

(i) $|P(v)|$, the size of $P(v)$, and

(ii) the (coordinates of the) vertices of the convex hull of $P(v)$ in circular order.

Given an open query half-plane h, we can determine $|S \cap h|$ as follows. We visit the root r of T_P, and check whether the boundary of h intersects $P(r)$. If it does, we pass down the question to the children of r. In general, when we visit a vertex v of T_P, we decide whether the boundary of h intersects $P(v)$. If it does not, we stop; otherwise, we visit the children of v. The vertices of T_P that have been visited during this process form a subtree $T'_P \subseteq T_P$. At each leaf v of T'_P we can decide whether $P(v) \subset h$ or $P(v) \cap h = \varnothing$. Starting at the leaves of T'_P and processing backward, we can easily compute the sum of $|P(v)|$ over all leaves for which $P(v) \subset h$. Obviously, this sum gives the number of points of S lying in h.

Notice that at each level of T_P, the number of vertices v for which h intersects $P(v)$ is at most the stabbing number of P. Since the number of levels in T_P is at most $\lceil \log_2 n \rceil + 1$, it is clear that for any query half-plane the number of vertices visited is

$$|V(T'_P)| \leq 3\lceil \log_2 n \rceil \sigma(P) = O(\sqrt{n} \log^2 n).$$

On the other hand, at each vertex $v \in T'_P$, the time spent is proportional to the time required to detect an intersection between a line ℓ and P_v. However, it is easy to see that ℓ intersects P_v if and only if ℓ intersects the convex hull of P_v, and that the intersection between a line and convex n-gon can be detected in $O(\log n)$ time (see Exercise 15.14).

Summarizing, we have obtained the following:

Corollary 15.15. *Given a set S of n points in the plane in general position, there exists a data structure using $O(n \log n)$ space that allows us to compute $|S \cap h|$ for any query half-plane in time $O(\sqrt{n} \log^3 n)$.*

Spanning trees with low stabbing number have been used to obtain fast algorithms for several problems in computational geometry. See Agarwal (1992), Edelsbrunner et al. (1989), and Welzl (1992) for related work.

Finally, we note that the following statement, which is a slightly weaker version of Theorem 11.6, can also be established using ε-nets (see Exercise 15.20).

Corollary 15.16. *Given a set L of n lines in general position and an integer $r \geq 3$, the plane can be decomposed into at most r^2 triangles (some of which are unbounded) such that the interior of each triangle intersects at most $c(n/r) \log r$ lines of L, where c is a suitable constant.*

EXERCISES

15.1 Prove (without using Theorem 15.1 or the duality theorem of linear programming) that $\nu^*(H) \le \tau^*(H)$ for every hypergraph H.

15.2 (König, 1936) Prove that if G is a bipartite graph, then $\nu(G) = \tau(G)$.

15.3 Prove that there exists no function $f: \mathbb{R} \to \mathbb{R}$ such that $\tau(H) \le f(\tau^*(H))$ for every hypergraph H.

15.4 (Erdős, 1964b; Lovász, 1975)

  (i) Prove that every r-uniform hypergraph H with at most 2^{r-1} edges is two-colorable; i.e., one can color its vertices by two colors so that every edge of H contains points of both colors.

  (ii) Prove that there exists a constant $c > 0$ such that for every r, one can find a non-two-colorable r-uniform hypergraph with at most $cr^2 2^r$ edges.

15.5 (Frankl and Pach, 1984) Let H be an r-uniform hypergraph with n vertices and VC-dimension d. Show that $|E(H)| \le \binom{n}{d}$.

15.6 (Vapnik and Chervonenkis, 1971) Given $\varepsilon > 0$ and a hypergraph H with vertex set $V = V(H)$ we say that a subset $A \subseteq V$ is an ε-*approximation* of H if

$$\left| \frac{|A \cap E|}{|A|} - \frac{|E|}{|V|} \right| \le \varepsilon \qquad \text{for every } E \in E(H);$$

  i.e., the relative number of points of A in any edge approximates the relative size of this edge with accuracy ε.

  (i)$\star$ Show that there is an absolute constant C such that for any $0 < \varepsilon < 1$, any hypergraph of VC-dimension d has an ε-approximation of at most $C(d/\varepsilon^2) \log (d/\varepsilon)$ elements.

  (ii) Show that if every edge of a hypergraph H has more than $\varepsilon|V(H)|$ points (for some $\varepsilon > 0$), then every ε-approximation A of H intersects all edges of H; i.e., A is a transversal of H.

15.7$\star$ Decide whether the following statement is true. Let H be a hypergraph of VC-dimension d, all of whose edges have at least $\varepsilon|V(H)|$ points for some $\varepsilon > d/|V(H)|$. Then there exists an $\lceil \varepsilon|V(H)| \rceil$-uniform hypergraph H' of VC-dimension d, with the property that for any $E \in E(H)$ one can find $E' \in E(H)$ such that $E' \subseteq E$.

15.8 (Pach and Woeginger, 1990; Komlós et al., 1992) Let H be a hypergraph of VC-dimension 1, all of whose edges have at least $\varepsilon|V(H)|$

points for some $0 < \varepsilon < 1$. Show that

$$\tau(H) \leq \max \{2, \lceil 1/\varepsilon \rceil - 1\},$$

and this bound cannot be improved.

15.9 (Komlós et al., 1992) Let H be a hypergraph of VC-dimension d, let $\varepsilon > 0$, and let μ be a probability measure on $V(H)$ such that $\mu(E) \geq \varepsilon$ for every $E \in E(H)$. Furthermore, let

$$\pi_H(T) = \max_{\substack{A \subseteq H \\ |A| = T}} |\{E \cap A \mid E \in E(H)\}|.$$

(i) Show that $\tau(H) \leq t(d, \varepsilon)$, where $t(d, \varepsilon)$ denotes the smallest positive integer t satisfying

$$2\pi_H(T)\left(1 - \frac{t}{T}\right)^{(T-t)\varepsilon - 1} < 1 \qquad \text{for some integer } T > t.$$

(ii) Show that for any $C > 2$,

$$\tau(H) \leq \frac{d}{\varepsilon}\left(\ln \frac{1}{\varepsilon} + 2 \ln \ln \frac{1}{\varepsilon} + C\right),$$

provided that $\varepsilon < \varepsilon_C$. In particular, if $d \geq 2$ and $C = 5$ (resp. 4 or 3), the assertion is true with $\varepsilon_C = 1/3$ (resp. $1/6$ or $1/50$).

15.10 (Schläfli, 1901)

(i) Show that any set of n hyperplanes in general position divides $\mathbf{R}^d$ into exactly $\sum_{i=0}^{d} \binom{n}{i}$ regions.

(ii) Show that any set of n hyperplanes in general position divides the projective d-space into exactly $\sum_{i=0}^{\lfloor d/2 \rfloor} \binom{n}{d-2i}$ regions.

(iii) Show that any set of n points in general position in $\mathbf{R}^d$ can be bisected by a hyperplane in exactly $\sum_{i=0}^{\lfloor d/2 \rfloor} \binom{n}{d-2i}$ different ways.

15.11 Determine the VC-dimension of the following range spaces:

(i) $X = \mathbf{R}^d$, $\mathcal{R} = \{$all open half-spaces in $\mathbf{R}^d\}$.

(ii) $X = \mathbf{R}^d$, $\mathcal{R} = \{$all open balls in $\mathbf{R}^d\}$.

15.12 (Naiman and Wynn, 1994) Given an n-element point set $S \subseteq \mathbf{R}^d$ in general position, show that

$$|\{S \cap B^d \mid B^d \text{ is a ball}\}| = \sum_{i=0}^{d+1} \binom{n}{i},$$

provided that $n \geq d + 3$.

15.13 Given a set S of n points in $\mathbb{R}^2$ in general position, show that there exists a spanning path whose stabbing number is $c\sqrt{n}$, where c is a suitable constant.

15.14 Show that for any convex polygon with n sides, one can test in $O(\log n)$ time whether it intersects a given line.

15.15 Given a set S of n points in $\mathbb{R}^2$ in general position, prove that there exists a data structure using only $O(n)$ space, which allows to calculate $|S \cap h|$ for any half-plane h in time $O(\sqrt{n}\log^3 n)$.

15.16 (i) Given a set of $4n$ points in $\mathbb{R}^2$ in general position and any line ℓ with the property that there are exactly $2n$ points in both open half-planes determined by ℓ, show that there is a line ℓ' crossing ℓ such that each quadrant formed by these two lines contains exactly n points.

   (ii) (Willard, 1982) Let S be a set of n points in $\mathbb{R}^2$ in general position. Use part (i) to construct a data structure requiring linear space that allows us to calculate $|S \cap h|$ for any half-plane h in $O(n^{\log 3/\log 4}) = O(n^{0.792\dots})$ time.

   (iii) (Edelsbrunner and Welzl, 1986b) Improve the exponent of the query time in part (ii) to $\log_2(\sqrt{5}+1)/2 \approx 0.695$.

15.17★ (Matoušek, 1992)   Let S be a set of n points in $\mathbb{R}^2$ in general position. A *geometric partitioning* of S is a collection $\{(S_1, \triangle_1), (S_2, \triangle_2), \dots, (S_m, \triangle_m)\}$, where $S_1, \dots, S_m$ is a partition of S and $\triangle_i$ is a triangle containing S_i. Show that there is a constant c such that for any r $(1 \le r \le n)$,

   (i) there exist a subset $S' \subseteq S$ of at least $n/2$ points and a geometric partitioning $\{(S_1', \triangle_1), (S_2', \triangle_2), \dots, (S_m', \triangle_m)\}$ of S' such that $|S_i'| = \lfloor n/r \rfloor$ $(1 \le i \le m)$ and every line intersects at most $c\sqrt{r}$ triangles;

   (ii) there exists a geometric partitioning of S of size at most cr such that $|S_i| \le n/r$ $(1 \le i \le m)$ and every line intersects at most $\sqrt{r}$ triangles.

15.18 Given any set S of n points in $\mathbb{R}^2$ in general position, $0 < \varepsilon < 1$, let $\mathcal{R}^\varepsilon$ be the family of all convex sets containing at least εn elements of S. A set of points $T \subseteq \mathbb{R}^2$ is said to be a "weak ε-net" for $\Sigma = (S, \mathcal{R}^\varepsilon)$ if every member of $\mathcal{R}^\varepsilon$ contains at least one element of T. (Note that the points of T do not necessarily belong to S.) Show that:

   (i) $\displaystyle\max_{|S|=n} \text{VC-dim}(\Sigma) = \lfloor n(1-\varepsilon) \rfloor$.

   (ii) There exists a constant c (independent of n, S, and ε) such that for every $0 < \varepsilon < 1$, Σ always has a weak ε-net consisting of at most $c/\varepsilon^{\log 4/\log(4/3)}$ points.

(iii) (Alon et al., 1992) The bound in part (ii) can be improved to c/ε^2.

(iv)⋆ (Chazelle et al., 1993) If the points of S are the vertices of a regular n-gon, then Σ has a weak ε-net whose size is at most c/ε, where c is a suitable constant.

(v) (Chazelle et al., 1993) If the points of S are the vertices of a convex n-gon, then Σ has a weak ε-net whose size is at most $(c/\varepsilon) \log^{\log 3}(1/\varepsilon)$, where c is a suitable constant.

15.19 (Aronov et al., 1994; Pach, 1991) We call two segments "independent" if neither of the lines supporting one of them crosses the other. Prove that n points in $\mathbb{R}^2$ in general position always determine at least $c\sqrt{n}$ mutually independent segments.

15.20 Use Corollary 15.6(i) to prove the following statement. Given a set L of n lines in general position and an integer $r \geq 3$, the plane can be decomposed into at most r^2 triangles (some of which are unbounded) such that the interior of each triangle intersects at most $c(n/r) \log r$ lines of L, where c is some fixed constant.

15.21 (Bárány and Lehel, 1987; Pach, 1994) Given two points $p, q \in \mathbb{R}^d$, let Box(p, q) denote the smallest box parallel to the axes which contains p and q. Show that from any finite set of points $S \subseteq \mathbb{R}^d$ one can choose $2^{2^{d+2}}$ or fewer points p_i $(1 \leq i \leq 2^{2^{d+2}})$ such that

$$\bigcup_{1 \leq i < j \leq 2^{2^{d+2}}} \text{Box}(p_i, p_j) \supseteq S.$$

15.22 (Ding et al., 1994) Solve Exercise 14.11 by using Theorem 15.9.

15.23 (Gyárfás and Lehel, 1969) Show that for any positive integers k and ν, there exists a number $f(k, \nu)$ with the following property. Let H be a finite family of subsets of $\mathbb{R}$, each a union of at most k intervals. If H has no $\nu + 1$ pairwise disjoint members, then one can find at most $f(k, \nu)$ points such that every member of H contains one of them.

16

Geometric Discrepancy

Let H be a hypergraph with vertex set $V(H)$ and edge set $E(H)$, respectively. Let c be a coloring of the vertices with two colors denoted by $+1$ and -1, that is, $c: V(H) \to \{+1, -1\}$. Given any edge E, let

$$c(E) = \sum_{x \in E} c(x),$$

and

$$\operatorname{disc}(H, c) = \max_{E \in E(H)} |c(E)|.$$

The *discrepancy* of H is defined as

$$\operatorname{disc}(H) = \min_{c} \operatorname{disc}(H, c),$$

where the minimum is taken over all two-colorings of the vertices. In other words, we want to find a two-coloring which is as "uniform" as possible, in the sense that every edge of H contains roughly the same number of points of color $+1$ and -1.

The theory of uniform distribution, especially discrepancy theory, has a long history, and it has several applications in number theory, geometry, and numerical analysis (see, e.g., Beck and Chen, 1987; Kuipers and Niederreiter, 1974; Hlawka, 1944; Beck and Sós, 1994; and Sós, 1989). It is not surprising that probabilistic and combinatorial methods play a central role in this field. However, it is intuitively less clear why harmonic analysis and integral geometry are instrumental in discrepancy theory. The aim of this chapter is to illustrate some of these techniques by concentrating on a few sample problems. In particular, we prove that the discrepancy of any hypergraph whose vertices are n points in the plane, and whose edges can be obtained by intersecting the vertex set with half-planes, is at most $cn^{1/4}$. We also show that this bound is asymptotically tight apart from the value of the constant c. These results are due to Matoušek (1995) and Alexander (1990), respectively, and improve some earlier estimates of Beck (1983b, 1991) by polylogarithmic factors.

We shall often use the following so-called *tail estimate* for the distribution of the sum of independent random variables.

Lemma 16.1 (Chernoff's inequality). *Let $\xi = \xi_1 + \xi_2 + \cdots + \xi_N$, where the ξ_i's are mutually independent random variables with $\Pr[\xi_i = +1] = \Pr[\xi_i = -1] = 1/2$. Then, for any $t > 0$,*

$$\Pr[\xi > t] < \exp\left(-\frac{t^2}{2N}\right).$$

Proof. Let c be a positive number to be fixed later. Then the expected value of $e^{c\xi_i}$,

$$\mathbf{E}[e^{c\xi_i}] = \tfrac{1}{2}e^c + \tfrac{1}{2}e^{-c} < e^{c^2/2}.$$

Since the ξ_i's are mutually independent,

$$\mathbf{E}[e^{c\xi}] = \prod_{i=1}^{N} \mathbf{E}[e^{c\xi_i}] < e^{Nc^2/2}.$$

Thus,

$$\Pr[\xi > t] = \Pr[e^{c\xi} > e^{ct}] < \frac{\mathbf{E}[e^{c\xi}]}{e^{ct}} = e^{N(c^2/2) - ct}.$$

Setting $c = t/N$, we obtain the desired upper bound on $\Pr[\xi > t]$.   □

METHOD OF FLOATING COLORS

Theorem 16.2. *Let H be hypergraph with n vertices and m hyperedges. Then*

$$\mathrm{disc}\,(H) \le \sqrt{2n \ln 2m}.$$

Proof. Take a random coloring c of the vertices. That is, assign $+1$ or -1 to every vertex of H independently with probability $\tfrac{1}{2}$.

By Lemma 16.1, for any $E \in E(H)$,

$$\Pr[|c(E)| > t] < 2\exp\left(-\frac{t^2}{2|E|}\right) \le 2\exp\left(-\frac{t^2}{2n}\right).$$

Thus,

$$\Pr[\mathrm{disc}\,(H, c) > t] \le \sum_{E \in E(H)} \Pr[|c(E)| > t]$$

$$< 2m \exp\left(-\frac{t^2}{2n}\right) = 1,$$

provided that $t = \sqrt{2n \ln 2m}$. Hence, there is a coloring c such that disc $(H, c) \le \sqrt{2n \ln 2m}$.   □

A simple probabilistic construction shows that this bound in general cannot be much improved (see Exercise 16.1). However, if we assume that every vertex of H is contained only in a small number of hyperedges, then H can be colored more uniformly, that is, with smaller discrepancy.

The number of edges $E \in E(H)$ containing a vertex x is called the *degree* of x in H and is denoted by $d_H(x)$ [or by $d(x)$].

Theorem 16.3 (Beck and Fiala, 1981). *If every vertex of a hypergraph H has degree at most D, then disc $(H) \le 2D$.*

Proof. The idea of the proof is that we relax the condition that a coloring can take only two values, $+1$ and -1. Instead, we consider a *quasi-coloring* $c: V(H) \rightarrow [-1, +1]$, whose values can lie anywhere in the closed interval $[-1, +1]$. We start with the trivial quasi-coloring $c \equiv 0$ having 0 discrepancy, and this will be gradually modified by setting some values to $+1$ or -1 until we obtain a true coloring. This procedure is widely known as the "method of floating colors" (cf. Spencer, 1987).

Assume that at some stage of this construction we have a quasi-coloring c. If $c(x) = +1$ or -1, we call this value *final* and we do not modify it later. If $-1 < c(x) < +1$, then x is said to have *floating colors*. A hyperedge $E \in E(H)$ is called *active* if more than D of its points have floating colors. Suppose further that all active edges have 0 discrepancy, i.e.,

$$c(E) = \sum_{x \in E} c(x) = 0 \qquad \text{for every active } E.$$

Suppose that not all edges of H are inactive, and let $E_1, E_2, \ldots, E_m$ denote the active edges. Renumber the vertices so that $x_1, x_2, \ldots, x_p$ have floating colors and the values of c are final on $x_{p+1}, \ldots, x_n$. Since every active edge has more than D points with floating colors, and every point belongs to at most D edges, we have $m < p$.

In the next step we modify the quasi-coloring c, as follows. Consider the system of m equations with p variables $(y_1, y_2, \ldots, y_p)$:

$$\sum_{\substack{x_j \in E_i \\ 1 \le j \le p}} y_j + \sum_{\substack{x_j \in E_i \\ p < j \le n}} c(x_j) = 0 \qquad (1 \le i \le m).$$

The locus of points $(y_1, y_2, \ldots, y_p)$ satisfying all of these equations is a k-dimensional flat in $\mathbb{R}^p$, for some $k \ge p - m \ge 1$, having nonempty intersection with the cube $C^p = [-1, +1]^p$. Thus, there exists at least one intersection point $(y_1', y_2', \ldots, y_p')$ lying on a facet $((p-1)$-dimensional face) of C^p; that is, $y_j' = +1$ or -1 for some j.

Setting

$$c'(x_j) = \begin{cases} y'_j & \text{if } 1 \le j \le p, \\ c(x_j) & \text{otherwise,} \end{cases}$$

we obtain a quasi-coloring c' with $c'(E) = 0$ for every active edge E. Besides, the number of final values of c' increased by at least one.

Proceeding like this, we end up with a quasi-coloring c^* having no active edges. If not all the values of c^* are +1's and −1's, then change all floating colors to, say, +1. Thus, we obtain a true coloring $\hat{c}: V(H) \to \{-1, +1\}$. We claim that $|\hat{c}(E)| \le 2D$ for every edge $E \in E(H)$.

Indeed, at the stage when E became inactive, we had $c'(E) = \sum_{x \in E} c'(x) = 0$, and after that we have changed the values of c' in at most D points of E. Hence,

$$|\hat{c}(E)| \le \sum_{x \in E} |\hat{c}(x) - c'(x)| + |c'(E)| \le 2D + 0. \qquad \square$$

An especially nice geometric consequence of the above result is the following theorem, which partially answers a question raised by G. Tusnády.

Theorem 16.4 (Beck, 1981a).   *Let S be a set of n points in $\mathbb{R}^d$ ($d \ge 2$) such that no two of them have the same x_i-coordinate for $1 \le i \le d$. Then there exists a coloring $c: S \to \{+1, -1\}$ with the property that for any box B whose sides are parallel to the axes,*

$$\left| \sum_{p \in B \cap S} c(p) \right| \le (2 \log 2n)^{2d}.$$

Proof.   For any $i \le d$, define a system $\mathcal{I}_i$ of *canonical intervals* on the x_i-axis, as follows. Let $x_i^1 < x_i^2 < \cdots < x_i^n$ denote the x_i-coordinates of the points of X, sorted in increasing order. Let $[x_i^1, x_i^n] \in \mathcal{I}_i$. Furthermore, for any element $[x_i^j, x_i^k] \in \mathcal{I}_i$ which is not a singleton ($k > j$), add two additional elements $[x_i^j, x_i^{\lfloor (j+k)/2 \rfloor}]$, $[x_i^{\lfloor (j+k)/2 \rfloor + 1}, x_i^k]$ to $\mathcal{I}_i$. Obviously, the elements of any contiguous subsequence of $x_i^1 < x_i^2 < \cdots < x_i^n$ can be covered by at most $2 \lceil \log n \rceil$ disjoint canonical intervals.

A box $I_1 \times I_2 \times \cdots \times I_d$ is called *canonical* if $I_i \in \mathcal{I}_i$ for every $1 \le i \le d$. Every point $p \in S$ is contained in at most $(\lceil \log n \rceil + 1)^d$ canonical boxes. Thus, it follows from Theorem 16.3 that there is a coloring $c: S \to \{+1, -1\}$ such that

$$\left| \sum_{p \in B' \cap S} c(p) \right| \le 2(\lceil \log n \rceil + 1)^d,$$

for any canonical box B'.

To complete the proof, it is sufficient to observe that for any box B whose sides are parallel to the axes, $B \cap S$ can be partitioned into at most $(2\lceil \log n \rceil)^d$ disjoint sets

$$B \cap S = (B^1 \cap S) \cup (B^2 \cap S) \cup \cdots \cup (B^k \cap S),$$

where each B^j is a canonical box. $\square$

DISCREPANCY AND VC-DIMENSION

The above argument made essential use of the fact that the boxes were in parallel position. Without this assumption one cannot obtain any polylogarithmic upper bound on the discrepancy. Nevertheless, even in this case we shall be able to exploit some structural information and derive nontrivial upper bounds (better than those following from Theorem 16.2).

We need some definitions. For any hypergraph $H = (V(H), E(H))$, we can define a *dual hypergraph* H^* by interchanging the roles of vertices and edges. That is, let

$$V(H^*) = E(H),$$
$$E(H^*) = \left\{ \{E \in E(H) \mid E \ni x\} \mid x \in V(H) \right\}.$$

Obviously, $(H^*)^* = H$.

The *shatter function* of H is defined as

$$\pi_H(m) = \max_{\substack{A \subseteq V(H) \\ |A| = m}} \left| \{E \cap A \mid E \in E(H)\} \right|.$$

Theorem 15.4 implies that if H has VC-dimension d, then

$$\pi_H(m) \le \sum_{i=0}^{d} \binom{m}{i} = O(m^d)$$

for all $m \le |V(H)| = n$.

The shatter function of H^* is often called the *dual shatter function* of H. For any $m \le |E(H)|$, $\pi_{H^*}(m)$ is the maximum number of equivalence classes into which m edges $E_1, \ldots, E_m \in E(H)$ can partition $V(H)$, where two vertices are considered equivalent if they belong to the same E_i's ($1 \le i \le m$).

Lemma 16.5. (i) *If* VC-dim$(H) < d$, *then* VC-dim$(H^*) < 2^d$.
(ii) *If* $\pi_H(m) \le Cm^d$ *for some* $C > 0$, $d > 1$ *and for all* $m \le |V(H)|$, *then*

$$\text{VC-dim}(H) \le 4d \log(Cd + 2).$$

Proof. (i) Suppose, in order to obtain a contradiction, that VC-dim(H^*) $\geq$ 2^d. This means that there is a 2^d-element subset $U \subseteq E(H)$ such that for any $W \subseteq U$, one can choose $x_W \in V(H)$ so that $\{E \in U \mid x_W \in E\} = W$. Label the elements of U by distinct subsets of $\{1, 2, \ldots, d\}$, and let W_i denote the set of all elements of U whose labels contain i ($1 \leq i \leq d$).

We claim that $\{x_{W_1}, x_{W_2}, \ldots, x_{W_d}\}$ is shattered in H. Indeed, for any subset $A \subseteq \{1, 2, \ldots, d\}$, there is an element $E \in U \subseteq E(H)$ whose label is A, and

$$E \cap \{x_{W_1}, \ldots, x_{W_d}\} = \{x_{W_i} \mid i \in A\}.$$

Thus, VC-dim(H) $\geq d$, a contradiction.

(ii) Observe that if VC-dim(H) = D, then $\pi_H(D) = 2^D \leq CD^d$. $\qquad\square$

In particular, Lemma 16.5(i) yields that if VC-dim(H) < d, then the shatter function of its dual hypergraph satisfies $\pi_{H^*}(n) = O(n^{2^d - 1})$. However, for most hypergraphs arising in geometric settings, one can derive significantly better upper bounds for $\pi_{H^*}(n)$.

Let H be a hypergraph and G be a graph on the same set of vertices, $V(G) = V(H)$. We say that an edge of G *stabs* a hyperedge $E \in E(H)$ if one of its endpoints is in E and the other is not. The *stabbing number* of G (with respect to H) is defined as the smallest number s such that every hyperedge of H is stabbed by at most s edges of G.

Chazelle and Welzl (1989) have noticed that the proofs of Theorem 15.12 and Corollary 15.14 can easily be extended to any hypergraph H whose dual shatter function satisfies $\pi_{H^*}(m) \leq Cm^d$ for all m. Moreover, Haussler (1991) managed to remove the logarithmic factors from the above bounds, even in this more general setting (see also Wernisch, 1994, and Exercise 16.7).

Thus, we have the following far-reaching generalization of Corollary 15.14.

Theorem 16.6. *For any $C > 0$, $d > 1$, there exists a constant $C'' = C''(C, d)$ with the following property. Let H be any hypergraph with n vertices whose dual shatter function satisfies $\pi_{H^*}(m) \leq Cm^d$ for all m. Then there is a spanning path P on $V(H)$, whose stabbing number with respect to H, $\sigma(P)$, satisfies*

$$\sigma(P) \leq C'' n^{1 - 1/d}.$$

The following important consequence of Theorem 16.6 has been pointed out by Matoušek, Welzl, and Wernisch (1993). In the case when H has small VC-dimension, it represents a definite improvement on the general bound given in Theorem 16.2. Moreover, for $d = 2$ and 3 it is known to be asymptotically best possible (see Matoušek, 1994).

Theorem 16.7. *For any $C > 0, d > 1$, there exists a constant $C'' = C''(C, d)$ with the following property. Let H be a hypergraph such that $\pi_{H^*}(m) \leq Cm^d$ for all m. Then*

$$\text{disc}(H) \leq C'' n^{(1/2)(1 - 1/d)} \sqrt{\log n}.$$

Proof. Apply Theorem 16.6 to find a spanning path $P = x_1 x_2 \cdots x_n$ whose stabbing number (with respect to H) is at most $C' n^{1-1/d}$.

Choose a random coloring $c = \{x_1, x_3, x_5, \ldots\} \to \{+1, -1\}$, where each point is colored $+1$ or -1 independently with probability $\frac{1}{2}$. Extend c to a coloring of the full vertex set by letting $c(x_{2i}) = -c(x_{2i-1})$ for every i ($1 \leq i \leq n/2$).

If a pair $x_{2i-1} x_{2i}$ does not stab an edge $E \in E(H)$, its contribution to $c(E) = \sum_{x \in E} c(x)$ is 0. The number of pairs $x_{2i-1} x_{2i}$ stabbing $E \in E(H)$ is at most $C' n^{1-1/d}$, and if n is odd and $x_n \in E$, then x_n also contributes to $c(E)$.

Thus, $c(E)$ can be written as the sum of at most $C' n^{1-1/d} + 1$ independent random variables taking values $+1$ and -1 with probability $\frac{1}{2}$. It follows by Chernoff's inequality (cf. Lemma 16.1) that for any fixed $E \in E(H)$,

$$\Pr[|c(E)| > t] < 2 \exp\left(-\frac{t^2}{2(C' n^{1-1/d} + 1)}\right).$$

Applying Lemma 16.5(ii) and (i), we have

$$\text{VC-dim}(H) \leq 2^{5d \log(Cd+2)} = (Cd + 2)^{5d}.$$

This, in turn, implies that

$$|E(H)| \leq \sum_{i=0}^{(Cd+2)^{5d}} \binom{n}{i} < n^{(Cd+2)^{5d}}.$$

Hence,

$$\Pr[\text{disc}(H, c) > t]$$
$$\leq \sum_{E \in E(H)} \Pr[|c(E)| > t]$$
$$< 2n^{(Cd+2)^{5d}} \cdot \exp\left(-\frac{t^2}{2(C' n^{1-1/d} + 1)}\right)$$
$$= 2 \exp\left((Cd + 2)^{5d} \ln n - \frac{t^2}{2(C' n^{1-1/d} + 1)}\right) < 1,$$

provided that $t = C'' n^{(1/2)(1-1/d)} \sqrt{\log n}$, where C'' is a sufficiently large constant, depending only on C and d. $\square$

Applying Theorem 16.7 to the hypergraph H whose vertex set is a set of n points in $\mathbb{R}^d$ and

$$E(H) = \{E \subseteq V(H) \mid \exists \text{ a ball } B \text{ such that } B \cap V(H) = E\},$$

we obtain the following result (see Exercise 16.8).

Corollary 16.8. *Let S be a set of n points in $\mathbb{R}^d, d \geq 2$. Then there exists a coloring $c: S \to \{+1, -1\}$ with the property that for any ball B,*

$$\left| \sum_{p \in B \cap S} c(p) \right| \leq Cn^{(1/2)(1-1/d)}\sqrt{\log n},$$

where C is a constant depending only on d.

A similar result holds for boxes instead of balls (see Exercise 16.9).

Corollary 16.9. *Let S be a set of n points in $\mathbb{R}^d$, $d \geq 2$. Then there exists a coloring $c: S \to \{+1, -1\}$ with the property that for any box B (not necessarily parallel to axes),*

$$\left| \sum_{p \in B \cap S} c(p) \right| \leq Cn^{(1/2)(1-1/d)}\sqrt{\log n},$$

where C is a constant depending only on d.

METHOD OF PARTIAL COLORING

In the preceding section we gave an upper bound on the discrepancy of a hypergraph H in terms of $\pi_{H^*}(n)$, the shatter function of its dual hypergraph (Theorem 16.7). In view of Lemma 16.5, this implies another nontrivial upper bound on disc (H) in terms of the "primal" shatter function $\pi_H(n)$. However, this bound can be improved substantially.

Theorem 16.10 (Matoušek, 1995). *For any $C > 0$, $d > 1$, there exists a constant $C'' = C''(C, d)$ with the following property. Let H be any hypergraph of n vertices whose shatter function satisfies $\pi_H(m) \leq Cm^d$ for all $m \leq n$. Then*

$$\text{disc}(H) \leq C''n^{(1/2)(1-1/d)}.$$

This result improves an earlier estimate of Matoušek et al. (1993) by a logarithmic factor, and it is asymptotically tight.

The proof presented below has two major ingredients. The first one is a *packing lemma* due to Haussler (1991), which generalizes an important geometric observation of Chazelle and Welzl (1989) to hypergraphs. The other major tool is the method of partial coloring developed by Beck (1981b) and

Spencer (1985), which is based on a clever combination of the probabilistic method and the pigeonhole principle.

Let $A \triangledown B$ denote the *symmetric difference* of the sets A and B, that is,

$$A \triangledown B = (A - B) \cup (B - A).$$

The *distance* of A and B (in the Hamming metric) is $|A \triangledown B|$. A set system is called *r-separated* if the distance between any two of its members is at least r.

Lemma 16.11 (Haussler).   *For any $C > 0, d > 1$, there exists a constant $C' = C'(C, d)$ with the following property. Let H be any hypergraph of n vertices, whose shatter function satisfies $\pi_H(m) \leq Cm^d$ for all $m \leq n$.*

Then, for every $r \geq 1$, any r-separated system of hyperedges of H has at most $C'(n/r)^d$ elements.

Proof (Sketch).   We give the proof of this lemma only in the special case when $V(H)$ is a set of points in $\mathbb{R}^d$ in general position, and $E(H)$ consists of all sets that can be obtained by intersecting $V(H)$ with half-spaces. Since the number of different ways in which a hyperplane can bisect an m-element subset of $V(H)$ is $\sum_{i=0}^{\lfloor d/2 \rfloor} \binom{m}{d-2i}$ [cf. Exercise 15.10(iii)], we have

$$\pi_H(m) = 2 \sum_{i=0}^{\lfloor d/2 \rfloor} \binom{m}{d - 2i} \leq m^d.$$

So in this case the condition of Lemma 16.11 is satisfied with $C = 1$.

Consider now an r-separated system of hyperedges E_i $(1 \leq i \leq I)$. For every i, choose a half-space S_i such that $S_i \cap V(H) = E_i$, and let h_i denote the boundary hyperplane of S_i. Assume without loss of generality that the intersection of S_i with the positive x-axis is unbounded for at least $I/2$ indices (say, for all $1 \leq i \leq I/2$). By geometric duality, each vertex $x \in V(H)$ corresponds to a hyperplane x^*, and each hyperplane h_i corresponds to a point h_i^* of $\mathbb{R}^d$ such that $|E_i \triangledown E_j|$ is equal to the number of hyperplanes x^* intersecting the segment $h_i^* h_j^*$ $(1 \leq i, j \leq I/2)$.

For any two points $p, q \in \mathbb{R}^d$, let dist(p, q) be defined as the number of hyperplanes x^* $(x \in V(H))$, intersecting the closed segment pq. (Notice that dist(p, q) is not a proper metric on $\mathbb{R}^d$, because it can vanish even if $p \neq q$, and dist$(p, p) \neq 0$ whenever p belongs to a hyperplane. Nevertheless, it satisfies the triangle inequality.) Furthermore, let $B^d(p, k)$ denote the set of all vertices v of the arrangement $\{x^* | x \in V(H)\}$, for which dist$(p, v) \leq k$. Since dist$(h_i^*, h_j^*) \geq r$, we know that the sets $B^d(h_i^*, r/2)$, $1 \leq i \leq I/2$, are pairwise disjoint. Taking into account that the total number of vertices of the arrangement is $\binom{n}{d}$, to verify Lemma 16.11 in our special case it is enough to prove that

$$|B^d(h_i^*, r/2)| \geq c_d(r/2)^d.$$

Claim: For every point not belonging to any hyperplane x^* ($x \in V(H)$), and for any integer k ($d \le k \le n/2$), we have $|B^d(p,k)| \ge c_d k^d$ for a suitable constant $c_d > 0$.

For $d = 1$, the assertion is trivially true. Let $d \ge 2$, and assume that the Claim has already been proved for all dimensions smaller than d. Choose any ray emanating from p which intersects at least $n/2$ hyperplanes of the arrangement in distinct points $p_1, \ldots, p_{\lceil n/2 \rceil}$ (in this order). Applying the induction hypothesis to the $(d-1)$-dimensional arrangement obtained by taking the intersection of the hyperplane containing p_i with the other $n-1$ hyperplanes ($1 \le i \le k-d+1$), we get that

$$|B^d(p,k)| \ge \frac{1}{d} \sum_{i=1}^{k-d+1} |B^{d-1}(p_i, k-i)|$$

$$\ge \frac{c_{d-1}}{d} \sum_{i=1}^{k-d+1} (k-i)^{d-1}$$

$$\ge c_d k^d,$$

as required. This proves the Claim and hence Lemma 16.11 in the special case we considered.   □

Proof of Theorem 16.10.   As in the proof of Theorem 16.4, we first define a *canonical decomposition* of every edge of H into relatively few "canonical" pieces. Then we show that there exists a good two-coloring of the family of these canonical subsets with small discrepancy. From this the result will follow immediately.

The canonical decomposition is based on Lemma 16.11. Suppose, for the sake of simplicity, that n is a power of 2, and $\emptyset \in E(H)$. For any i, $0 \le i < \log n$, let H_i be a hypergraph on the vertex set $V(H) = V$, whose edges are chosen from $E(H)$ so that they form a *maximal* 2^i-*separated system*. Then $H_0 = H$ and by Lemma 16.11, we have

$$|E(H_i)| < C' \left(\frac{n}{2^i} \right)^d \qquad \text{for every } i. \tag{16.1}$$

For notational convenience, let $H_{\log n}$ denote the hypergraph consisting of only one edge, the empty set.

For every i, $1 \le i \le \log n$, define two other hypergraphs H_i' and H_i'', as follows. For any $E \in E(H_{i-1})$, fix an edge $E^* \in E(H_i)$ with $|E \triangledown E^*| \le 2^i$, and let

$$E(H_i') = \{E - E^* \mid E \in E(H_{i-1})\},$$
$$E(H_i'') = \{E^* - E \mid E \in E(H_{i-1})\}.$$

By our conventions, $H'_{\log n} = H_{\log n-1}$ and $E(H''_{\log n}) = \{\varnothing\}$. Now each $E \in E(H) = E(H_0)$ can be written as

$$E = (E^* \cup (E - E^*)) - (E^* - E).$$

In other words,

$$E = (E_1 \cup E'_1) - E''_1$$

for some $E_1 \in E(H_1)$, $E'_1 \in E(H'_1)$, and $E''_1 \in E(H''_1)$. Similarly, we can express E_1 as

$$E_1 = (E_2 \cup E'_2) - E''_2$$

for some $E_2 \in E(H_2)$, $E'_2 \in E(H'_2)$, and $E''_2 \in E(H''_2)$. Proceeding in this way, after making $\log n$ steps, we obtain a canonical decomposition of E as

$$E = ((((\cdots (E_{\log n} \cup E'_{\log n}) \cdots) \cup E'_2) - E''_2) \cup E'_1) - E''_1,$$

where $E_{\log n} = \varnothing$, $E'_i \in E(H'_i)$, and $E''_i \in E(H''_i)$. The crucial property of this decomposition is that one always takes unions of *disjoint* sets or subtracts from a set one of its *subsets*. Thus, for any coloring $c: V \to \{+1, -1\}$, we have

$$|c(E)| \le \sum_{i=1}^{\log n} (|c(E'_i)| + |c(E''_i)|). \qquad (16.2)$$

Let $\hat{H}_i$ denote the hypergraph with $V(\hat{H}_i) = V$ and $E(\hat{H}_i) = E(H'_i) \cup E(H''_i)$, $1 \le i \le \log n$. It follows from (16.1) and the definitions that

$$|E(\hat{H}_i)| \le 2|E(H_{i-1})| \le 2C'(n/2^{i-1})^d,$$
$$|E| \le 2^i \qquad \text{for every } E \in E(\hat{H}_i). \qquad (16.3)$$

It is sufficient to establish the following.

Lemma 16.12. *There exists a three-coloring* $c: V \to \{+1, -1, 0\}$ *with the following properties.*

(i) $\sum_{i=1}^{\log n} \max_{E \in E(\hat{H}_i)} |c(E)| \le Kn^{(1/2)(1-1/d)}$, *where* $K = K(C, d)$ *is a suitable constant;*

(ii) $c(x) \ne 0$ *for at least* $n/3$ *points* $x \in V$.

Indeed, Theorem 16.10 easily follows by repeated application of the lemma. Let c be a coloring satisfying the above conditions, and let

$$V_1 = \{x \in V \mid c(x) = 0\}.$$

At the second iteration, we repeat the whole procedure for the subhypergraph

$H[V_1]$ of H *induced* by V_1, which is defined by

$$V(H[V_1]) = V_1,$$
$$E(H[V_1]) = \{E \cap V_1 \mid E \in E(H)\}.$$

Now we obtain a coloring $c_1: V_1 \to \{+1, -1, 0\}$, and let $V_2 = \{x \in V_1 \mid c_1(x) = 0\}$, and so on. We can stop this process when we obtain a coloring with fewer than $n^{(1/2)(1 - 1/d)}$ zeros, and color all these points with $+1$.

Let $\bar{c}: V \to \{+1, -1\}$ be the "superposition" of the above colorings, i.e.,

$$\bar{c}(x) = \begin{cases} c(x) & \text{if } x \in V - V_1, \\ c_1(x) & \text{if } x \in V_1 - V_2, \\ c_2(x) & \text{if } x \in V_2 - V_3 \\ \vdots & \vdots \\ +1 & \text{otherwise.} \end{cases}$$

It follows from (16.2) and Lemma 16.12(i) that for every $E \in E(H)$,

$$|\bar{c}(E)| < |c(E)| + |c_1(E \cap V_1)| + \|c_2(E \cap V_2)| + \cdots + n^{(1/2)(1 - 1/d)}$$

$$< 2K \sum_{i=0}^{\infty} \left(\left(\frac{2}{3} \right)^i n \right)^{(1/2)(1 - 1/d)} + n^{(1/2)(1 - 1/d)}$$

$$< \frac{2K + 1}{1 - \left(\frac{2}{3} \right)^{(1/2)(1 - 1/d)}} n^{(1/2)(1 - 1/d)},$$

which completes the proof of the theorem.   □

Hence, it remains to establish Lemma 16.12. For this we need some preparation.

Definition 16.13.   Let ξ be a random variable whose values are taken from a finite set S. The *entropy* of ξ is defined as

$$H(\xi) = - \sum_{s \in S} p_s \log p_s,$$

where $p_s = \Pr[\xi = s]$, and we sum over all values for which $p_s > 0$.

We leave to the reader the easy proofs of the following properties of the entropy that can be derived directly from the definition:

(a) $H(\xi) \le - \sum_{s \in S} \frac{1}{|S|} \log \frac{1}{|S|} = \log |S|$;

(b) if $H(\xi) \le h$, then there is a value $s \in S$ with $p_s \ge 2^{-h}$;

(c) for any random variable $\xi = (\xi_1, \xi_2)$ consisting of two components, we have $H(\xi) \leq H(\xi_1) + H(\xi_2)$

(see Exercise 16.10).
   Now we can prove Lemma 16.12.

Proof of Lemma 16.12. Let M be a large positive constant to be specified later, and assume for simplicity that

$$m = \log(M^{1/d} n^{1 - 1/d})$$

is an integer. For any i, $1 \leq i \leq \log n$, set

$$D_i = M n^{(1/2)(1 - 1/d)} 2^{-|i - m|/4}.$$

Let $c: V \rightarrow \{+1, -1\}$ be a random two-coloring, where the vertices are colored with $+1$ or -1 independently with probability $\frac{1}{2}$. For any i and any $E \subseteq E(\hat{H}_i)$, define a random variable $b_{i,E} = b_{i,E}(c)$, as

$$b_{i,E} = \left[\frac{c(E)}{D_i} \right],$$

where $[x]$ denotes a nearest integer to x. Finally, let

$$b = b(c) = (b_{i,E} \mid 1 \leq i \leq \log n, \ E \in E(\hat{H}_i));$$

i.e., b is a vector with $\sum_{i=1}^{\log n} |E(\hat{H}_i)|$ integer coordinates.

Claim: If M is sufficiently large, then $H(b) \leq n/20$.

   Suppose that the Claim is true. Then, by property (b) of the entropy function, there exists a particular vector b_0, such that

$$\Pr[b(c) = b_0] \geq 2^{-n/20}.$$

In other words, the number of different colorings $c: V \rightarrow \{+1, -1\}$ with $b(c) = b_0$ is at least $2^{19n/20}$. Pick any one of them, say c_1. Clearly, the total number of two-colorings that differ from c_1 in at most $n/3$ points is

$$\sum_{i=0}^{n/3} \binom{n}{i} < 2^{19n/20}.$$

Thus, there exists a two-coloring c_2 with $b(c_2) = b(c_1) = b_0$ such that c_1 and c_2 differ for more than $n/3$ vertices. This yields that $\bar{c} = (c_1 - c_2)/2$ is a $(+1, -1, 0)$-coloring of V, which satisfies condition (ii) in Lemma 16.12. Notice that $\bar{c}$ also satisfies condition (i). Indeed, for any $E \in E(\hat{H}_i)$ we have $b_{i,E}(c_1) = b_{i,E}(c_2)$, whence

$$\left| \frac{c_1(E)}{D_i} - \frac{c_2(E)}{D_i} \right| \leq 1,$$

$$|\bar{c}(E)| = \left| \frac{c_1(E) - c_2(E)}{2} \right| \leq \frac{D_i}{2}.$$

Therefore,

$$\sum_{i=1}^{\log n} \max_{E \in E(\hat{H}_i)} |\bar{c}(E)| \leq \sum_{i=1}^{\log n} \frac{D_i}{2} < \frac{M}{1 - 2^{-1/4}} n^{(1/2)(1 - 1/d)},$$

and Lemma 16.12 holds.

Finally, we turn to the proof of the Claim. It follows from property (c) of the entropy function that

$$H(b) \leq \sum_{i=1}^{\log n} \sum_{E \in E(\hat{H}_i)} H(b_{i,E}),$$

so it is enough to bound the terms $H(b_{i,E})$.

Fix an index i and an edge $E \in E(\hat{H}_i)$, and set $p_s = \Pr[b_{i,E} = s]$ for any integer s. Since $c(E)$ is the sum of $|E|$ mutually independent random variables with values $+1$ and -1, in view of (16.3), we obtain by Chernoff's inequality (Lemma 16.1) that

$$p_s \leq \Pr[c(E) \geq (s - 1/2)D_i]$$

$$< \exp\left(-s^2 \frac{M^{2 - 1/d}}{8} 2^{m - i - |i - m|/2} \right).$$

If $i \leq m$, this yields

$$H(b_{i,E}) = \sum_{s = -\infty}^{\infty} p_s \log p_s$$

$$\leq - \log p_0 + 2 \sum_{s=1}^{\infty} \exp\left(-s^2 M 2^{(m - i)/2} \right)(\log e)s^2 M 2^{(m - i)/2}$$

$$\leq - \log p_0 + \exp\left(-2^{(m - i)/2} \right)$$

$$\leq - \log(1 - 2\Pr[c(E) \geq D_i/2]) + \exp\left(-2^{(m - i)/2} \right)$$

$$\leq - \log\left(1 - 2\exp\left(-M 2^{(m - i)/2} \right) \right) + \exp\left(-2^{(m - i)/2} \right)$$

$$\leq 2\exp\left(-2^{(m - i)/2} \right),$$

provided that M is sufficiently large. Thus, using (16.3) to bound $|E(\hat{H}_i)|$, we obtain

$$\sum_{i=1}^{m} \sum_{E \in E(\hat{H}_i)} H(b_{i,E}) \leq \sum_{i=1}^{m} 4C' \left(\frac{n}{2^{i-1}} \right)^d \exp\left(-2^{(m-i)/2}\right)$$

$$= \frac{4C'}{M} n \sum_{i=1}^{m} 2^{d(m-i+1)} \exp\left(-2^{(m-i)/2}\right)$$

$$\leq \frac{n}{40}.$$

If $i > m$, then we can use property (a) of the entropy. In this case,

$$H(b_{i,E}) = -\sum_{|s| \leq 2^{3(i-m)/4}} p_s \log p_s - \sum_{|s| > 2^{3(i-m)/4}} p_s \log p_s$$

$$\leq \log\left(2 \cdot 2^{3(i-m)/4} + 1\right) +$$

$$2 \sum_{|s| > 2^{3(i-m)/4}} \exp\left(-s^2 M 2^{-3(i-m)/2}\right)(\log e) s^2 M 2^{-3(i-m)/2}$$

$$\leq \left(\frac{3(i-m)}{4} + 2 \right) + 1.$$

Thus, by (16.3),

$$\sum_{i=m+1}^{\log n} \sum_{E \in E(\hat{H}_i)} H(b_{i,E}) \leq \sum_{i=m+1}^{\log n} 2C' \left(\frac{n}{2^{i-1}} \right)^d (i - m + 3)$$

$$= \frac{2C'}{M} n \sum_{i=m+1}^{\log n} \frac{i - m + 3}{2^{(i-m-1)d}}$$

$$\leq \frac{n}{40},$$

provided that M is large enough. This completes the proof of the Claim and hence Lemma 16.12. $\qquad\square$

It follows immediately from Theorem 16.10 that the logarithmic factors of the bounds in Corollaries 16.8 and 16.9 can be eliminated in certain special cases.

Corollary 16.14. *Let S be a set of n points in $\mathbb{R}^d$ for $d \geq 2$. Then there exists a coloring $c: S \to \{+1, -1\}$ with the property that for any unit ball B,*

$$\left| \sum_{p \in B \cap S} c(p) \right| \leq C n^{(1/2)(1 - 1/d)},$$

where C is a constant depending only on d. Consequently, the same result holds for half-spaces instead of unit balls.

The bound in Theorem 16.10 cannot be improved in general. It is well known, for example, that

$$\operatorname{disc}(H) \geq C_d n^{(1/2)(1 - 1/d)}$$

for a hypergraph H_d with $n = q^d + q^{d-1} + \cdots + 1$ vertices, whose edge set consists of all hyperplanes of a finite d-dimensional projective space of order q.

We include a simple proof of this fact, due to Di Paola, for $d = 2$. Notice that $H = H_2$ satisfies the condition for Theorem 16.10, because $\pi_H(m) \leq \binom{m}{2}$ for all $m \leq n$. Consider now any coloring $c: V(H) \to \{+1, -1\}$. Clearly,

$$\sum_{E \in E(H)} c^2(E) = \sum_{E \in E(H)} \left(\sum_{x \in E} c(x) \right)^2$$

$$= \sum_{E \in E(H)} \sum_{x \in E} c^2(x) + \sum_{x \in V(H)} \sum_{\substack{y \in V(H) \\ y \neq x}} c(x)c(y)$$

$$= (q + 1) \sum_{x \in V(H)} c^2(x) + \sum_{x \in V(H)} \sum_{y \in V(H)} c(x)c(y) - \sum_{x \in V(H)} c^2(x)$$

$$= q \sum_{x \in V(H)} c^2(x) + \left(\sum_{x \in V(H)} c(x) \right)^2$$

$$\geq qn.$$

Since $|E(H)| = n$, we obtain from here that there exists an edge $E \in E(H)$ with $|c(E)| \geq \sqrt{q} > (n - \sqrt{n} - 1)^{1/4}$, as required. The argument generalizes to higher dimensions.

DISCREPANCY AND INTEGRAL GEOMETRY

Improving a slightly weaker result of Beck (1983b), Alexander (1990) has shown that the bound in Corollary 16.14 is asymptotically tight. Of course, this also implies that Corollaries 16.8 and 16.9 are not far from being optimal. In this section we focus on the planar version of the problem. We are going to prove that for any two-coloring of the integer points $\{(i, j) | 1 \leq i, j \leq \sqrt{n}\}$, there exists a half-plane h within which one of the colors outnumbers the other by at least $cn^{1/4}$.

The following simple proof of Chazelle (1993), as well as the original argument of Alexander, is based on some elementary observations in integral geometry. This discipline originated with the famous *Buffon needle problem* and a series of other paradoxes involving geometric probabilities (see Buffon, 1770). The foundations of an exact theory were laid by Santaló (1952, 1976).

We start with an important special case of Cauchy's formula (1850) (see also Crofton, 1868, 1885; Blaschke, 1955). Let K be a convex disc (body) in the plane, containing $\mathbf{0}$ in its interior. For any $0 \leq \varphi < 2\pi$, let $h(\varphi)$ be the distance from $\mathbf{0}$ to the supporting line of K orthogonal to the direction φ, that is,

$$h(\varphi) = \sup_{P \in K} \langle P, (\cos \varphi, \sin \varphi) \rangle.$$

The function $h(\varphi)$ is called the *support function* of K.

Theorem 16.15 (Cauchy). *Let $K \subseteq \mathbb{R}^2$ be a convex disc with support function $h(\varphi)$, and let* Per K *denote its perimeter. Then*

$$\text{Per } K = \int_0^{2\pi} h(\varphi)\, d\varphi.$$

Proof. Let $\mathbf{0} \in K$, and without loss of generality assume that $h''(\varphi)$, the second derivative of the support function of K, exists and is continuous. For any given $0 \leq \varphi < 2\pi$, let $\ell(\varphi)$ denote the supporting line orthogonal to the direction φ. Furthermore, let $P(\varphi) = (x(\varphi), y(\varphi))$ and $H(\varphi) = (h(\varphi) \cos \varphi, h(\varphi) \sin \varphi)$ denote the intersection point of $\ell(\varphi)$ with the boundary of K and the foot of the normal dropped from $\mathbf{0}$ to $\ell(\varphi)$, respectively (see Figure 16.1).

Then we have

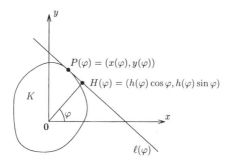

Figure 16.1. Illustration for the proof of Theorem 16.15.

$$h(\varphi) = \langle P(\varphi), (\cos\,\varphi, \sin\,\varphi)\rangle$$
$$= x(\varphi)\cos\,\varphi + y(\varphi)\sin\,\varphi,$$
$$h'(\varphi) = \langle P'(\varphi), (\cos\,\varphi, \sin\,\varphi)\rangle + \langle P(\varphi), (-\sin\,\varphi, \cos\,\varphi)\rangle$$
$$= 0 - x(\varphi)\sin\,\varphi + y(\varphi)\cos\varphi.$$

Equivalently,

$$x(\varphi) = h(\varphi)\cos\,\varphi - h'(\varphi)\sin\,\varphi,$$
$$y(\varphi) = h(\varphi)\sin\,\varphi + h'(\varphi)\cos\,\varphi.$$

Thus,

$$
\begin{aligned}
\operatorname{Per} K &= \int_0^{2\pi} \left((x'(\varphi))^2 + (y'(\varphi))^2\right)^{1/2} d\varphi \\
&= \int_0^{2\pi} \left((-h(\varphi)\sin\,\varphi - h''(\varphi)\sin\,\varphi)^2 + (h(\varphi)\cos\,\varphi + h''(\varphi)\cos\,\varphi)^2\right)^{1/2} d\varphi \\
&= \int_0^{2\pi} (h(\varphi) + h''(\varphi))\,d\varphi \\
&= \int_0^{2\pi} h(\varphi)\,d\varphi,
\end{aligned}
$$

as asserted (see Exercise 16.11).   □

We mention two immediate corollaries of Cauchy's formula.

(i) Let $K' \subseteq K$ be convex discs in the plane. Then

$$\operatorname{Per} K' \le \operatorname{Per} K.$$

(ii) (Barbier, 1860) Let K be a convex disc of *constant width* 1; that is, let $h(\varphi) + h(\varphi + \pi) = 1$ for every φ. Then $\operatorname{Per} K = \pi$.

Every line ℓ not passing through the origin is uniquely determined by its closest point to **0**. Let $\rho(\ell)$ and $\varphi(\ell)$ denote the polar coordinates of this point, that is, distance from **0** to ℓ, and the angle between the x-axis and the normal to ℓ, respectively. In this way, any set of lines L can be characterized by a subset of the (ρ, φ)-plane, and we define $\mu(L)$ to be the Lebesgue measure of this subset. Then μ is invariant under the group of motions of $\mathbb{R}^2$, and the measure of the set of lines intersecting a given convex set $K \subseteq \mathbb{R}^2$ is equal to $\operatorname{Per} K$ (by Theorem 16.15).

In the sequel let $U = [0, 1/4]^2$ denote a square of unit perimeter, and let us restrict our attention to the set L of all lines intersecting U. Then $\mu(L) = 1$, so μ is a *probability measure* on L. For a subset $L' \subseteq L$ and a *random line* ℓ

intersecting U, we can say that

$$\Pr[\ell \in L'] = \mu(L') \le 1.$$

Similarly, we can talk about a *random slab* of width w, which is a parallel strip centered along a random line $\ell \in L$:

$$\text{slab}(\ell, w) = \{p \in \mathbb{R}^2 \mid \text{dist}(p, \ell) \le w/2\}.$$

These definitions provide a possible framework for asking (and answering) exact questions about geometric probabilities. For example, let p and p' be two points of U at distance $s \ge w$ from each other, and suppose that their distances from the boundary of U are greater than $w/2$. Then the probability that a random slab of width w will contain both points,

$$\Pr[p, p' \in \text{slab}(\ell, w)] = \int_{-\arcsin(w/s)}^{+\arcsin(w/s)} (w - s|\sin \varphi|)\, d\varphi$$

$$= 2w \arcsin(w/s) + 2s\sqrt{1 - (w/s)^2} - 2s. \tag{16.4}$$

Now we turn to Chazelle's proof for the main result of this section.

Theorem 16.16 (Alexander). *There exists a set S of n points in the plane such that for any two-coloring $c: S \to \{+1, -1\}$, one can find a half-plane h with*

$$\big|\,|\{p \in S \cap h \mid c(p) = +1\}| - |\{p \in S \cap h \mid c(p) = -1\}|\,\big| \ge Cn^{1/4},$$

where C is a positive constant.

Proof. Assume without loss of generality that $n = m^2$ for some integer $m > 0$, and let

$$S = \left\{\left(\frac{i}{4(m+1)}, \frac{j}{4(m+1)}\right) \in \mathbb{R}^2 \,\Big|\, 1 \le i, j \le m\right\} \subseteq U.$$

Let $w = \varepsilon/m$, where $\varepsilon > 0$ is a small constant which will be specified later. Let $c: S \to \{+1, -1\}$ be a two-coloring of the points of S. In the rest of the proof we regard the values $+1, -1$ as "blue" and "red," respectively.

Let ℓ be any line intersecting U, and let ℓ^+ (resp. ℓ^-) denote the open half-plane bounded by ℓ that contains (resp. does not contain) the origin. Further, let

$$R_\ell^+ = \{p \in S \mid p \in \ell^+ \cap \text{slab}(\ell, w) \text{ and } p \text{ is red}\},$$
$$R_\ell^- = \{p \in S \mid p \in \ell^- \cap \text{slab}(\ell, w) \text{ and } p \text{ is red}\},$$
$$B_\ell^+ = \{p \in S \mid p \in \ell^+ \cap \text{slab}(\ell, w) \text{ and } p \text{ is blue}\},$$
$$B_\ell^- = \{p \in S \mid p \in \ell^+ \cap \text{slab}(\ell, w) \text{ and } p \text{ is blue}\}.$$

Clearly, it is sufficient to show that the absolute value of

$$\Delta(\ell) = (|R_\ell^+| - |B_\ell^+|) - (|R_\ell^-| - |B_\ell^-|)$$

is at least $Cn^{1/4}$, for some $\ell \in L$. Actually, we will prove that the expected value of $\Delta^2(\ell)$, in the probability space defined above, is at least $C'n^{1/2}$. The idea of estimating this quantity is somewhat reminiscent of the argument described at the end of the preceding section, which gives a similar lower bound for the discrepancy of a finite projective plane.

Note that the probability that a random line passes through some point of S is 0. Given a line ℓ avoiding all elements of S, a segment $p_i p_j$ ($p_i \neq p_j \in S$) is said to be *crossing* if p_i and p_j are on different sides of ℓ.

$$\Delta^2(\ell) = \left(\sum_{p_i \in R_\ell^+} 1 - \sum_{p_i \in B_\ell^+} 1 - \sum_{p_i \in R_\ell^-} 1 + \sum_{p_i \in B_\ell^-} 1 \right)^2$$

$$= |S \cap \text{slab}(\ell, w)|$$
$$+ 2|\{p_i p_j \subset \text{slab}(\ell, w) : p_i p_j \text{ is monochromatic noncrossing}\}|$$
$$- 2|\{p_i p_j \subset \text{slab}(\ell, w) : p_i p_j \text{ is monochromatic crossing}\}|$$
$$+ 2|\{p_i p_j \subset \text{slab}(\ell, w) : p_i p_j \text{ is bichromatic crossing}\}|$$
$$- 2|\{p_i p_j \subset \text{slab}(\ell, w) : p_i p_j \text{ is bichromatic noncrossing}\}|.$$

By integrating over all lines $\ell \in L$, we obtain

$$\mathbf{E}[\Delta^2(\ell)] \geq \sum_{i=1}^{n} \Pr[p_i \in \text{slab}(\ell, w)] - 2 \sum_{1 \leq i < j \leq n} P_{ij}, \qquad (16.5)$$

where

$$P_{ij} = |\Pr[p_i p_j \subseteq \text{slab}(\ell, w) \text{ crossing}] - \Pr[p_i p_j \subseteq \text{slab}(\ell, w) \text{ noncrossing}]|.$$

By Cauchy's formula, the probability that p_i is contained in a random slab of width w is equal to the perimeter of a circle of radius $w/2$. Let ℓ' and ℓ'' denote the parallel lines at distance $w/4$ from ℓ, and let $d_{ij} = |p_i - p_j|$. It follows from (16.4) that

$$P_{ij} = \left| \Pr[p_i p_j \subseteq \text{slab}(\ell, w)] - 2\left(\Pr\left[p_i p_j \subseteq \text{slab}(\ell', \frac{w}{2})\right] + \Pr\left[p_i p_j \subseteq \text{slab}\left(\ell'', \frac{w}{2}\right)\right]\right) \right|$$

$$= 2w\left(\arcsin\frac{w}{d_{ij}} - 2\arcsin\frac{w/2}{d_{ij}} \right) + 2d_{ij}\left(3 + \sqrt{1 - \left(\frac{w}{d_{ij}}\right)^2} - 4\sqrt{1 - \left(\frac{w/2}{d_{ij}}\right)^2} \right).$$

Since

$$w = \frac{\varepsilon}{m} < \frac{1}{4(m+1)} \le d_{ij},$$

we find (e.g., by Taylor's expansion around $w = 0$) that the last expression can be bounded from above by $w^4/(2d_{ij}^3)$, provided that ε is sufficiently small. Hence, it follows from (16.5) that

$$\mathbf{E}[\Delta^2(\ell)] \ge n\pi w - 2 \sum_{1 \le i < j \le n} \frac{w^4}{2d_{ij}^3}.$$

Using the fact that S forms a grid of side length $\frac{1}{4(m+1)}$, for any $p_i \in S$ there are at most $8k$ other points $p_j \in S$ at Manhattan distance (L_1-distance) $\frac{k}{4(m+1)}$ from p_i. Therefore, if ε is sufficiently small,

$$\mathbf{E}[\Delta^2(\ell)] \ge n\pi w - \frac{n}{2} \sum_{1 \le k < m} \frac{8kw^4}{\left(\frac{1}{\sqrt{2}} \frac{k}{4(m+1)}\right)^3}$$

$$\ge n\pi w - 10^4 n(m+1)^3 w^4$$

$$\ge n\pi w/2 = \frac{\varepsilon\pi}{2} n^{1/2}.$$

completing the proof.   □

The above argument remains valid for any set $S = \{p_1, \ldots, p_n\} \subseteq \mathbb{R}^2$ with the property that

$$\max_{i \ne j} |p_i - p_j| \Big/ \min_{i \ne j} |p_i - p_j| < C\sqrt{n}.$$

The proof can also be generalized to higher dimensions (see Chazelle et al., 1995).

DISCREPANCY AND ε-APPROXIMATION

As we have pointed out before, the original approach to transversal problems suggested by Vapnik and Chervonenkis (1971) was the starting point of many investigations in hypergraph theory, combinatorial and computational geometry, discrepancy theory, and so on. The particular question addressed in that paper was a statistical one: Given a set system (hypergraph) H with n vertices, we want to take a relatively small "sample" T from its vertex set $V(H)$ so that the size of any set (edge) $E \in E(H)$ is approximately $n|E \cap T|/|T|$. We close this chapter and our book by returning to this problem and improving the original results of Vapnik and Chervonenkis in the light of the new developments (cf. Exercise 15.6).

Definition 16.17.   Let $H = (V(H), E(H))$ be a hypergraph with n vertices, and let $\varepsilon > 0$. A subset $T \subseteq V(H)$ is called an ε-*approximation* for H if

$$\left| \frac{|E \cap T|}{|T|} - \frac{|E|}{n} \right| \leq \varepsilon$$

for every edge $E \in E(H)$.

Clearly, an ε-approximation for H is also an ε'-net for any $\varepsilon' > \varepsilon$ (see Definition 15.10).

Lemma 16.18.   *Let $H = (V(H), E(H))$ be a hypergraph with n vertices, and suppose that $V(H) \in E(H)$. Let $c: V(H) \rightarrow \{+1, -1\}$ be a coloring of the vertex set with* disc $(H, c) = D \leq n/2$, *and set $T = \{x \in V(H) \mid c(x) = +1\}$. Then*

   (i) $n/4 \leq |T| \leq 3n/4$,

  (ii) T *is a $(4D/n)$-approximation for H.*

  Proof.   Since $|c(E)| \leq D$, we obtain that

$$\left| 2|E \cap T| - |E| \right| \leq D \leq \frac{n}{2} \qquad \text{for every } E \in E(H).$$

In particular, for $E = V(H)$,

$$\left| 2|T| - n \right| \leq D \leq \frac{n}{2},$$

which yields (i).

    On the other hand, for any hyperedge E,

$$\begin{aligned}
\left| \frac{|E \cap T|}{|T|} - \frac{|E|}{n} \right| &\leq \left| \frac{2|E \cap T| - |E|}{2|T|} \right| + \left| \frac{|E|}{2|T|} - \frac{|E|}{n} \right| \\
&\leq \frac{D}{2|T|} + \frac{|E|}{n} \left| \frac{n - 2|T|}{2|T|} \right| \\
&\leq \frac{D}{|T|} \leq \frac{4D}{n}. \qquad\qquad \square
\end{aligned}$$

In this chapter we have established various upper bounds for the discrepancy of a hypergraph H. Combining these results with Lemma 16.18, we obtain very good ε-approximations for H in the sense that the "margin of error," ε, is extremely small. The problem is that the sizes of these "samples" are very big. We can overcome this difficulty by iterating the process, i.e., by approximating a good sample with a smaller one.

Theorem 16.19 (Matoušek et al., 1993). *Let H be a hypergraph with n vertices, let $0 < \varepsilon < 1$, and let $\pi_H(m)$ and $\pi_{H^*}(m)$ denote the shatter function and the dual shatter function of H, respectively. For any $C > 0, d > 1$, there exists a constant $C' = C'(C, d)$ such that*

(i) *if $\pi_H(m) \leq Cm^d$, then H has an ε-approximation T with*

$$|T| \leq C'\left(\frac{1}{\varepsilon}\right)^{2d/(d+1)}$$

(ii) *if $\pi_{H^*}(m) \leq Cm^d$, then H has an ε-approximation T with*

$$|T| \leq C'\left(\frac{1}{\varepsilon}\right)^{2d/(d+1)}\left(\log\frac{1}{\varepsilon}\right)^{d/(d+1)}.$$

Proof. Assume without loss of generality that $V(H) \in E(H)$. To establish part (i), first apply Theorem 16.10 to the hypergraph $H_0 = H$. If $n_0 = n$ is sufficiently large, we obtain a coloring $c_0 : V(H_0) \to \{+1, -1\}$ with

$$\operatorname{disc}(H_0, c_0) = D_0 \leq C'' n_0^{(1/2)(1-1/d)} \leq \frac{n_0}{2}.$$

By Lemma 16.18, the set $T_1 = \{x \in V(H_0) \mid c_0(x) = +1\}$ is a $(4D_0/n_0)$-approximation for H_0. Let $H_1 = H_0[T_1]$ be the subhypergraph of H_0 induced by T_1, i.e.,

$$E(H_1) = \{E \cap T_1 \mid E \in E(H_0)\}.$$

Setting $n_1 = |V(H_1)| = |T_1|$, we have $n_0/4 \leq n_1 \leq 3n_0/4$.

If n_1 is still large enough, there is a coloring $c_1 : V(H_1) \to \{+1, -1\}$ with

$$\operatorname{disc}(H_1, c_1) = D_1 \leq C'' n_1^{(1/2)(1-1/d)} \leq \frac{n_1}{2}.$$

Then $T_2 = \{x \in V(H_1) \mid c_1(x) = +1\}$ is a $(4D_1/n_1)$-approximation for H_1. Set $H_2 = H_1[T_2]$, $n_2 = |V(H_2)| = |T_2|$, and so on. We stop this procedure when

$$|T_k| \leq C'\left(\frac{1}{\varepsilon}\right)^{2d/(d+1)} = t,$$

where C' is chosen so large as to satisfy

$$t^{-(d+1)(2d)} < \frac{1 - \sqrt{3/2}}{4C''}\varepsilon.$$

Now, for every $0 \leq i < k$,

$$\text{disc}\,(H_i, c_i) = D_i \le C'' n_i^{(1/2)(1 - 1/d)}$$

$$\le C'' n_i t^{-(d+1)/(2d)}$$

$$\le \frac{(1 - \sqrt{3}/2)\varepsilon}{4} n_i$$

$$< \frac{n_i}{2},$$

so that the conditions of Lemma 16.18 are satisfied.

To complete the proof of part (i), we have to note only that T_k is a D-approximation for H with

$$D = \frac{4D_{k-1}}{n_{k-1}} + \frac{4D_{k-2}}{n_{k-2}} + \cdots + \frac{4D_0}{n_0}$$

$$\le 4C''(n_{k-1}^{-(d+1)/(2d)} + n_{k-2}^{-(d+1)/(2d)} + \cdots + n_0^{-(d+1)/(2d)})$$

$$\le 4C'' t^{-(d+1)/(2d)} \left(1 + \left(\frac{3}{4}\right)^{(d+1)/(2d)} + \left(\frac{3}{4}\right)^{2(d+1)/(2d)} + \cdots\right)$$

$$< \varepsilon,$$

provided that C' (and hence every n_i) is sufficiently large.

Part (ii) can be deduced from Theorem 16.6 in exactly the same way. □

The conditions in parts (i) and (ii) of Theorem 16.19 are satisfied for any hypergraph H with VC-dim$(H) = d$ and VC-dim$(H^*) = d$, respectively.

EXERCISES

16.1 Prove that if m is sufficiently large in terms of n, there exists a hypergraph with n vertices and m edges such that disc $(H) > C\sqrt{n \log m}$, where $C > 0$ is some suitable constant.

16.2 Prove that the vertices of any r-uniform hypergraph with at most 2^{r-1} edges can be colored by two colors so that no edge is monochromatic.

16.3★ (Erdős and Lovász, 1975) Prove that if every edge of an r-uniform hypergraph intersects at most 2^{r-3} other edges, then there is a two-coloring of the vertices such that no edge is monochromatic.

16.4 (Las Vergnas and Lovász, 1972) Let H be a hypergraph with the property that for every k, the union of any k edges has at least $k + 1$ vertices. Prove that the points of H can be colored by two colors so that no edge is monochromatic.

16.5 Prove that under the assumptions of Theorem 16.3, disc $(H) \le 2D - 1$.

16.6 (International Mathematics Olympiad, 1986)   Let S be a finite set of points in the plane. Prove that there exists a two-coloring $c\colon S \to \{+1, -1\}$ so that on every horizontal and vertical line the sum of the values of c is $+1, -1$, or 0.

16.7 (Wernisch, 1994)   Deduce Theorem 16.6 from Lemma 16.11.

16.8 Let H be a hypergraph whose vertices are n points in $\mathbb{R}^d$, $d \le 2$, and let

$$E(H) = \{E \subseteq V(H)\,|\,\text{there is a ball or a halfspace } B \text{ such that } B \cap V(H) = E\}.$$

Prove that $\pi_{H^*}(n) \le C_d n^d$, where C_d is a constant depending only on d.

16.9 Prove that for any $C > 0$, $d \ge 1$, and any positive integer k, there is a constant $C' = C'(C, d, k)$ with the following property.

Let H be any hypergraph of n vertices such that $\pi_{H^*}(m) \le Cm^d$ for all m, and let

$$E(H') = \{\varphi(E_1, E_2, \ldots, E_k)\,|\,E_1, E_2, \ldots, E_k \in E(H)\},$$

where φ is a fixed set-theoretic formula of k variables. Then

$$\operatorname{disc}(H') \le C' n^{(1/2)(1 - 1/d)}\sqrt{\log n}.$$

16.10 Let ξ and $H(\xi)$ be as in Definition 16.13. Prove that

(i) $H(\xi) \le -\displaystyle\sum_{s \in S} \frac{1}{|S|} \log \frac{1}{|S|} = \log |S|$;

(ii) if $H(\xi) \le h$, then there is a value $s \in S$ with $p_s \ge 2^{-h}$;

(iii) for any random variable $\xi = (\xi_1, \xi_2)$ consisting of two components, we have $H(\xi) \le H(\xi_1) + H(\xi_2)$.

16.11 Let $K \subseteq \mathbb{R}^2$ be a convex disc with a twice differentiable support function $h(\varphi)$. Prove that $h(\varphi) + h''(\varphi) \ge 0$.

16.12 (Hansen, 1978; Lutwak, 1979)   Let γ be a simple closed Jordan curve of length π in the plane. Prove that

(i) γ is contained in a rectangle of unit area;

(ii) γ is contained in a triangle of area $3\sqrt{3}/4$.

16.13 (Spencer, 1985)   Let H be a hypergraph with n vertices and n edges. Prove that $\operatorname{disc}(H) \le C\sqrt{n}$, for a suitable constant C.

16.14★ (Beck, 1981b; Matoušek and Spencer, 1994)   Let H_n denote the hypergraph on the vertex set $V = \{1, 2, \ldots, n\}$, whose edge set consists of all arithmetic progressions within V. Prove that $\operatorname{disc}(H_n) \le Cn^{1/4}\log n$, for a suitable constant C.

16.15 Let π_1 and π_2 be two permutations of $V = \{1, 2, \ldots, n\}$, where n is even. Let H_n denote the hypergraph on the vertex set V, whose edges are all sets of the form $\{\pi_i(j), \pi_i(j + 1), \ldots, \pi_i(k)\}$, $i = 1, 2$, $1 \leq j < k \leq n$.

(i) Prove that disc $(H_n) \leq 2$.

(ii) What happens if we have three permutations?

Hints for Exercises

In this appendix we provide hints for most of the exercises. The problems marked with a star are more challenging. The wise and conscientious reader is encouraged to try them unaided before consulting the papers cited in the exercises or evaluating the hints we have offered. The exercises that are neither marked with a star nor have a hint are usually quite straightforward.

CHAPTER 1

[1.1] Show that each u_i can be written as $\sum_{j=1}^{d} a_{ij} v_j$, where $a_{ij} \in \mathbb{Z}$.

[1.2] Apply Minkowski's theorem (Theorem 1.7) to the lattice

$$\Lambda = \{(n, \alpha n - m) \in \mathbb{R}^2 \mid m \text{ and } n \text{ are integers}\}$$

and to the rectangle C bounded by the straight lines $x = k$, $x = -k$, $y = 1/k$, $y = -1/k$.

[1.3] Apply Corollary 1.6 in the form

$$|n\alpha - m| \le \frac{1}{\sqrt{3n}}.$$

[1.4] Apply Theorem 1.8 to $S = \frac{1}{2}C$.

[1.5] Observe that the set of all d-tuples $(n_1, n_2, \ldots, n_d)$ of reals satisfying the inequalities $|l_i(n_1, \ldots, n_d)| \le b_i$ is a parallelepiped in $\mathbb{R}^d$. Apply Minkowski's theorem (Theorem 1.7).

[1.6] To show (ii), notice that $\mathbb{Z}_p^+$ must contain an element b of order $4 = (p - 1)/m$. Then $b^2 \equiv -1 \pmod{p}$, because -1 is the only element of the cyclic group, whose order is 2.

[1.7] To prove that a full-dimensional discrete subgroup $X \subseteq \mathbb{R}^d$ is a lattice, pick d linearly independent vectors $x_1, \ldots, x_d$. Letting H denote the subspace induced by $\{x_1, \ldots, x_{d-1}\}$, we can assume by induction on d that $X \cap H$ is a $(d - 1)$-dimensional lattice generated by some vectors $u_1, \ldots, u_{d-1}$. Show that there exists $u_d \in X - H$ whose distance from H is minimal, and that $X = \Lambda(u_1, \ldots, u_d)$.

[1.8] Use Exercise 1.7 and show that

$$\det \Lambda = \lim_{R \to \infty} \frac{|\mathbb{Z}^d \cap B^d(R)|}{|\Lambda \cap B^d(R)|} = k,$$

where $B^d(R)$ is the ball of radius R centered at origin.

[1.9]   Try a number of the form $8m + 7$.

[1.10] Combine Lemma 1.10 and Wilson's theorem [equation (1.2)].

CHAPTER 2

[2.1]   Use the idea of the proof of Theorem 2.1.

[2.2]   Let P be a convex n-gon of minimum area circumscribed about C, and let P_i denote the image of P under rotation through $2i\pi/k$, $0 \le i < k$. If there are no i and j such that a cap of P_i is strictly contained in the some cap of P_j, then P has k-fold rotational symmetry. Otherwise, apply the same argument as in the proof of Theorem 2.5.

[2.3]   No, try to modify an ellipse.

[2.4]   (i) Determine how $\text{dev}_A\,(C, Q_n)$ changes during infinitesimal translations and rotations of the sides of Q_n.

[2.5]   (i) It follows by a variational argument using small perturbations.
(ii) Apply (i) and Theorem 2.4.

[2.6]   Substitute equations (2.1) and (2.2) in

$$A(C) = \left| \int_0^\pi y(\phi)x'(\phi)d\phi \right| + \left| \int_\pi^{2\pi} y(\phi)x'(\phi)d\phi \right|.$$

[2.7]   Use the Fourier series of $y(\phi)$.

[2.8]   Let Q_n be a circumscribing n-gon all of whose angles are $(1 - 2/n)\pi$, and let q_n denote the n-gon induced by those points where Q_n touches C.

[2.9]   Assume that C is centrally symmetric with respect to the common center of B and B', and it contains a regular simplex inscribed in B'. Use the fact that a centrally symmetric body of width W always contains a ball of diameter W, and follow the proof of Corollary 2.12

[2.10] (i) Prove the claim for $k = d - 1$ by showing that the radius of the largest ball inscribed in S is at most $1/d$. Prove the claim for $k < d - 1$ by induction on $d - k$.

(ii) Consider the set of k-dimensional simplices F_k induced by the vertices of p_n^d for $k \approx d/(2 \ln n)$. By (i), every point of p_n^d is close to at least one of them, so

$$\text{Vol}\,(p_n^d) \le \binom{n}{k+1} \left(\max_{F_k} \text{Vol}_k(F_k) \right) \text{Vol}_{d-k}(B^{d-k}(\rho)),$$

where $\rho = \sqrt{(d - k)/(dk)}$ and $B^{d-k}(\rho)$ is the $(d - k)$-dimensional ball of radius ρ. Use that

$$\mathrm{Vol}_i\,(B^i) = \frac{2\sqrt{\pi}}{i\Gamma(i/2)},$$

where the Γ-*function* is defined by $\Gamma(x+1)=x\Gamma(x), \Gamma(1)=1, \Gamma(1/2)=\sqrt{\pi}$.

CHAPTER 3

[3.2] To prove the first statement, construct a packing C_n of congruent copies of C in the disc $D(n)$ with density

$$d(C_n, D(n)) \geq \delta(C) - \frac{O(1)}{n}$$

for every n. Choose a subsequence $n_1, n_2, \ldots$ such that C_{n_i} converges when restricted to $D(1)$, as $i \to \infty$. From this, choose a subsequence for which C_n converges when restricted to $D(2)$, and so on. Show that the limit packing meets the requirements.

[3.4] Let C be the densest packing of unit discs in a cone. Choose P and P' to be a circular disc and a regular triangle, respectively.

[3.5] Follow the proof of Theorem 3.7.

[3.6] Try a regular octagon.

[3.7] Use a compactness argument.

[3.8] (ii) Use (i) and Dowker's theorem (Theorem 2.1) in exactly the same way as in the proof of Theorem 3.2.

(iii) It follows from (ii), using the observation that for a centrally symmetric C, P_4 can be chosen to be a parallelogram.

[3.9] Generalize the proof of Theorem 3.14.

CHAPTER 4

[4.2] To see that $D(C)$ is convex, notice that for any pair of points $p_1 = q_1 - q_1', p_2 = q_2 - q_2' \in D(C)$ and for any $0 \leq \lambda \leq 1$,

$$\lambda p_1 + (1 - \lambda)p_2 = (\lambda q_1 + (1 - \lambda)q_2) - (\lambda q_1' + (1 - \lambda)q_2').$$

[4.3] Show first that for any direction v, there is an affinely regular hexagon $p_1 p_2 p_3 p_4 p_5 p_6$ all of whose vertices except p_6 lie on the boundary of the disc, and $p_1 p_2$ is parallel to v.

[4.4] Assume without loss of generality that the half-length parallelogram $p_1 p_2 p_3 p_4$ in the direction of the x-axis is a rectangle, and let

$q_1 q_2$ (and $r_1 r_2$) denote the longest horizontal (resp. vertical) chord. Assume further that the intersection of $q_1 q_2$ and $r_1 r_2$ is at the origin. Let $p_1 = (x_1, y_1)$, $p_2 = (x_2, y_1)$, $p_3 = (x_2, y_2)$, $p_4 = (x_1, y_2)$, $q_1 = (x_1^*, 0)$, $q_2 = (x_2^*, 0)$, $r_1 = (0, y_1^*)$, and $r_2 = (0, y_2^*)$ for some $x_1^* < x_1 < 0 < x_2 < x_2^*$ and $y_1^* < y_1 < 0 < y_2 < y_2^*$, where $x_2^* - x_1^* = 2(x_2 - x_1)$. By convexity, p_3 is above the chord $q_2 r_2$, which implies that $y_2 \geq y_2^*(1 - x_2/x_2^*)$. Similarly, we obtain three other inequalities, and combining them, we obtain $y_2^* - y_1^* \leq 2(y_2 - y_1)$.

[4.5] Let $D(C)$ denote the difference region of C. For any set $P \subseteq \mathbb{R}^2$, show that $C = \{C + p \mid p \in P\}$ is a packing if and only if

$$\mathcal{D} = \left\{ \frac{D(C)}{2} + p \;\middle|\; p \in P \right\}$$

is a packing. Thus, the statement follows from Corollary 3.6.

[4.6] Show that in any densest double-lattice packing, a copy of C touches four copies of $-C$, and that these four intersection points form an extensive parallelogram.

CHAPTER 5

[5.2] Use the fact that the intersection of two convex sets is convex.

[5.4] Use the idea of the proof of Theorem 5.5.

CHAPTER 6

[6.1] (i) Prove by induction on d, as follows. Let B be a ball of radius $\sqrt{2}$ containing all c_i's. First show that the points can be moved to the boundary of B without decreasing their pairwise distances. Assume that $c_1 = (\sqrt{2}, 0, \ldots, 0)$. Let B' be the intersection of B and the plane $x_1 = 0$. Show that each c_i $(i > 1)$ can be mapped to a point c_i' on the boundary of B' such that $|c_i' - c_j'| > 2$.

[6.2] Let $p, q \in C_{-\rho}$. Using the convexity of C, show that the Minkowski sum of the ball ρB^d and the segment pq is contained in C.

[6.3] Let $\{\varphi_i \mid i = 1, 2, \ldots\}$ be a collection of isometries of $\mathbb{R}^d$ such that $\{\varphi_i(C) \mid i = 1, 2, \ldots\}$ is a packing. Notice that it is sufficient to prove that

$$\sum_{i=1}^{\infty} f(\varphi^{-1}(x)) \leq 1 \qquad \text{for any } x \in \mathbb{R}^d.$$

[6.4] Apply Theorem 6.5 for the density function used in the proof of Corollary 6.3, and let $\rho = r(C)$.

[6.6] Show that if p is a common point of the closed Dirichlet cells corresponding to the elements of C', there is a sphere centered at p which passes through all points of C'. Prove that this sphere cannot contain any element of C in its interior.

[6.8] Let S^d be a regular simplex in $\mathbb{R}^d$ with side length 2. Show that the distance between a vertex and the centroid of S^d is $\sqrt{2d/(d+1)}$.

[6.9] Start with a packing of (unequal) circles in the plane, whose density is very close to 1, and turn it into a covering of density at most $1 + \varepsilon$. This covering can be extended to a covering of $\mathbb{R}^3$ with equal balls.

CHAPTER 7

[7.1] Notice that C is the unit ball in the corresponding Minkowski space, and follow the proof of Proposition 7.1.

[7.2] Use a compactness argument.

[7.3] (iii) It follows from (ii).

(iv), (v) They follow from (iii).

(vi) Analyze the product $\left(\sum_{i=1}^{\infty} \mu(i)\,\frac{1}{i^d} \right)\left(\sum_{j=1}^{\infty} \frac{1}{j^d} \right)$.

[7.4] (i) Follow the proof of Theorem 7.7.

(ii) Consider the intersection of C with a large ball of radius R around the origin **0**.

[7.5] Integrate by parts.

[7.6] Apply Theorem 7.11 and use Stirling's formula,

$$d! \approx \sqrt{2\pi d}\left(\frac{d}{e} \right)^d.$$

[7.7] Assume that C is strictly convex. Consider the intersection of C with the plane induced by **0**, $(x_1, \ldots, x_d)$, and $(1, 0, \ldots, 0)$, and apply the planar version of the statement.

[7.8] Assume without loss of generality that $\|x_1\|_C = \|x_2\|_C = \|x_3\|_C = 1$. By using a linear transformation, if necessary, one can also assume that

$$\{(x, y)\,||x| + |y| \le 1\} \subseteq C \subseteq \{(x, y)\,||x|, |y| \le 1\}.$$

If there exist $i \neq j$ such that the angle between x_i and x_j is at most $\pi/2$, then $\|x_i + x_j\|_C > 1$. Otherwise, observe that $x_i + x_j$ is on the boundary of $C + x_1 + x_2 + x_3$, for every $i \neq j$.

[7.9] Let $a = \|x\|_{C,p}$, $b = \|y\|_{C,p}$. Choose $u, v \in p\mathbb{Z}^d$ such that $(x - u)/a$, $(y - v)/b \in C$. Show that $(x + y - u - v)/(a + b) \in C$. Observe that we are not completely done because $x + y$ is not necessarily in $\mathbb{Z}_p^d$.

[7.10] Apply Exercise 1.7.

[7.11] (ii)–(iv) Use induction on n.
(v) Use (iv) and Exercise 7.10.

[7.12] Consider a cross section of the packing parallel to the bases of cylinders.

[7.13] (ii) Notice that the complement of the union of balls in the densest lattice packing contains infinite cylinders.

[7.14] Consider the thinnest lattice covering of $\mathbb{R}^3$ with unit balls. Replace each ball by a ball of radius $1 - \varepsilon$. The small uncovered holes can be covered by a discrete set of parallel circular cylinders, which, in turn, can be covered (efficiently) with translates of a long ellipsoid whose volume is the same as the volume of a ball of radius $1 - \varepsilon$. Now use part (i) of Exercise 7.13.

CHAPTER 8

[8.1] Suppose that r_i increases while r_j and r_k remain the same. By the law of cosines,

$$\cos(\angle v_j v_i v_k) = 1 - \frac{2 r_j r_k}{(r_i + r_j)(r_i + r_k)},$$

$$\cos(\angle v_i v_j v_k) = 1 - \frac{2 r_k}{r_j + r_k}\left(1 - \frac{r_j}{r_i + r_j}\right).$$

Thus, $\cos(\angle v_j v_i v_k)$ increases, and the cosines of the other two angles of the triangle $v_i v_j v_k$ decrease.

[8.2] Let $f(0) = x$, and for the sake of contradiction assume that there exists a point $y \in B^d$ which does not belong to $f(B^d)$. Choose a monotone increasing sequence of positive reals r_n ($n = 1, 2, \ldots$) tending to 1, and let $S(r) = \{p \in \mathbb{R}^d \mid |p - \mathbf{0}| = r\}$. Let y_n be an intersection point of $f(S(r_n))$ and the segment xy. Then $y_n = f(p_n)$ for a suitable $p_n \in S(r_n)$ ($n = 1, 2, \ldots$), and one can choose an infinite subsequence of $\{p_1, p_2, \ldots\}$ converging to a boundary point of B^d, which violates our assumption.

[8.3] Let $\overline{F}$ denote the collection of faces all of whose vertices belong to $\overline{U} = V(G) - U$. Since these vertices do not form a perfect triangulation on the vertex set $\overline{U}$, by (8.1), $|\overline{F}| < 2|\overline{U}| - 4$. That is,

$$2n - 4 - |F(U)| < 2(n - U|) - 4.$$

[8.4] Suppose, for the sake of contradiction, that $|A(C)| \geq \frac{2}{3}n$. Let G' denote the subgraph of G induced by the vertices inside and on C. Show first that for any two vertices of C there is a shortest path in G', all of whose edges belong to C. Use this fact and the minimality condition to prove that C has exactly $2k$ vertices, where $k = \lfloor \sqrt{2n} \rfloor$. Denote them by $v_0, v_1, \ldots, v_{2k} = v_0$, in order. Use Menger's theorem (see Lovász, 1979) to show that there are $k+1$ vertex disjoint paths in G' between $\{v_0, v_1, \ldots, v_k\}$ and $\{v_k, v_{k+1}, \ldots, v_{2k}\}$. Denote these paths by P_i ($0 \leq i \leq k$), where P_i connects v_i to v_{2k-i}. Since P_i has at least $\min\{2i+1, 2k-2i+1\}$ vertices, we obtain that the number of vertices of G' is larger than n, a contradiction.

[8.5] (i) Generalize Lemma 8.4 in the following way. Let $P \subseteq \mathbb{R}^d$ be a finite set whose elements have some nonnegative weights $w(p)$ such that $\sum_{p \in P} w(p) = 1$. Show that there exists a point $q \in \mathbb{R}^d$ with the property that any half-space H that does not contain q satisfies

$$\sum_{p \in H \cap P} w(p) \leq \frac{d}{d+1}.$$

Then follow the proof of Theorem 8.3.

(ii) Construct a graph G', all of whose vertices have degree at most 4, by replacing each vertex v of G, whose degree is d_i, with a $\lceil d_i/4 \rceil \times \lceil d_i/4 \rceil$ grid. Assign weight 0 to every internal vertex of the grid corresponding to v_i, and distribute $w(v_i)$ evenly among its boundary vertices. Apply part (i) to the weighted graph G'.

[8.6] Prove and use the following lemma of de Fraysseix et al. (1990). Let G be a triangulated planar graph with exterior face uvw. Then we can label the vertices of G by $v_1 = u$, $v_2 = v$, $v_3, \ldots, v_n = w$, so that for any $k \geq 3$,

(a) the exterior face of the subgraph G_k of G induced by $v_1, \ldots, v_k$ is bounded by a cycle C_k containing the edge uv;

(b) v_{k+1} is in the exterior of C_k, and its neighbors form a subinterval of the path $C_k - uv$, having at least two elements.

[8.8] Use Euler's formula for planar graphs to prove that there exists a vertex of degree at most 5.

[8.9] By Lemma 5.6 it is sufficient to prove the claim for centrally symmetric convex bodies. Show that for a centrally symmetric disc C^*, all translates of C^* that touch C are contained in $3C^*$.

[8.10] Let $C_1, \ldots, C_k$ be mutually touching translates of a centrally symmetric convex body C with centers $O_1, \ldots, O_k$. Let K denote the convex hull of $\{O_1, \ldots, O_k\}$, and m_{ij} the midpoint of $O_i O_j$. Show that

$$K_i = \mathrm{conv}\,(O_i \cup \{m_{ij} \mid 1 \leq j \neq i \leq k\}) \subseteq K \cap C_i,$$

and K_i is similar to K for every i.

[8.11] Apply Koebe's theorem (Theorem 8.1). Let C be a disc packing that realizes a triangulated planar graph G. Show that C can be selected so that for any two adjacent discs in C the ratio of the smaller radius to the larger radius is bounded from below by a constant, depending only on the maximum degree of G.

[8.12] Let G be the original graph. Define a graph G' on the same vertex set whose two vertices are connected if and only if they are adjacent or have a common neighbor in G. Show that the chromatic number of G' is at most $d^2 + 1$, and embed each color class in a short arc of a circle.

[8.13] Let $2\alpha_i$ denote the angle between two rays tangent to C_i emanating from the center of C ($1 \leq i \leq k$). Observe that

$$r(C_i) = r(C)\,\frac{\sin \alpha_i}{1 - \sin \alpha_i}.$$

Show that $\left(\dfrac{\sin \alpha}{1 - \sin \alpha}\right)^n$ is a convex function in $(0, \pi/2)$, and apply Jensen's inequality.

[8.14] Show that the sum $\sum_{j=1}^{n} 1/e_j$ is minimized when $\sum_{j=1}^{i} e_j = \gamma i^2$ for every $1 \leq i \leq n$.

[8.15] Apply Theorem 8.9.

[8.16] (i) Consider a circle whose center is a vertex of the convex hull of the centers.

(ii) Consider the stereographic projection of the incircles of the faces of a regular dodecahedron.

[8.17] (i) Let C be a unit square. Consider the packing

$$\{C + (i,j) \mid i = 0, 1, \text{ and } j \in \mathbb{Z}\}.$$

[8.18] Construct a regular polytope with 120 facets in $\mathbb{R}^4$ (see Coxeter, 1962). The images of the balls inscribed in these facets under a stereographic projection into $\mathbb{R}^3$ will form a 12-neighbored packing with 120 incongruent balls.

CHAPTER 9

[9.1] Show that any two cycles have at most one common vertex, and that the cycles of the graph are connected in a cactus-like fashion (as in Figure 9.2).

[9.2] Assume that G does not contain a Hamiltonian cycle but is maximal in the sense that G becomes Hamiltonian if a new edge is added to it. Modifying the proof of Theorem 9.1, prove that any vertex of degree $\geq (n-1)/2$ is connected to every vertex of degree $\geq n/2$, and then arrive at a contradiction by showing that G has a Hamiltonian cycle.

[9.3] Use Exercise 9.2.

[9.4] Define a graph G, as follows. Let $V(G) = V_1 \cup V_2$, where $V_1 = \{1, 2, \ldots, \lfloor n^{2/3} \rfloor\}$, $V_2 = \{p \mid n^{2/3} \leq p \leq n, p \text{ is a prime}\}$. For each n_i that is not the square of a prime, select a pair $x, y \in V_1 \cup V_2$ such that $x \neq y$ and $xy = n_i$, and connect them by an edge. Prove that G has no cycles.

[9.5] Let V_1 and V_2 be two classes of complete $(r-1)$-partite graphs, and assume that $|V_1| \geq |V_2| + 2$. Show that by moving a vertex from V_1 to V_2, the number of edges increases.

[9.6] (i) It follows from Turán's theorem (Theorem 9.3) and Exercise 9.5.
(ii) Apply (i) to the complement of G.

[9.7] (i) Suppose that among any k members of H with a point in common, there are two whose intersection has at most $d - 1$ elements. For any x, let H_x denote the subfamily of H consisting of all members that contain x. By Exercise 9.6 there are at least $|H_x|^2/(2k-2) - |H_x|/2$ pairs $\{E_i, E_j\} \subset H_x$, $i \neq j$, with $1 \leq |E_i \cap E_j| \leq d - 1$. On the other hand, any such pair belongs to at most $d - 1$ subfamilies H_x. Thus,

$$(d-1)\binom{m}{2} \geq \sum_x \left(\frac{|H_x|^2}{2k-2} - \frac{|H_x|}{2} \right),$$

and the result follows by using Jensen's inequality.
(ii) Let $N_G(x)$ denote the set of all neighbors of $x \in V(G)$. Apply (i) to the family $H = \{N_G(x) \mid x \in V(G)\}$ with

$$m = n = |V(G)|, s = 2|E(G)|/n, \text{ and } d = \binom{k}{2} + k - 1.$$

[9.8] Apply Exercise 7.8 and Exercise 9.6.

[9.9] Use Exercise 9.8.

[9.10] Let $T \subseteq V(G)$ be a smallest subset of vertices such that any edge of

G is incident with at least one element of T. Observe that the degree of any vertex is at most $\alpha(G)$.

[9.11] If $|E(G)| > n^2/4$, then, by Turán's theorem (Theorem 9.3), G contains a triangle. If $|E(G)| = n^2/4$, then the triangle-free graph is a complete bipartite graph.

[9.12] For $n \geq r + 2$, pick a vertex $x \in V(G)$ of minimum degree. Suppose $d_G(x) \leq \lfloor (1 - 1/(r-1))n \rfloor$. Show that $|E(G - x)| \geq |E(T_{r-1}(n-1))| + 1$, and apply induction on n.

[9.13] Let (a_i, b_i, c_i), $1 \leq i \leq k$, be a maximal set of vertex disjoint triangles. Let G_i be the subgraph induced by $V(G) - \bigcup_{j \leq i} \{a_j, b_j, c_j\}$. First show that there exists an i such that $|E(G_{i-1})| - |E(G_i)| \geq cn$, for some constant $c > 1$, and then show that there are $\approx (c-1)n/3$ triangles in G_{i-1} incident with one of the edges in $\{(a_i, b_i), (b_i, c_i), (c_i, a_i)\}$.

[9.14] Let (v_1, v_2, v_3) be a triangle in G. Prove that if there are $n + k$ edges, each incident with at least one of v_1, v_2, and v_3, then G has at least k triangles of the form (v_i, v_j, v_l), $1 \leq i, j \leq 3 < l$. Consider the subgraph G' induced by $V(G) - \{v_1, v_2, v_3\}$. If G' has at most $\lfloor (n-3)^2/4 \rfloor$ edges, use the above claim; otherwise, show that G' has at least $\lfloor (n-3)/2 \rfloor$ triangles.

[9.15] Use Turán's theorem (Theorem 9.3) twice (for n and $n-2$) to show that among all triangle-free graphs with n vertices, $T_2(n)$ has the maximum number of pairs of disjoint edges. Observe that any pair of disjoint edges can be contained in at most one cycle of length 4.

[9.16] Let $|V(G)| = n$. Choose randomly an $\lfloor n/2 \rfloor$-element subset $V_1 \subseteq V(G)$. Show that the expected number of edges connecting V_1 and V_2 is

$$\sum_{xy \in E(G)} 2 \binom{n-2}{\lfloor n/2 \rfloor - 1} \bigg/ \binom{n}{\lfloor n/2 \rfloor} \geq |E(G)|/2.$$

[9.17] Use the estimate found in the proof of Theorem 9.5.

[9.20] Assume, for example, that $c = 4$, and fix a partition of the set of all k-tuples of X into four classes C_i ($1 \leq i \leq 4$). If $|X| \geq n_0(n_0(r, k), k)$, then Theorem 9.13 yields that there exists an $n_0(r, k)$-element subset $X' \subseteq X$, all of whose k-tuples belong either to $C_1 \cup C_2$ or to $C_3 \cup C_4$. In either case, the result follows by another application of Theorem 9.13.

[9.21] (i) Let $G_0 = G$. Assume that we have already defined an infinite subgraph $G_i \subseteq G$. Pick any $x_i \in V(G_i)$. Let G_{i+1} be the subgraph of G_i induced by the vertices adjacent (resp. nonadjacent) to x_i, according to whether x_i has infinitely many neighbors in G_i.

(ii) Assume, for a contradiction, that for every n there exists a graph G_n on the vertex set $\{1, \ldots, n\}$, which contains no complete or empty subgraphs with r vertices. Define a nested sequence of infinite subsets of $\mathbb{N}$, $N_1 \supseteq N_2 \supseteq N_3 \supseteq \cdots$ such that the restriction of G_n to $\{1, \ldots, i\}$ is the same for every $n \in N_i$. These restrictions "converge" to a counterexample to part (i).

[9.22] Let $V_1 = \{v_1, v_2, \ldots\}$, $V_2 = \{w_1, w_2, \ldots\}$, and $E(G) = \{v_i w_j \mid i \leq j\}$.

[9.23] Apply Theorem 9.13 for $k = 2$, where $X = \{x_1, x_2, \ldots\}$ and $\{x_i, x_j\}$, $i < j$, belongs to the first (second) class if $x_i \leq x_j$ (resp. $x_i > x_j$).

[9.24] Let $R_k(r, s)$ be defined as the smallest integer n with the property that for any coloring of all k-tuples of $\{1, 2, \ldots, n\}$ with red and blue, there is an r-element set, all of whose k-tuples are red, or an s-element set, all of whose k-tuples are blue. Show that

$$R_k(r, s) \leq R_{k-1}\big(R_k(r-1, s), R_k(r, s-1)\big) + 1$$

for any $r, s > k$.

[9.25] (i) Let U_i ($1 \leq i \leq k_r(G)$) and V_j ($1 \leq j \leq k_{r-1}(G)$) denote those r-element resp. $(r-1)$-element subsets of $V(G)$, which induce complete subgraphs of G. Let u_i denote the number of K_{r+1}'s containing U_i, and let v_j denote the number of K_r's containing V_j. Let N denote the number of pairs of the form (U_i, W), where W is an r-element subset of $V(G)$, which does not induce a K_r, and $|U_i \cap W| = r - 1$. Show that

$$\sum_{i=1}^{k_r(G)} (r-1)(n-r-u_i) \leq N \leq \sum_{j=1}^{k_{r-1}(G)} v_j(n-r+1-v_j).$$

(ii) Using (i), prove by induction that

$$\frac{k_r(G)}{k_{r-1}(G)} \geq \frac{x-r+1}{xr} n.$$

[9.26] The proof is almost identical to that of Theorem 9.19.

[9.27] Use Exercise 9.26 for $r = 3$. To show that $d_4 = 1/\sqrt{2}$, prove that any set of four points determines a nonacute triangle.

[9.28] For any $c < c_r^*(P)$, construct a graph G_c with $V(G_c) = P$ by connecting two points if and only if their distance is larger than c. Apply Exercise 9.19 to G_c.

[9.29] It is sufficient to prove the assertion for $r = 4$.

CHAPTER 10

[10.1]   (i) Assume that $|p_i - p_j| = 1$ for all $0 \leq i < j \leq k$. Show that the vectors $p_0 p_i$ $(1 \leq i \leq k)$ are linearly independent.

(ii) No. Apply Ramsey's theorem (Theorem 9.13) and part (i).

[10.2]   (i) For any fixed $p_1, p_2 \in P$, all other points $p \in P$ have the property that $|p - p_1| - |p - p_2|$ is an integer, whose absolute value is at most $|p_1 - p_2|$. Hence, all points of $P - \{p_1, p_2\}$ lie on finitely many hyperbolas or lines. Observe that two hyperbolas with different axes have at most four points in common.

(ii) Let α be such that $\tan \alpha$ is rational, but α/π is irrational. Let p_i denote the point with polar coordinates $\rho(p_i) = 1$ and $\varphi(p_i) = 2i\alpha$ $(i = 1, 2, \ldots)$.

(iii) It follows from (ii).

[10.3]   The unit distance graph consists of paths, edges, and isolated vertices.

[10.4]   Generalize the proof of Theorem 10.5.

[10.5]   Notice that any two disjoint compact convex sets have at most four common tangents.

[10.6]   (ii) It is sufficient to show that $p_i - p_j$ is orthogonal to $q_g - q_h$, for all choices of the indices. Observe that if $i \neq j$, then q_g and q_h lie on the perpendicular bisector hyperplane of p_i and p_j.

[10.7]   Use a random decomposition into k parts, and show that the expected number of k-tuples with the required property is $(k!/k^k)|E(G)|$.

[10.8]   (i) Use Theorem 10.12.

(ii) Use (i) and Theorem 10.2 to establish the assertion for $k = 4$.

[10.10]   Let P be a set of n points in the plane. For any $1 \leq l \leq k + 2$, define a graph G_l with $V(G_l) = P$, whose two vertices are joined by an edge if and only if their distance belongs to $[d_l, d_l + c\sqrt{n}\,]$. Use the following generalization of Szemerédi's regularity lemma (Theorem 9.16): For any $0 < \delta < 1$ and any natural number m, there exist n_0 and M with the property that P can be partitioned into m' classes $P_1, \ldots, P_{m'}$ as equally as possible such that $m \leq m' \leq M$ and all but at most $\delta m'^2$ pairs are δ-regular in every G_l $(1 \leq l \leq k + 2)$. Then use Lemma 9.17 and follow the steps of the proof of Theorem 10.14.

[10.11]   See Claim B in the proof of Theorem 10.14.

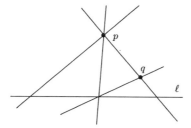

Figure H 11.1. Illustration for the solution of Exercise 11.1(i).

CHAPTER 11

[11.1] (i) Assume that there is no ordinary crossing. Choose a line $\ell \in L$ and an intersection point $p \notin \ell$ of lines in L that is closest to ℓ (see Figure H 11.1). Arrive at a contradiction by showing that there exists another intersection point $q \neq p$ such that $d(\ell,p) > d(\ell,q)$.

(ii) Use induction on n.

[11.2] (i) We can assume that every line $\ell \in L$ has at least three distinct points of intersection (crossings) with the other lines. Two such points belonging to the same line ℓ are called *neighbors* if either all other intersection points on ℓ lie between them or none of them do. Associate an ordinary crossing p with a line ℓ (not necessarily containing p) if ℓ contains exactly two neighbors of p. Notice that any ordinary crossing is associated with at most six lines. Show that if fewer than three ordinary crossings are associated with a line ℓ, then ℓ contains exactly two ordinary crossings. If there are fewer than $3n/7$ lines containing exactly two ordinary crossings, double-count the number of pairs (p,ℓ) such that p is associated with ℓ.

[11.3] For each side s_i of C, let F_i denote the set of faces of $\mathcal{A}(L) \cap C$ intersected by s_i, and let $L_i \subseteq L$ denote the set of lines that intersect s_i. Call an edge of an element of F_i *red* if it lies on a line of L_i, and *blue* otherwise. Bound the number of red and blue edges separately. Use Turán's theorem (Theorem 9.3) to bound the number of blue edges.

[11.4] Use Lemma 11.1.

[11.5] Use Jensen's inequality.

[11.6] Prove that the summand is equal to the probability that $\triangle$ is a trapezoid that meets exactly one line $\ell \in \mathcal{R}$, and $\triangle = \triangle_{\mathcal{R}-\{\ell\}}$. Hence, the sum is equal to the expected number of such trapezoids. Show that the *actual* number of such trapezoids is at most four for *any* choice of $\mathcal{R} \subseteq L$.

[11.7] (i) A line intersecting the interior of the edge $q_{(i-1)d} q_{id}$ is incident with a vertex q_j, $(i-1)d < j < id$.

(ii) Show that if a segment pq whose endpoints are not vertices of the arrangement intersects m lines of L, then $|\text{level}(p) - \text{level}(q)| \leq m$. Also, use the fact that the level of points in a sufficiently small neighborhood of a vertex of the k-level is between $k-1$ and $k+2$.

[11.8] (i), (ii) Use the Möbius inversion formula (see Exercise 7.3).

(iii) Show that

$$\sum_{r=1}^{k} \frac{\phi(r)}{r^2} \geq \log k \sum_{r=1}^{k} \frac{\mu(r)}{r^2} - c_1 \quad \text{and} \quad \sum_{r=1}^{k} \frac{\mu(r)}{r^2} \geq c_2,$$

where c_1, c_2 are positive constants.

[11.9] Perturb the lines slightly so that the arrangement of resulting lines becomes simple.

[11.10] For each ℓ_i, create $\lceil nw_i / \sum_{j=1}^{n} w_j \rceil$ copies of ℓ_i and use Theorem 11.6 to construct a partitioning of the resulting multiset of lines.

[11.11] Choose a random subset $\mathcal{R} \subseteq L$ of $r \approx n/k$ lines. Give a lower bound on the expected value of $H_0(\mathcal{R})$ in terms of $H_{\leq k}(L)$, and use the fact that $H_0(\mathcal{R}) \leq r$.

[11.12] Let $D_k(L)$ denote the number of discs satisfying the required property. First show that $D_0(L) = \binom{n-1}{2}$ is the same as the number of bounded cells in the arrangement $\mathcal{A}(L)$. Then prove that for every k,

$$\sum_{j=0}^{k} \binom{n-j-3}{k-j} D_j(L) = \binom{n}{k} \binom{n-k-1}{2}.$$

[11.13] Use Corollary 11.8.

[11.15] (ii) Let $f(n,d)$ denote the maximum number of rich cells in an arrangement of n hyperplanes in $\mathbb{R}^d$. Prove the recurrence relation

$$f(n,d) \leq f(n-1,d) + f(n-1,d-1).$$

To prove the tightness of this bound, take n points (vertices) of the $(d+1)$-dimensional *moment curve* $M(t) = (t, t^2, \ldots, t^{d+1})$, and consider the d-dimensional hyperplanes through the origin perpendicular to these vectors. Cut this arrangement by another d-dimensional hyperplane.

CHAPTER 12

[12.1] Let $k = \lfloor \sqrt{\log n} \rfloor$. Consider a $\lceil n/k \rceil \times k$ piece of the integer grid. Show that there are cn^2 triangles of area $k!/2$ having a horizontal side. Taking into account all other triangles with the same area (just like in the proof of the second part of Theorem 12.3), we obtain $cn^2 \log \log n$ as a lower bound.

[12.2] Let S be a set of n points in the plane. Fix a point p in S. For any other point q, let ℓ_q denote the perpendicular bisector of pq. Use Corollary 11.8 to give an upper bound on the number of incidences between the lines ℓ_q ($q \in S - \{p\}$) and the points of $S - \{p\}$.

[12.3] (i) Define a four-uniform hypergraph H on the vertex set $V(H) = S_n$, whose edges are those quadruples which either contain three collinear points or consist of two disjoint pairs which determine parallel lines. Using Exercise 11.8(iii), show that the number of quadruples of both kinds can be bounded from above by $cn^3 \log n$, and then apply Theorem 10.11.

(ii) Let $P \subseteq S_n$ be a set of m points with the desired property. For any integers k and $r \leq \sqrt{n}/k$, S_n can be covered with roughly n/k^2 disjoint translates of the "subgrid"

$$S^{k,r} = \{(rx, ry) \mid 0 \leq x, y < k\}.$$

Denoting the number of elements of P in these translates by m_i ($i = 1, 2, \ldots$), show that the total number of slopes determined by those pairs of points of P which belong to the same translate is

$$\sum \binom{m_i}{2} \geq \frac{c(mk)^2}{n},$$

and vary r.

[12.4] For any pair of points $\{p, q\}$, consider the locus of those points r for which the triangle pqr has the required property. Show that these curves belong to a family of 2 degrees of freedom, and apply Theorem 12.6.

[12.5] (i) Count the number of simplices homothetic to the simplex spanned by the points

$$(0, 0, \ldots, 0), (1, 0, \ldots, 0), (0, 1, \ldots, 0), \ldots, (0, 0, \ldots, 1).$$

Delete from P all elements for which $x_1 + x_2 + \cdots + x_d$ is maximum, and apply induction.

[12.6] Let $T = \triangle abc$, and let ϕ denote an orientation-preserving similarity transformation of the plane with $\phi(a) = a$ and $\phi(b) = c$. Let P_1 (resp.

P_2) be the set of points dividing the edge bc (resp. ac) into m equal parts. Consider the set

$$\{a\} \cup P_1 \cup P_2 \cup \{\phi(p) \mid p \in P_1\}.$$

[12.7] Fix $s > 2/\varepsilon$, $K = \lfloor (n^{1/s} - 1)/2 \rfloor$, and consider the set

$$P = \left\{ \sum_{i=0}^{s-1} k_i \alpha^i \;\middle|\; k_i \in \mathbb{Z}, |k_i| \le K \text{ for every } i \right\}.$$

[12.8] Construct a system of square grids

$$\Lambda^i = \{n_1 v_1^i + n_2 v_2^i \mid n_1, n_2 \in \mathbb{Z}\}$$

for $1 \le i \le k = \lceil 4\pi/\varepsilon \rceil$, where v_1^i is a vector of length $\frac{1}{2}$ in direction $e^{-\sqrt{-1}(i-1)\varepsilon/2}$, and v_2^i is a vector of length $\frac{1}{2}$ orthogonal to v_1. For each i count the number of points $p \in P$ such that the ray emanating from p in direction v_1^i intersects at most c/ε nonempty squares of the grid Λ^i, and for each point p count the number of indices i that satisfy the above property.

[12.9] (i) Apply Theorem 10.12.
(ii) Let P_1 be a set of $\lfloor n/3 \rfloor$ points on the unit circle in the $x_1 x_2$-plane, $P_2 = \{(0,0,i,0) \mid 1 \le i \le n/3\}$, and $P_3 = \{(-1,0,0,i) \mid 1 \le i \le n/3\}$. Set $P = P_1 \cup P_2 \cup P_3$.

[12.10] Let $C_i' = C_i + \frac{1}{2}B^2$, where B^2 denotes the unit disc around $\mathbf{0}$. Prove that the boundaries of two expanded sets C_i', C_j' ($i \ne j$) intersect in at most two points.

[12.11] Prove that the maximum number of incidences between m points and n unit spheres in $\mathbb{R}^3$ is at most $c(m^{5/7}n^{6/7} + n\sqrt{m} + n)$. To this end, let σ_i denote the number of unit spheres containing the ith point, and use Theorem 11.10 to obtain an upper bound for $\sum_{i=1}^m \binom{\sigma_i}{2}$.

[12.12] Take two congruent copies of P lying on parallel lines in the plane, and denote them by P_1 and P_2. Notice that an $\lfloor \varepsilon n \rfloor$-element subset of P_1 is similar to a subset of P_2 if and only if they can be projected onto each other from an appropriate center. Use Corollary 11.8 to prove that the number of such centers cannot exceed $C'n$.

CHAPTER 13

[13.1] Use Reuleaux triangles as shown in Figure H 13.1.

[13.2] Let $c(q)$ denote the circle centered at p and passing through a point $q \in C$. Observe that the tangent line to $c(q)$ at q intersects pr.

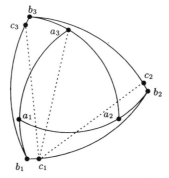

Figure H 13.1. Each vertex of the polygon $a_1 b_1 c_1 a_2 b_2 c_2 a_3 b_3 c_3$ has three vertices equidistant from it.

[13.4] Assume that the center of C is at origin. Put a unit disc on every point $(2i, 0)$ $(i \geq 1)$, and extend this to a densest hexagonal packing of unit discs. Let C be the portion of this packing lying inside a cone of angle $2\pi/n$ which is symmetric about the x-axis and whose apex is at the origin. Rotate C n times by angle $2\pi/n$ around the origin, and take the union of the rotated packings. Extend this to a maximal packing, and let n tend to infinity.

[13.5] Let $P(n)$ denote a configuration of n points satisfying the required properties. Let $P(2n)$ be defined as the union of $P(n)$ and a translate of $P(n)$ by unit distance in some generic direction.

[13.6] (i) For any $p \in C$, let p^+ (resp. p^-) denote the vertex of conv C adjacent to p in the clockwise (resp. counterclockwise) order. Let $pq \in E(G_k)$. If $|p^+ - q|$ (resp. $|p - q^-|$) is larger than $|p - q|$, then the edge p^+q (resp. pq^-) is said to *cover* pq. Select a maximal sequence $p_1 q_1 = pq, p_2 q_2, \dots, p_j q_j$ such that $p_{i+1} q_{i+1}$ covers $p_i q_i$ for every $i < j$. Clearly, $j \leq k$. The edge $p_j q_j \in E(G_k)$ is called a *majorant* of pq. For any $p \in C$, let $p_{\min}$ and $p_{\max}$ denote the first and the last neighbors of p in G_k in the clockwise order, and choose p that maximizes $\|p_{\max} p\|$, the number of edges of conv C lying after $p_{\max}$ but before p in the clockwise order. Let uv be a majorant of $p_{\min} p$. Let q be the kth vertex from v in the counterclockwise direction, and let $v'u'$ be a majorant of $qq_{\max}$. (See Figure H 13.6.) Then

$$\|p_{\min} p_{\max}\| = \|p_{\min} u\| + \|u q_{\max}\| + \|q_{\max} p_{\max}\|$$

$$\leq (k - 1 - \|vp\|) + (k - 1) + (\|vp\| + k)$$

$$= 3k - 2.$$

(ii), (iii) Delete from C any vertex of $G_k(C)$ whose degree is at most $3k - 1$, and use induction.

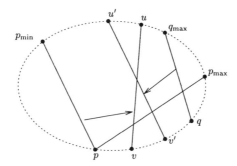

Figure H 13.6. Illustration for the hint of Exercise 13.6.

[13.7] (i) Use the fact that a planar graph is four-colorable.

(ii) See Figure H 13.7.

[13.8] Use the centers of a regular hexagonal packing of equal circles (see Figure 13.4.).

[13.9] (i) Call a pair (i,j) *farthest neighbor pair* if p_j is a farthest neighbor of p_i. Let d_i denote the distance between p_i and its farthest neighbors, and let D_i (resp. C_i) denote the disc (resp. circle) of radius d_i centered at p_i. Assume that $d_i \le d_{i+1}$ for all i. Mark a farthest neighbor pair (i,j) if there is another farthest neighbor pair (i',j) for some $i' < i$. If (i,j) is a marked pair, then p_j is an intersection point of C_i and the boundary of $\bigcup_{k<i} D_k$. Show that there are at most two such intersection points. In fact, with some extra care, one can show that

$$\Phi(n) = \begin{cases} 3n - 3 & \text{if } n \text{ is even,} \\ 3n - 4 & \text{if } n \text{ is odd.} \end{cases}$$

(ii) Show that if $p_{i_1}, p_{i_2}, \ldots, p_{i_m}$ are the farthest neighbors of p_i, ordered in clockwise order, then $p_{i_2}, \ldots, p_{i_{m-1}}$ cannot be farthest neighbors of any other point of P.

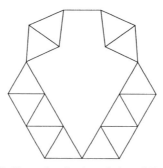

Figure H 13.7. Illustration for the solution of Exercise 13.7(ii).

[13.11] Use Lenz's construction (Figure 10.4).

[13.12] Clearly, every distance $d_i(n)$ occurs between $(0,0)$ and some point of the triangular region

$$T_n = \{(i,j) \mid 0 \le j \le i < n\}.$$

Let π denote the orthogonal projection of $\mathbb{R}^2$ onto a line passing through the origin, whose slope is $1 - \varepsilon$ for some very small $\varepsilon > 0$. Order the points $p \in T_n$ according to the distance between the origin and $\pi(p)$. Show that for every $i < n$, the distance between $(0,0)$ and the ith largest element in this ordering is $d_i(n)$.

[13.13] Let I denote the number of isosceles triangles spanned by triples of P. From the proof of Theorem 13.7, we obtain the result that

$$\sum_{i=1}^{k} n \binom{2s_i/n}{2} \le I \le 2 \binom{n}{2},$$

and the statement follows by Jensen's inequality.

[13.14] Define a cell decomposition of P by assigning to each vertex v of P the set of all points of P which are closer to v than to any other vertex. Apply Corollary 13.15 to the vertex set of P.

[13.15] Apply the linear algebra method. Let $V(H) = \{v_1, \ldots, v_n\}$, and let $E(H) = \{E_1, \ldots, E_m\}$. For any $1 \le j \le m$, let $x_j = (x_{j1}, \ldots, x_{jn})$, where $x_{ji} = 1$ if $v_i \in E_j$, and $x_{ji} = 0$ if $v_i \notin E_j$. Then $\langle x_j, x_k \rangle \equiv 0 \pmod 2$, unless $j = k$. Show that the vectors x_j are linearly independent over the field $GF(2)$.

[13.16] Apply the linear algebra method. Let $V(H) = \{v_1, \ldots, v_n\}$, and let $E(H) = \{E_1, \ldots, E_m\}$, $|E_1| \le |E_2| \le \cdots \le |E_m|$. For any $1 \le j \le m$, let $a_j = (a_{j1}, \ldots, a_{jn})$, where $a_{ji} = 1$ if $v_i \in E_j$, and $a_{ji} = 0$ if $v_i \notin E_j$. Let f_j be the polynomial in n variables $(x_1, \ldots, x_m) = x$, defined by

$$f_j(x) = \prod_{\substack{t \in T \\ t < |E_j|}} (\langle a_j, x \rangle - t).$$

Observe that $f_j(a_k) = 0$ if $k < j$, but $f_j(a_j) \ne 0$. Let $\hat{f}_j$ denote the multilinear polynomial obtained from f_j by repeated application of the relations $x_1^2 = x_1, \ldots, x_n^2 = x_n$. Show that the polynomials $\hat{f}_j$ $(1 \le j \le m)$ are linearly independent over $\mathbb{R}$.

[13.17] To establish the lower bound, show that there is no vertex of the minimum-distance graph, whose degree is 5 and all of whose neighbors and second neighbors are also of degree 5. For the upper bound, deform a square lattice.

CHAPTER 14

[14.1] (ii) Show that two segments representing the largest distance within a point set cannot be disjoint.

[14.2] If G has a vertex x incident with only one edge $e \in E(G)$, then let $f(e) = x$ and delete x from the graph. If every vertex has degree at least 2, then assign each pointed vertex to the leftmost edge incident with it. (See the proof of Theorem 14.3 for the terminology.)

[14.3] Generalize the construction shown in Figure H 14.3.

[14.4] Let $x_1, \ldots, x_n$ denote the vertices of a convex n-gon in clockwise order. For any $i < j$, connect x_i and x_j by an edge if and only if $j \le 2k + 1$ or i is an even number not exceeding $2k$.

[14.5] Decompose the vertex set of a convex n-gon into $k - 1$ subsets of consecutive vertices, $V_1, V_2, \ldots, V_{k-1}$, as equally as possible, and connect two vertices by a segment if and only if they belong to different subsets.

[14.6] Let $\binom{X}{r}$ denote the family of all r-element subsets of X. To prove (i), it is sufficient to show that for every $r < n/2$ there is a bijection (matching) f between $\binom{X}{r}$ and a subset of $\binom{X}{r+1}$ such that $f(A) \supseteq A$ for every $A \in \binom{X}{r}$. To see this one can use, for example, *Hall's marriage theorem* (1935): Let G be a bipartite graph on the vertex set $V_1 \cup V_2$. There is a matching between V_1 and a subset of V_2 if and only if the following condition holds for every $U \subseteq V_1$. The number of vertices in V_2 adjacent to at least one element of U is at least $|U|$.

[14.7] (i) Follow the proof of Theorem 14.10.

(ii) Construct five unit segments $x_i y_i$ $(0 \le i \le 4)$ with the property that $\max_{i,j} (|x_i - x_j|, |y_i - y_j|) < \varepsilon$ for some small $\varepsilon > 0$, and $x_i y_i$ intersects

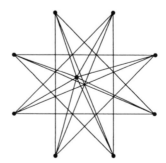

Figure H 14.3. $n = 9$.

$x_j \, y_j$ if and only if $|j - i| \equiv \pm 1$ (mod 5). Replace each segment with a similar construction of five segments, and proceed by induction.

[**14.8**] Use induction on k based on Corollary 14.13.

[**14.9**] (i) If all edges along the boundary of the convex n-gon are of the same color, then the assertion is true. Otherwise, delete a point incident with two sides of different colors, and apply induction.

(ii) Use the same argument as in (i), except that now delete the point together with its two neighbors.

[**14.10**] (i) Let $f(G)$ denote the number of $(k + 1)$-tuples of pairwise crossing edges in a convex geometric graph G. Observe that

$$(n - 2k - 2)f(G) = \sum_{v \in V(G)} f(G - v),$$

and apply induction on n, as in the proof of Theorem 14.12.

[**14.11**] Define a small number of partial orderings on the ground set $\mathcal{B}$ such that any two members of $\mathcal{B}$, whose intersection contains no element of P, are comparable by at least one of them. Each of these orderings should satisfy the condition that if there is a chain of length $n + 1$, then (a) holds. Use the idea of the proof of Theorem 14.10.

[**14.12**] Assume that no three arcs have an interior point in common. Let p and q be two consecutive intersection points of the arcs γ_1 and γ_2. Eliminate these intersection points by replacing the piece of γ_1 (or γ_2) between p and q by a curve running very close to γ_2 (resp. γ_1). One of these replacements will reduce the total number of intersection points.

[**14.13**] Let $\{e_i, f_i\}$ ($1 \leq i \leq m$) be an enumeration of all pairs of independent edges of G. Assign to each drawing of G that satisfies (a) the $(0,1)$-vector $\mathbf{x} = (x_1, \ldots, x_m)$, where $x_i = 1$ if and only if e_i and f_i intersect each other an odd number of times. Furthermore, for each pair (v, e), such that $v \in V(G)$, $e \in E(G)$, and $v \notin e$, let $\mathbf{x}_{(v, e)}$ denote the $(0,1)$-vector, all of whose coordinates are 0, except those which correspond to a pair $\{e, f\}$ with $v \in f$. Let L be the subspace modulo 2 generated by the vectors $\mathbf{x}_{(v, e)}$. Show that for any two drawings of G, which satisfy (a), the difference between the corresponding vectors, $\mathbf{x} - \mathbf{x}'$ (mod 2), belongs to L.

[**14.14**] (ii) Consider the image of an odd cycle C in a thrackle which contains no three arcs with a common interior point. This closed curve will divide the plane into a finite number of cells, which can be colored with black and white so that no two cells that share a boundary segment have the same color. Observe that the image of any edge of $G - C$ connects two points of different colors.

(iii) By Kuratowski's theorem it is sufficient to show that G contains no subdivision of K_5 or $K_{3,3}$. To prove that G has no subdivision of K_5, observe the following. If C and C' are two cycles of G with exactly one common vertex, then the closed curves representing C and C' do not cross each other at this vertex.

CHAPTER 15

[15.1] Let $t : V(H) \to \mathbb{R}^+$ (resp. $p : E(H) \to \mathbb{R}^+$) be an optimal fractional transversal (resp. packing) of H. Estimate $\sum t(x)p(E)$ over all pairs (x, E) with $x \in E$.

[15.2] Apply the max-flow min-cut theorem for networks.

[15.3] Calculate $\tau(H)$ and $\tau^*(H)$ for the complete r-uniform hypergraph with $|V(H)| = n$, $E(H) = \{E \subseteq V(H) \mid |E| = r\}$, and vary the parameter r.

[15.4] (i) Color the vertices of H with two colors randomly.

(ii) Let S be a $2r^2$-element set. Define a hypergraph H whose vertices are the r-element subsets of S. Assign to each partition $S = S_1 \cup S_2$ a hyperedge E_s consisting of all r-tuples contained entirely in S_1 or in S_2. Apply Theorem 15.2 to show that $\tau(H) \leq cr^2 2^r$.

[15.5] Use the idea of the second proof of Theorem 15.4.

[15.6] (i) Use the idea of the proof of Theorem 15.5.

[15.8] Remove every vertex which is not contained in any edge. For every pair $x, y \in X$, write $x \prec y$ (resp. $y \prec x$) if there is no edge E such that $E \cap \{x, y\} = \{x\}$ (resp. $\{y\}$). Call x and y equivalent if $x \prec y$ and $y \prec x$. Show that $\prec$ defines a partial order on the equivalence classes of X, which has at most $\lceil 1/\varepsilon \rceil - 1$ maximal elements.

[15.10] (i) For any fixed d, use induction on n.

(ii) Let $G_d(n)$ (resp. $G'_d(n)$) denote the number of regions into which $\mathbb{R}^d$ (resp. the projective d-space) is divided by n hyperplanes in general position. Then

$$G'_d(n) = G_d(n) - G'_{d-1}(n).$$

(iii) By duality, the statement is equivalent to (ii).

[15.11] (i) Use the following theorem of Radon. Any $(d + 2)$-element set $A \subseteq \mathbb{R}^d$ can be partitioned into two parts $A = A_1 \cup A_2$ such that the convex hull of A_1 and the convex hull of A_2 have at least one point in common (see Eckhoff, 1993).

[15.12] For any $p_j \in S$, let $p'_j = (p_j, \|p_j\|^2)$ denote the projection of p_j into the $(d + 1)$-dimensional paraboloid $\Pi = \{(p, r) \in \mathbb{R}^d \times \mathbb{R} \mid \|p\|^2 = r\}$. By Exercise 15.10(iii), there is a collection of

$$\sum_{i=0}^{\lfloor (d+1)/2 \rfloor} \binom{n+1}{d+1-2i} = \sum_{i=0}^{d+1} \binom{n}{i}$$

hyperplanes in $\mathbb{R}^{d+1}$ such that each of them bisects the set consisting of the points p'_i and a faraway point on the r-axis in a different way. Consider the system of spheres obtained by projecting the intersection of Π and each of these hyperplanes into $\mathbb{R}^d$.

[15.13]   Improve the bound in Lemma 15.13 to $c'm/\sqrt{n}$ by using Theorem 11.6 instead of an ε-net argument.

[15.14]   Let P be a convex n-gon, and let l be a line. By a binary search on the slope of the edges of P, compute the two tangents of P that are parallel to l, and check whether l lies between them.

[15.15]   Use the same construction as in the proof of Corollary 15.15, except that now store at each vertex v only those vertices of the convex hull of $P(v)$ that do not belong to the convex hull of $P(w)$, where w is the parent of v.

[15.16]   (i) The line l bisects S into two $2n$-element sets, S_1 and S_2. For any direction θ, let $L_i(\theta)$ denote the set of all lines l' parallel to this direction that bisect S_i into two n-element sets ($i = 1, 2$). Use a continuity argument to show that $L_1(\theta) \cap L_2(\theta) \neq \varnothing$ for some θ.

(ii) Construct a four-way tree T each of whose nodes v is associated with a subset $S_v \subseteq S$, as follows. Partition S_v into four subsets $S_v^1, \ldots, S_v^4$ by two lines l, l' using part (i). Store $|S_v|$ and the equations of l, l' at v. Create four children $w_1, \ldots, w_4$ for v, and associate S_v^i with w_i.

(iii) Using (i) construct a binary tree, so that for each node v visited by a query, one of the children of v and only one child of the other child of v are visited.

[15.18]   (ii) Use Exercise 15.16(i).

(iii) Let l be a vertical line dividing S into two roughly equal parts. Recursively, construct two weak $(3\varepsilon/2)$-nets T_1 and T_2 for the $(n/2)$-element point sets on the left-hand side and on the right-hand side of l, respectively. Notice that $T_1 \cup T_2$ will cover all convex sets except those which contain more than $\varepsilon n/4$ points on both sides of l. However, all such sets can be covered by roughly

$$\frac{n^2/4}{(\varepsilon n/4) \cdot (3\varepsilon n/4)}$$

points of l, chosen from the set of intersection points of l with the segments connecting the elements of S.

(v) Let $S = \{p_0, \ldots, p_{n-1}\}$. For a fixed k, let $S_i = \{p_j \mid (i-1)k < j < ik\}$, $1 \leq i \leq n/k$. Find recursively a small weak $(\varepsilon n/3k)$-net for each S_i, and add to them the points p_0, p_{ik}, and all points that can be obtained as the intersection of $p_0 p_{ik}$ and $p_{jk} p_{(i+1)k}$ $(1 \leq j \leq i < n/k)$. Show that these points form a weak ε-net for S. Set $k = \lfloor n\sqrt{\varepsilon/3} \rfloor$.

[15.19] Apply Theorem 15.12 and inequality (10.1).

[15.20] For any triangle $\triangle$, let $R_\triangle$ be the set of all lines intersecting $\triangle$. The range space $X = \{\text{all lines in } \mathbb{R}^2\}$, $\mathcal{R} = \{R_\triangle \mid \triangle \text{ is a triangle}\}$ has bounded VC-dimension. Apply Corollary 15.6(i) with $\varepsilon = (c \log r)/r$ to find an ε-net L' for the subspace of $(X, \mathcal{R})$ induced by L, and triangulate the arrangement $\mathcal{A}(L')$.

[15.21] Define a hypergraph H by setting $V(H) = \{\text{Box}(p, q) \mid p, q \in P\}$, $E(H) = \{E_r \mid r \in P\}$, where $E_r = \{\text{Box}(p, q) \mid r \in \text{Box}(p, q)\}$. Show that $\lambda(H) \leq 2^{2^{d-1}}$ by using the following lemma of Erdős and Szekeres (1935). Any sequence of $k^2 + 1$ real numbers contains a monotone subsequence of length $k + 1$. Apply Theorem 15.9.

[15.22] Construct a hypergraph H with $V(H) = P$, $E(H) = \{B \cap P \mid B \in \mathcal{B}\}$. Show that $\lambda(H)$ is bounded by a constant depending only on d, and apply Theorem 15.9.

[15.23] Apply Theorem 15.9. To bound $\lambda(H)$, choose some members (edges) of H, $E_1, \ldots, E_\lambda$, such that for any $i < j$, $E_i \cap E_j$ has a point x_{ij} which does not belong to any other E_h $(h \neq i, j)$. Write each E_i $(1 \leq i \leq \lambda)$ as the union of k intervals, $E_i = I_{i1} \cup \cdots \cup I_{ik}$. If $x_{ij} \in I_{ip} \cap I_{jq}$ for some $i < j$, then call (E_i, E_j) a pair of type (p, q). Show that there are no four edges such that all six pairs determined by them are of the same type, and use Theorem 9.13.

CHAPTER 16

[16.1] Try a random hypergraph.

[16.2] Use a random coloring. For an ingenious algorithmic proof, see Erdős and Selfridge (1973).

[16.3] Use the following statement, known as *Lovász local lemma* (see Alon and Spencer, 1992). Let $A_1, \ldots, A_n$ be events in a probability space, and let G be a graph on the vertex set $\{1, 2, \ldots, n\}$ with the properties:

(i) each A_i is independent of the collection of all events A_j for which j is not adjacent to i in G.

(ii) $\Pr[A_i] \le 1/4D$ for every i, where D denotes the maximum degree of the vertices in G.

Then $\Pr[\bar{A}_1\bar{A}_2\ldots\bar{A}_n] > 0$, that is, with positive probability none of the events will occur.

[16.4] Construct a tree T on $V(H)$ such that every $E \in E(H)$ contains at least one edge of T.

[16.5] Analyze the original proof.

[16.6] Eliminate all closed polygons whose vertices belong to S and whose edges are alternately horizontal and vertical, and use induction.

[16.7] Deduce from Lemma 16.11 that there is a pair of points $p, q \in V(H)$ such that the edge $\{p,q\}$ stabs at most $Cn^{1-1/d}$ hyperedges of H for a suitable constant C. Now follow the proof of Theorem 15.12.

[16.8] Use the fact that a family of n spheres or half-spaces in $\mathbb{R}^d$ partitions the space into $C_d n^d$ cells.

[16.9] Modify the proof of Theorem 16.7. Notice that every edge of a spanning path P that stabs an edge $\varphi(E_1,\ldots,E_k) \in E(H')$ must also stab at least one E_i, $1 \le i \le k$, and that $\pi_{H'}(m) \le (\pi_H(m))^k$ for every m.

[16.11] Show that using the notation in the proof of Theorem 16.15,

$$h(\varphi) + h''(\varphi) = \langle P'(\varphi), u(\varphi)\rangle,$$

where $u(\varphi)$ is a unit vector parallel to the line $\ell(\varphi)$.

[16.12] (i) Apply Theorem 16.15 and the Cauchy-Schwarz inequality to bound, for example,

$$\int_0^{2\pi} \left(h(\varphi) + h(\varphi + \pi)\right)^{1/2}\left(h(\varphi + \pi/2) + h(\varphi + 3\pi/2)\right)^{1/2} d\varphi,$$

where γ encloses a convex region with support function h.

[16.13] Let $c: V(H) \rightarrow \{+1,-1\}$ be a random two coloring, and set $b_E = [c(E)/(100\sqrt{n})]$, where $[x]$ denotes a nearest integer to x. Furthermore, let $b = b(c) = (b_E | E \in E(H))$. As in the proof of Lemma 16.12, show that $H(b) \le n/20$. Hence, there is a three-coloring $\bar{c}: V(H) \rightarrow \{+1,-1,0\}$ such that $\bar{c}(x) \ne 0$ for at least $n/3$ points, and $|\bar{c}(E)| \le 50\sqrt{n}$ for every E. Use this statement iteratively.

[16.14] An arithmetic progression of the form $\{a + kd \mid j2^i \le k < (j + 1)2^i\}$, $1 \le a \le d$ and $i, j \ge 0$, is called a *canonical* progression. Clearly, any arithmetic progression can be decomposed into canonical ones. Let H_i denote the hypergraph formed by all canonical progressions of length 2^i. Use the same argument as in the proof of Lemma 16.12 to obtain a three-coloring $c: V(H) \to \{+1, -1, 0\}$ such that $c(x) \neq 0$ for at least $n/3$ points, and

$$\sum_{i=1}^{\log n} \text{disc}(H_i, c) \leq C n^{1/4}.$$

Iterate this result. (This is now somewhat trickier!)

[16.15] (i) For $i = 1, 2$, partition V into $n/2$ disjoint pairs $(p_i^1, q_i^1), \ldots, (p_i^{n/2}, q_i^{n/2})$ such that $\pi_i(p_i^m) = q_i^m$, $1 \leq m \leq n/2$. Color the points with $+1$ and -1 so that the points of each pair will be colored by different colors.

(ii) As in the proof of Theorem 16.4, divide each permutation into "canonical intervals" and apply Theorem 16.3 to obtain a $c \log^2 n$ bound on the discrepancy of H. Improve it to $c \log n$.

BIBLIOGRAPHY

P. AGARWAL (1992), Ray shooting and other applications of spanning trees with low stabbing number, *SIAM J. Computing* 22, 540–570.

P. AGARWAL, B. ARONOV, J. PACH, R. POLLACK, AND M. SHARIR (1995), Quasi-planar graphs have a linear number of edges, manuscript.

A. AHO, J. HOPCROFT, AND J. ULLMAN (1974), *Design and Analysis of Algorithms*, Addison-Wesley, Reading, MA.

M. AJTAI, V. CHVÁTAL, M. NEWBORN, AND E. SZEMERÉDI (1982), Crossing-free subgraphs, *Ann. Discrete Mathematics* 12, 9–12.

J. AKIYAMA AND N. ALON (1989), Disjoint simplices and geometric hypergraphs, in: *Combinatorial Mathematics* (G. Bloom et al., eds.), *Ann. New York Academy of Sciences*, vol. 555, New York Academy of Sciences, New York, 1–3.

R. ALEXANDER (1990), Geometric methods in the study of irregularities of distribution, *Combinatorica* 10, 115–136.

N. ALON AND P. ERDŐS (1989), Disjoint edges in geometric graphs, *Discrete and Computational Geometry* 4, 287–290.

N. ALON AND E. GYŐRI (1986), The number of small semispaces of a finite set of points in the plane, *J. Combinatorial Theory, Series A* 41, 154–157.

N. ALON AND J. SPENCER (1992), *The Probabilistic Method*, Wiley-Interscience, New York.

N. ALON, L. BABAI, AND H. SUZUKI (1991), Multilinear polynomials and Frankl Ray–Chaudhuri Wilson type intersection theorems, *J. Combinatorial Theory, Series A* 58, 165–180.

N. ALON, P. SEYMOUR, AND R. THOMAS (1990), A separator theorem for nonplanar graphs, *J. American Mathematical Society* 3, 801–808.

N. ALON, P. SEYMOUR, AND R. THOMAS (1994), Planar separators, *SIAM J. Discrete Mathematics* 7, 184–193.

N. ALON, I. BÁRÁNY, Z. FÜREDI, AND D. KLEITMAN (1992), Point selection and weak ϵ-nets for convex hulls, *Combinatorics, Probability and Computing* 1, 189–200.

N. ALON, R. DUKE, H. LEFMANN, AND V. RÖDL (1994), The algorithmic aspect of the regularity lemma, *J. Algorithms* 16, 80–109.

E. ALTMAN (1963), On a problem of Erdős, *American Mathematical Monthly* 70, 148–157.

E. ALTMAN (1972), Some theorems on convex polygons, *Canadian Mathematical Bull.* 15, 329–340.

E. ANDREEV (1970a), On convex polyhedra in Lobačevskiĭ spaces, *Matematicheskiĭ Sbornik, Nov. Ser.* 81, 445–478. [English translation appears in *Mathematics of the USSR, Sbornik* 10 (1970), 413–440.]

E. ANDREEV (1970b), On convex polyhedra of finite volume in Lobačevskiĭ spaces, *Matematicheskiĭ Sbornik, Nov. Ser.* 83, 256–260. [English translation appears in *Mathematics of the USSR, Sbornik* 12 (1970), 255–259.]

H. ANNING AND P. ERDŐS (1945), Integral distances, *Bull. American Mathematical Society* 51, 598–600.

R. ANSTEE (1985), General forbidden configuration theorems, *J. Combinatorial Theory, Series A* 40, 108–124.

R. ANSTEE AND Z. FÜREDI (1986), Forbidden submatrices, *Discrete Mathematics* 62, 225–243.

M. ANTHONY AND N. BIGGS (1992), *Computational Learning Theory*, Cambridge University Press, New York.

D. APPLEGATE AND R. KANNAN (1991), Sampling and integration of near log–concave functions, *Proc. 23rd Annual ACM Symposium on Theory of Computing*, 156–163.

B. ARONOV, D. NAIMAN, J. PACH, AND M. SHARIR (1993), An invariant property of balls in arrangements of hyperplanes, *Discrete and Computational Geometry* 10, 421–425.

B. ARONOV, B. CHAZELLE, H. EDELSBRUNNER, L. GUIBAS, M. SHARIR, AND R. WENGER (1991), Points and triangles in the plane and halving planes in space, *Discrete and Computational Geometry* 6, 435–442.

B. ARONOV, P. ERDŐS, W. GODDARD, D. KLEITMAN, M. KLUGERMAN, J. PACH, AND L. SCHULMAN (1994), Crossing families, *Combinatorica* 14, 127–134.

D. AVIS (1984), The number of furthest neighbour pairs of a finite planar set, *American Mathematical Monthly* 91, 417–420.

D. AVIS, P. ERDŐS, AND J. PACH (1988), Repeated distances in the space, *Graphs and Combinatorics* 4, 207–217.

D. AVIS, P. ERDŐS, AND J. PACH (1991), Distinct distances determined by subsets of a point set in space, *Computational Geometry: Theory and Applications* 1, 1–11.

S. AVITAL AND H. HANANI (1966), Graphs, *Gilyonot Lematematika* 3, 2–8.

L. BABAI AND P. FRANKL (1988), Linear algebra methods in combinatorics, Part I, Tech. Rept., University of Chicago, July 1988. Expanded edition, 1992.

K. BALL (1992), A lower bound for the optimal density of lattice packings, *Duke J. Mathematics* 68, 217–221.

R. BAMBAH AND H. DAVENPORT (1952), The covering of *n*-dimensional space by spheres, *J. London Mathematical Society* 27, 224–229.

R. BAMBAH AND C. ROGERS (1952), Covering the plane with convex sets, *J. London Mathematical Society* 27, 304–314.

R. BAMBAH, C. ROGERS, AND H. ZASSENHAUS (1964), On coverings with convex domains, *Acta Arithmetica* 9, 191–207.

E. BARANOVSKIĬ (1964), On packing *n*-dimensional Euclidean space by equal spheres I, *Izvestija Visših Učebnyh Zavedeniĭ Matematika* (2) 39, 14–24.

I. BÁRÁNY AND Z. FÜREDI (1987), Computing the volume is difficult, *Discrete and Computational Geometry* 2, 319–326.

I. BÁRÁNY AND J. LEHEL (1987), Covering with Euclidean boxes, *European J. Combinatorics* 8, 113–119.

I. BÁRÁNY, Z. FÜREDI, AND L. LOVÁSZ (1990), On the number of halving planes, *Combinatorica* 10, 175–183.

I. BÁRÁNY, Z. FÜREDI, AND J. PACH (1984), Discrete convex functions and proof of the six circle conjecture of Fejes Tóth, *Canadian J. Mathematics* 36, 569–576.

I. BÁRÁNY, H. BUNTING, D. LARMAN, AND J. PACH (1993), Rich cells in an arrangement of hyperplanes, manuscript.

E. BARBIER (1860), Note sur le probléme de l'aiguille et le jeu du joint couvert, *J. Mathématiques Pures et Appliquées* (2) 5, 273–286.

P. BATEMAN AND P. ERDŐS (1951), Geometric extrema suggested by a lemma of Besicovitch, *American Mathematical Monthly* 58, 306–314.

J. BECK (1981a), Balanced two-colourings of finite sets in the square I, *Combinatorica* 1, 327–335.

J. BECK (1981b), Roth's estimate of the discrepancy of integer sequences is nearly sharp, *Combinatorica* 1, 319–325.

J. BECK (1983a), On the lattice property of the plane and some problems of Dirac, Motzkin and Erdős in combinatorial geometry, *Combinatorica* 3, 281–297.

J. BECK (1983b), On a problem of K. F. Roth concerning irregularities of distribution, *Inventiones Mathematicae* 74, 477–487.

J. BECK (1991), Quasi-random 2-colorings of point sets, *Random Structures and Algorithms* 2, 289–302.

J. BECK AND W. CHEN (1987), *Irregularities of Distribution*, Cambridge University Press, New York.

J. BECK AND T. FIALA (1981), Integer making theorems, *Discrete Applied Mathematics* 3, 1–8.

J. BECK AND V. T. SÓS (1994), Discrepancy theory, to appear in *Handbook of Combinatorics*.

J. BECK AND J. SPENCER (1984), Unit distances, *J. Combinatorial Theory, Series A* 37, 231–238.

C. BENSON (1966), Minimal regular graphs of girth eight and twelve, *Canadian J. Mathematics* 18, 1091–1094.

E. BERLEKAMP (1969), On subsets with intersections of even cardinality, *Canadian Mathematical Bull.* 12, 363–366.

M. BERN, D. EPPSTEIN, P. PLASSMANN, AND F. YAO (1991), Horizon theorems for lines and polygons, in: *Discrete and Computational Geometry: Papers from DIMACS Special Year*, DIMACS Series, vol. 6 (J. Goodman et al., eds.), American Mathematical Society, Providence, RI, 45–66.

A. BEZDEK (1984), On the section of lattice-coverings of balls, *Rend. Circ. Math. Palermo, Ser. II, Suppl. 3*, 23–45.

A. BEZDEK AND G. KERTÉSZ (1987), Counter-examples to a packing problem of L. Fejes Tóth, in: *Intuitive Geometry, Colloquia Mathematica Societatis János Bolyai*, vol. 48 (K. Böröczky and G. Fejes Tóth, eds.), North-Holland, Amsterdam, 29–36.

A. BEZDEK AND W. KUPERBERG (1991), Packing Euclidean space with congruent cylinders and with congruent ellipsoids, in: *Applied Geometry and Discrete Mathematics: The Victor Klee Festschrift* (P. Gritzmann and B. Sturmfels, eds.), DIMACS Series, vol. 4, American Mathematical Society, Providence, RI, 71–80.

D. BIENSTOCK AND E. GYŐRI (1991), An extremal problem on sparse 0-1 matrices, *SIAM J. Discrete Mathematics* 4, 17–27.

W. BLASCHKE (1923), *Vorlesungen über Differentialgeometrie II*, Berlin.

W. BLASCHKE (1955), *Vorlesungen über Integralgeometrie*, 3rd ed., Deutscher Verlag der Wissenschaften, Berlin.

H. BLICHFELDT (1914), A new principle in the geometry of numbers with some applications, *Trans. American Mathematical Society* 15, 227–235.

H. BLICHFELDT (1921), Note on geometry of numbers, *Bull. American Mathematical Society* 27, 152–153.

H. BLICHFELDT (1929), The minimum value of quadratic forms, and the closest packing of spheres, *Mathematische Ann.* 101, 605–608.

G. BLIND (1969), Über Unterdeckungen der Ebene durch Kreise, *J. reine und angewandte Mathematik* 236, 145–173.

G. BLIND (1983), Problem No. 34, Research problems, *Periodica Mathematica Hungarica* 14, 309–312.

A. BLUMER, A. EHRENFEUCHT, D. HAUSSLER, AND M. WARMUTH (1989), Learnability and the Vapnik–Chervonenkis dimension. *J. Association for Computing Machinery* 36, 929–965.

B. BOLLOBÁS (1976), On complete subgraphs of different orders, *Mathematical Proc. Cambridge Philosophical Society* 79, 19–24.

B. BOLLOBÁS (1978), *Extremal Graph Theory*, Academic Press, New York.

B. BOLLOBÁS (1979), *Graph Theory: An Introductory Course*, Graduate Texts in Mathematics 63, Springer-Verlag, New York.

B. BOLLOBÁS (1985), *Random Graphs*, Academic Press, London.

B. BOLLOBÁS (1986), *Combinatorics*, Cambridge University Press, Cambridge.

V. BOLTYANSKY AND I. GOHBERG (1985), *Results and Problems in Combinatorial Geometry*, Cambridge University Press, New York.

J. BONDY (1971), Large cycle graphs, *Discrete Mathematics* 1, 121–132.

J. BONDY AND M. SIMONOVITS (1974), Cycles of even lengths in graphs, *J. Combinatorial Theory, Series B* 16, 97–105.

T. BONNESEN AND W. FENCHEL (1934), *Theorie der konvexen Körper, Ergebnisse der Mathematik 3*, Vol. 1, Springer-Verlag, New York.

K. BÖRÖCZKY (1986), Closest packing and loosest covering of the space with balls, *Studia Scientiarum Mathematicarum Hungarica* 21, 79–89.

K. BORSUK (1933), Drei Sätze über die n-dimensionale euklidische Sphäre, *Fundamenta Mathematicae* 20, 177–190.

P. BORWEIN AND W. MOSER (1990), A survey of Sylvester's problem and its generalizations, *Aequationes Mathematicae* 40, 111–135.

J. BOURGAIN AND J. LINDENSTRAUSS (1991), On covering a set in $\mathbb{R}^d$ by balls of the same diameter, in: *Geometric Aspects of Functional Analysis, Lecture Notes in Mathematics*, vol. 1469 (J. Lindenstrauss and V. Milman, eds.), Springer-Verlag, New York, 138–144.

P. BRASS (1992a), The maximum number of second smallest distances in finite planar sets, *Discrete and Computational Geometry* 7, 371–379.

P. BRASS (1992b), *Beweis einer Vermutung von Erdős und Pach aus der kombinatorischen Geometrie,* Ph.D. Thesis, Department of Discrete Mathematics, Technical University Braunschweig, Braunschweig.

P. BRASS (1992c), On the maximum densities of the jth smallest distances, in: *Sets, Graphs and Numbers, Colloquia Mathematica Societatis János Bolyai*, vol. 60, North-Holland, Amsterdam, 119–126.

G. BRIGHTWELL AND E. SCHEINERMAN (1993), Representation of planar graphs, *SIAM J. Discrete Mathematics* 6, 214–229.

W. BROWN (1966), On graphs that do not contain a Thomsen graph, *Canadian Mathematical Bull.* 9, 281–285.

W. BROWN (1983), On an open problem of Paul Turán concerning 3-graphs, in: *Studies in Pure Mathematics: To the Memory of Paul Turán* (P. Erdős, ed.), Birkhäuser Verlag, Basel, 91–93.

W. BROWN AND F. HARARY (1970), Extremal digraphs, in: *Combinatorial Theory and Its Applications, Colloquia Mathematica Societatis János Bolyai*, vol. 10 (P. Erdős et al., eds.), North-Holland, Amsterdam, 135–198.

W. BROWN AND M. SIMONOVITS (1984), Digraph extremal problems, hypergraph extremal problems, and the densities of graph structures, *Discrete Mathematics* 48, 147–162.

W. BROWN, P. ERDŐS, AND M. SIMONOVITS (1973), Extremal problems for directed graphs, *J. Combinatorial Theory, Series B* 15, 77–93.

G. BUFFON (1770), Essai d'arithmétique morale, *Supplément á l'Histoire Naturelle*, vol. 4.

D. DE CAEN (1983a), Extension of a theorem of Moon and Moser on complete hypergraphs, *Ars Combinatoria* 16, 5–10.

D. DE CAEN (1983b), A note on the probabilistic approach to Turán's problem, *J. Combinatorial Theory, Series B* 34, 340–349.

D. DE CAEN (1983c), Linear constraints related to Turán's problem, *Proc. 6th Southeastern Conference on Combinatorics, Graph Theory and Computing, Congressus Numerantium* 39, 291–303.

R. CANHAM (1969), A theorem on arrangements of lines in the plane, *Israel J. Mathematics* 7, 393–397.

V. CAPOYLEAS AND J. PACH (1992), A Turán-type theorem on chords of a convex polygon, *J. Combinatorial Theory, Series B* 56, 9–15.

J. CASSELS (1953), A short proof of the Minkowski–Hlawka theorem, *Proc. Cambridge Philosophical Society* 49, 165–166.

J. CASSELS (1959), *An Introduction to the Geometry of Numbers*, Springer-Verlag, New York.

A. CAUCHY (1850), Mémoire sur la rectification des courbes et la quadrature des surfaces courbes, *Mem. Academie des Sciences, Paris* 22, 3–15.

B. CHAZELLE (1993), Geometric discrepancy revisited, *Proc. 34th Annual IEEE Symposium on Foundations of Computer Science*, 392–399.

B. CHAZELLE AND J. FRIEDMAN (1990), A deterministic view of random sampling and its use in geometry, *Combinatorica* 10, 229–249.

B. CHAZELLE AND E. WELZL (1989), Quasi optimal range searching in spaces of finite VC-dimension, *Discrete and Computational Geometry* 4, 467–489.

B. CHAZELLE, L. GUIBAS, AND D. T. LEE (1985), The power of geometric duality, *BIT* 25, 76–90.

B. CHAZELLE, J. MATOUŠEK, AND M. SHARIR (1995), An elementary approach to lower bounds in geometric discrepancy, *Discrete and Computational Geometry* 13, 363–381.

B. CHAZELLE, H. EDELSBRUNNER, M. GRIGNI, L. GUIBAS, M. SHARIR, AND E. WELZL (1993), Improved bounds on weak ϵ-nets for convex sets, *Proc. 25th Annual ACM Symposium on Theory of Computing*, 495–504. Also in *Discrete and Computational Geometry* 13 (1995), 1–15.

H. CHERNOFF (1952), A measure of asymptotic efficiency for tests of a hypothesis based on the sum of observations, *Ann. Mathematical Statistics* 23, 493–507.

F. CHUNG (1984), The number of different distances determined by n points in the plane, *J. Combinatorial Theory, Series A* 36, 342–354.

F. CHUNG (1989), Sphere-and-point incidence relations in higher dimensions with applications to unit distances and furthest neighbor pairs, *Discrete and Computational Geometry* 4, 183–190.

F. CHUNG (1990), Separator theorems and their applications, in: *Paths, Flows, and VLSI-Layout* (B. Korte et al., eds.), Springer-Verlag, New York, 17–34.

F. CHUNG, R. GRAHAM, AND J. PACH (1987), see P. Erdős, Some combinatorial and metric problems in geometry, in: *Intuitive Geometry, Colloquia Mathematica Societatis János Bolyai*, vol. 48 (K. Böröczky and G. Fejes Tóth, eds.), North-Holland, Amsterdam, 167–177.

F. CHUNG, E. SZEMERÉDI, AND W. TROTTER, JR. (1992), The number of distinct distances determined by a set of points in the Euclidean plane, *Discrete and Computational Geometry* 7, 1–11.

V. CHVÁTAL (1979), A greedy heuristic for the set covering problem, *Mathematical Operations Research* 4, 233–235.

V. CHVÁTAL (1983), *Linear Programming*, Freeman, New York.

V. CHVÁTAL AND E. SZEMERÉDI (1981), On the Erdős–Stone theorem, *J. London Mathematical Society* (2) 23, 207–214.

K. CLARKSON (1987), New applications of random sampling in computational geometry, *Discrete and Computational Geometry* 2, 195–222.

K. CLARKSON AND P. SHOR (1989), Applications of random sampling in computational geometry, II, *Discrete and Computational Geometry* 4, 387–421.

K. CLARKSON, H. EDELSBRUNNER, L. GUIBAS, M. SHARIR, AND E. WELZL (1990), Combinatorial complexity bounds for arrangements of curves and surfaces, *Discrete and Computational Geometry* 5, 99–160.

Y. COLIN DE VERDIÈRE (1989), Empilements de cercles: Convergence d'une méthode de point fixe, *Forum Mathematicum* 1, 395–402.

J. CONWAY AND N. SLOANE (1988), *Sphere Packings, Lattices and Groups*, Springer-Verlag, New York.

J. CONWAY, H. CROFT, P. ERDŐS, AND M. GUY (1979), On the distribution of values of angles determined by coplanar points, *J. London Mathematical Society* (2) 19, 137–143.

T. COREMAN, C. LEISERSON, AND R. RIVEST (1990), *Introduction to Algorithms*, McGraw-Hill, New York.

J. VAN DER CORPUT (1936), Verallgemeinerung einer Mordellschen Beweismethode in der Geometrie der Zahlen II, *Acta Arithmetica* 1, 62–66.

R. COURANT (1965), The least dense lattice packing of two-dimensional convex bodies, *Communications on Pure and Applied Mathematics* 18, 339–343.

H. COXETER (1948), A problem of collinear points, *American Mathematical Monthly* 55, 26–28.

H. COXETER (1962), *Regular Polytopes*, 3rd ed., Dover, New York.

H. COXETER (1969), *Introduction to Geometry*, 2nd ed., Wiley, New York.

H. COXETER, L. FEW, AND C. ROGERS (1959), Covering space with equal spheres, *Mathematika* 6, 147–157.

H. CROFT (1967), Some geometrical thoughts II, *Mathematical Gazette* 51, 125–129.

M. CROFTON (1868), On the theory of local probability, *Philosophical Trans. Royal Society of London* 158, 181–199.

M. CROFTON (1885), Probability, in: *Encyclopedia Britannica*, 9th ed., vol. 19, 768–788.

J. CSIMA AND E. SAWYER (1993), There exist $6n/13$ ordinary points, *Discrete and Computational Geometry* 9, 187–202.

J. CSIMA AND E. SAWYER (1995), The $6n/13$ theorem revisited, *Proc. Seventh International Conference in Graph Theory, Combinatorics, Algorithms, and Applications, Kalamazoo, MI* (Y. Alavi and A. Schenk, eds.), Wiley, New York, 235–249.

I. CSISZÁR AND J. KÖRNER (1981), *Information Theory: Coding Theorems for Discrete Memoryless Systems*, Academic Press, New York, and Akadémiai Kiadó, Budapest.

G. CSIZMADIA (1995), Furthest neighbors in space, *Discrete Mathematics*, to appear.

L. DANZER (1987), see P. Erdős, Some combinatorial and metric problems in geometry, in: *Intuitive Geometry, Colloquia Mathematica Societatis János Bolyai*, vol. 48 (K. Böröczky and G. Fejes Tóth, eds.), North-Holland, Amsterdam, 167–177.

L. DANZER AND B. GRÜNBAUM (1962), Über zwei Probleme bezüglich konvexer Körper von P. Erdős und von V. Klee, *Mathematische Zeitschrift* 79, 95–99.

L. DANZER, B. GRÜNBAUM, AND V. KLEE (1963), Helly's theorem and its relatives, in: *Convexity, Proc. Symposia Pure Applied Mathematics*, vol. 7 (V. Klee, ed.), American Mathematical Society, Providence, RI, 101–181.

H. DAVENPORT (1952), The covering of space by spheres, *Rendiconti del Circolo Matematico di Palermo* (2) 1, 92–107.

H. DAVENPORT AND C. ROGERS (1947), Hlawka's theorem in the geometry of numbers, *Duke J. Mathematics* 14, 367–375.

B. DELAUNAY (1934), Sur la sphère vide, *Bull. Académie des Sciences USSR (VII), Classe des Sciences Mathématiques et Naturelles*, 793–800.

T. DEY AND H. EDELSBRUNNER (1993), Counting triangle crossings and halving planes, *Proc. 9th Annual Symposium on Computational Geometry*, 270–273. Also: *Discrete and Computational Geometry* 12 (1994), 281–289.

M. DEZA, V. GRISHUKHIN, AND M. LAURENT (1993), Hypermetrics in geometry of numbers, Tech. Rept. 93–4, Rapport de Recherche du Laboratoire d'Informatique de l'Ecole Normale Supérieure.

R. DILWORTH (1950), A decomposition theorem for partially ordered sets, *Annals of Mathematics* 51, 161–166.

G. DING, P. SEYMOUR, AND P. WINKLER (1994), Bounding the vertex cover number of a hypergraph, *Combinatorica* 14, 23–34.

G. DIRAC (1951), Collinearity properties of sets of points, *Quarterly J. Math.* (2) 2, 221–227.

G. DIRAC (1952), Some theorems on abstract graphs, *Proc. London Mathematical Society* (3) 2, 69–81.

G. DIRAC (1963), Extensions of Turán's theorem on graphs, *Acta Mathematica Academiae Scientiarum Hungarica* 14, 417–422.

H. DJIDJEV (1982), On the problem of partitioning planar graphs, *SIAM J. Discrete Mathematics* 3, 229–240.

C. H. DOWKER (1944), On minimum circumscribed polygons, *Bull. American Mathematical Society* 50, 120–122.

P. DOYLE, Z.-X. HE, AND B. RODIN (1994), The asymptotic value of the circle–packing rigidity constants s_n, *Discrete and Computational Geometry* 12, 105–116.

P. DOYLE AND J. SNELL (1984), *Random Walks and Electric Networks*, The Carus Mathematical Monographs, vol. 22, Mathematical Association of America, Washington DC.

M. DYER AND A. FRIEZE (1988), On the complexity of computing the volume, *SIAM J. Computing* 17, 967–974.

M. DYER, A. FRIEZE, AND R. KANNAN (1991), A random polynomial time algorithm for estimating volumes of convex bodies, *J. Association for Computing Machinery* 38, 1–17.

J. ECKHOFF (1993), Helly, Radon and Carathéodory type theorems, in: *Handbook of Convex Geometry*, vol. A (P. Gruber and J. Wills, eds.), North-Holland, Amsterdam, 389–448.

H. EDELSBRUNNER (1987), *Algorithms in Combinatorial Geometry*, Springer-Verlag, New York.

H. EDELSBRUNNER AND P. HAJNAL (1991), A lower bound on the number of unit distances between the points of a convex polygon, *J. Combinatorial Theory, Series A* 56, 312–316.

H. EDELSBRUNNER AND S. SKIENA (1989), On the number of furthest neighbor pairs in a point set, *American Mathematical Monthly* 96, 614–618.

H. EDELSBRUNNER AND E. WELZL (1986a), On the maximal number of edges of many faces in arrangements, *J. Combinatorial Theory, Series A*, 41, 159–166.

H. EDELSBRUNNER AND E. WELZL (1986b), Half-planar range searching in linear space and $O(n^{0.695})$ query time, *Information Processing Letters* 23, 289–293.

H. EDELSBRUNNER AND E. WELZL (1986c), Constructing belts in two-dimensional arrangements with applications, *SIAM J. Computing* 15, 271–284.

H. EDELSBRUNNER, J. O'ROURKE, AND R. SEIDEL (1986), Constructing arrangements of lines and hyperplanes, *SIAM J. Computing* 15, 317–340.

H. EDELSBRUNNER, D. ROBISON, AND X. SHEN (1990), Covering convex sets with non-overlapping polygons, *Discrete Mathematics* 81, 153–164.

H. EDELSBRUNNER, L. GUIBAS, J. HERSHBERGER, R. SEIDEL, M. SHARIR, J. SNOEYINK, AND E. WELZL (1989), Implicitly representing arrangements of lines and of segments, *Discrete and Computational Geometry* 4, 433–466.

C. EDWARDS (1975), Triangles in simple graphs and some related problems, *Proc. 5th British Columbia Combinatorial Conference*, 159–164.

H. EGGLESTON (1957a), Approximation to plane convex curves I. Dowker-type theorems, *Proc. London Mathematical Society* (3) 7, 351–377.

H. EGGLESTON (1957b), *Problems in Euclidean Space: Application of Convexity*, Pergamon Press, London.

G. ELEKES (1986), A geometric inequality and the complexity of computing the volume, *Discrete and Computational Geometry* 1, 289–292.

G. ELEKES (1992), Irregular pairs are necessary in Szemerédi's regularity lemma, manuscript.

G. ELEKES AND P. ERDŐS (1994), Similar configurations and pseudo grids, in: *Intuitive Geometry, Colloquia Mathematica Societatis János Bolyai* (K. Böröczky and G. Fejes Tóth, eds.), North-Holland, Amsterdam, 85–104.

N. ELKIES, A. ODLYZKO, AND N. RUSH (1991), On the packing densities of superballs and other bodies, *Inventiones Mathematicae* 105, 613–639.

V. ENNOLA (1961), On the lattice constant of a symmetric convex domain, *J. London Mathematical Society* 36, 135–138.

P. ERDŐS (1938), On sequences of integers none of which divides the product of two others, and related problems, *Mitteilungen des Forschungsinstituts für Mathematik und Mechanik, Tomsk* 2, 74–82.

P. ERDŐS (1943), Problem for solution, no. 4065, *American Mathematical Monthly* 50, 65.

P. ERDŐS (1945), Integral distances, *Bull. American Mathematical Society* 51, 996.

P. ERDŐS (1946), On sets of distances of *n* points, *American Mathematical Monthly* 53, 248–250.

P. ERDŐS (1947), Some remarks on the theory of graphs, *Bull. American Mathematical Society* 53, 292–294.

P. ERDŐS (1950), Some remarks on set theory, *Proc. American Mathematical Society* 1, 127–141.

P. ERDŐS (1955a), Some theorems on graphs, *Riveon Lematematika* 10, 13–16.

P. ERDŐS (1955b), Aufgabe, *Elemente der Mathematik* 10, 114.

P. ERDŐS (1960), On sets of distances of *n* points in Euclidean space, *Magyar Tudományos Akadémia Matematikai Kutató Intézet Közleményei* 5, 165–169.

P. ERDŐS (1962a), On a theorem of Rademacher–Turán, *Illinois J. Mathematics* 6, 122–127.

P. ERDŐS (1962b), Remarks on a paper of Pósa, *Magyar Tudományos Akadémia Matematikai Kutató Intézet Közleményei* 7, 227–228.

P. ERDŐS (1962c), On the number of complete subgraphs contained in certain graphs, *Magyar Tudományos Akadémia Matematikai Kutató Intézet Közleményei* 7, 459–464.

P. ERDŐS (1963a), On the structure of linear graphs, *Israel J. Mathematics* 1, 156–160.

P. ERDŐS (1963b), Extremal problems in graph theory, in: *Theory of Graphs and Its Applications: Proc. Symposium on the Theory of Graphs and Its Applications*, Smolenice, Czechoslovakia, 29–36.

P. ERDŐS (1964a), On the number of triangles contained in certain graphs, *Canadian Mathematical Bull.* 7, 53–56.

P. ERDŐS (1964b), On a combinatorial problem II, *Acta Mathematica Academiae Scientiarum Hungarica* 15, 445–447.

P. ERDŐS (1964c), On some problems of graphs and generalized graphs, *Israel J. Mathematics* 2, 183–190.

P. ERDŐS (1965), On some extremal problems in graph theory, *Israel J. Mathematics* 3, 113–116.

P. ERDŐS (1966), On cliques in graphs, *Israel J. Mathematics* 4, 233–234.

P. ERDŐS (1967a), On some applications of graph theory to geometry, *Canadian J. Mathematics* 19, 968–971.

P. ERDŐS (1967b), On bipartite subgraphs of a graph (in Hungarian), *Matematikai Lapok* 18, 283–288.

P. ERDŐS (1970), On the graph theorem of Turán (in Hungarian), *Matematikai Lapok* 21, 249–251.

P. ERDŐS (1974), Some remarks on the theory of graphs, *Bull. American Mathematical Society* 53, 292–294.

P. ERDŐS (1975), On some problems in elementary and combinatorial geometry, *Ann. Mathematics, Series 4*, 130, 99–108.

P. ERDŐS (1978), Some applications of graph theory and combinatorial methods to number theory and geometry, in: *Algebraic Methods in Graph Theory, Colloquia Mathematica Societatis János Bolyai*, vol. 13 (Lovász and V. T. Sós, eds.), North-Holland, Amsterdam, 137–148.

P. ERDŐS (1980), Some combinatorial problems in geometry, in: *Geometry and Differential Geometry*, Lecture Notes in Mathematics, vol. 792 (R. Artzy and I. Vaisman, eds.), Springer-Verlag, New York, 46–53.

P. ERDŐS (1981), On the combinatorial problems which I would most like to see solved, *Combinatorica* 1, 25–42.

P. ERDŐS (1982), Personal reminiscences and remarks of the mathematical work of Tibor Gallai, *Combinatorica* 2, 207–212.

P. ERDŐS (1983), Extremal problems in number theory, combinatorics and geometry, in: *Proc. International Congress of Mathematics*, 1–3.

P. ERDŐS (1984), On some problems in graph theory, combinatorial analysis and combinatorial number theory, in: *Graph Theory and Combinatorics* (B. Bollobás, ed.), Academic Press, New York, 1–17.

P. ERDŐS (1986), On some metric and combinatorial geometry problems, *Discrete Mathematics* 60, 147–153.

P. ERDŐS (1987), Some combinatorial and metric problems in geometry, in: *Intuitive Geometry, Colloquia Mathematica Societatis János Bolyai*, vol. 48 (K. Böröczky and G. Fejes Tóth, eds.), North-Holland, Amsterdam, 167–177.

P. ERDŐS AND T. GALLAI (1959), On maximal paths and circuits of graphs, *Acta Mathematica Academiae Scientiarum Hungarica* 10, 337–356.

P. ERDŐS AND R. GUY (1970), Distinct distances between lattice points, *Elemente der Mathematik* 25, 121–123.

P. ERDŐS AND D. KLEITMAN (1968), On coloring graphs to maximize the proportion of multicolored k-edges, *J. Combinatorial Theory* 5, 164–169.

P. ERDŐS AND L. LOVÁSZ (1975), Problems and results on 3-chromatic hypergraphs and some related problems, in: *Infinite and Finite Sets, Colloquia Mathematica Societatis János Bolyai*, vol. 10 (A. Hajnal et al., eds.), North-Holland, Amsterdam, 609–617.

P. ERDŐS AND L. MOSER (1959), Problem 11, *Canadian Mathematical Bull.* 2, 43.

P. ERDŐS AND L. MOSER (1970), An extremal problem in graph theory, *Australian J. Mathematics*, 11, 42–47.

P. ERDŐS AND J. PACH (1990), Variations on the theme of repeated distances, *Combinatorica* 10, 261–269.

P. ERDŐS AND L. PÓSA (1962), On the maximal number of disjoint circuits of a graph, *Publ. Mathematicae Debrecen* 9, 3–12.

P. ERDŐS AND G. PURDY (1971), Some extremal problems in geometry, *J. Combinatorial Theory, Series A* 10, 246–252.

P. ERDŐS AND G. PURDY (1975), Some extremal problems in geometry III, *Proc. 6th Southeastern Conference on Combinatorics, Graph Theory and Computing, Congressus Numerantium* 14, 291–308.

P. ERDŐS AND G. PURDY (1976), Some extremal problems in geometry IV, *Proc. 7th Southeastern*

Conference on Combinatorics, Graph Theory and Computing, Congressus Numerantium 17, 307–322.

P. ERDŐS AND G. PURDY (1989), Combinatorics of geometric configurations, to appear in: *Handbook of Combinatorics*, (M. Grötschel et al., eds.), Elsevier Science, Amsterdam.

P. ERDŐS AND A. RÉNYI (1962), On a problem in the theory of graphs, *Magyar Tudományos Akadémia Matematikai Kutató Intézet Közleményei* 7A, 623–641.

P. ERDŐS AND C. ROGERS (1953), The covering of *n*-dimensional space by spheres, *J. London Mathematical Society* 28, 287–293.

P. ERDŐS AND C. ROGERS (1962), Covering space with convex bodies, *Acta Arithmetica* 7, 281–285.

P. ERDŐS AND J. SELFRIDGE (1973), On a combinatorial game, *J. Combinatorial Theory, Series A* 14, 298–301.

P. ERDŐS AND M. SIMONOVITS (1966), A limit theorem in graph theory, *Studia Scientiarum Mathematicarum Hungarica* 1, 51–57.

P. ERDŐS AND M. SIMONOVITS (1983), Supersaturated graphs, *Combinatorica* 3, 181–192

P. ERDŐS AND J. SPENCER (1974), *Probabilistic Methods in Combinatorics*, Academic Press, London, and Akadémiai Kiadó, Budapest.

P. ERDŐS AND A. STONE (1946), On the structure of linear graphs, *Bull. American Mathematical Society* 52, 1087–1091.

P. ERDŐS AND G. SZEKERES (1935), A combinatorial problem in geometry, *Compositio Mathematica* 2, 463–470.

P. ERDŐS AND G. SZEKERES (1961), On some extremum problems in elementary geometry, *Ann. Universitatis Scientiarum Budapestinensis, Eötvös, Sectio Mathematica* 3–4, 53–62.

P. ERDŐS, P. GRUBER, AND J. HAMMER (1989), *Lattice Points*, Longman Scientific Technological, Harlow, Essex, England.

P. ERDŐS, D. HICKERSON, AND J. PACH (1989), A problem of Leo Moser about repeated distances on the sphere, *American Mathematical Monthly* 96, 569–575.

P. ERDŐS, L. LOVÁSZ, AND K. VESZTERGOMBI (1989), On the graph of large distances, *Discrete and Computational Geometry* 4, 541–549.

P. ERDŐS, E. MAKAI, AND J. PACH (1993), Nearly equal distances in the plane, *Combinatorics, Probability and Computing* 2, 401–408.

P. ERDŐS, A. RÉNYI, AND V. T. SÓS (1966), On a problem of graph theory, *Studia Scientiarum Mathematicarum Hungarica* 1, 215–235.

P. ERDŐS, Z. FÜREDI, J. PACH, AND I. RUZSA (1993), The grid revisited, *Discrete Mathematics* 111, 189–196.

P. ERDŐS, R. GRAHAM, I. RUZSA, AND H. TAYLOR (1992), Bounds for arrays of dots with distinct slopes or lengths, *Combinatorica* 12, 39–44.

P. ERDŐS, A. HAJNAL, A. MÁTÉ, AND R. RADO (1984), *Combinatorial Set Theory: Partition Relations for Cardinals*, North-Holland, Amsterdam.

P. ERDŐS L. LOVÁSZ, A. SIMMONS, AND E. STRAUS (1973), Dissection graphs of planar point sets, in: *A Survey of Combinatorial Theory* (G. Srivastava, ed.), North-Holland, Amsterdam, 139–149.

P. ERDŐS, E. MAKAI, J. PACH, AND J. SPENCER (1991), Gaps in difference sets, and the graph of nearly equal distances, in: *Applied Geometry and Discrete Mathematics: The Victor Klee Festschrift* (P. Gritzmann and B. Sturmfels, eds.), DIMACS Series, vol. 4, American Mathematical Society, Providence, RI, 265–273.

P. ERDŐS, A. MEIR, V. SÓS, AND P. TURÁN (1971), On some applications of graph theory II (presented to R. Rado), in: *Studies in Pure Applied Mathematics*, (L. Mirksy, ed.), Academic Press, London, 89–93.

P. Erdős, A. Meir, V. Sós, and P. Turán (1972a), On some applications of graph theory I, *Discrete Mathematics* 2, 207–228. [Corrigendum, *Discrete Mathematics* 4 (1973), 90.]

P. Erdős, A. Meir, V. Sós, and P. Turán (1972b), On some applications of graph theory III, *Canadian Mathematical Bull.* 15, 27–32.

I. Fáry (1948), On straight line representation of planar graphs, *Acta Scientiarum Mathematicarum (Szeged)* 11, 229–233.

I. Fáry (1950), Sur la densité des réseaux de domaines convexes, *Bull. Société Mathématique de France* 78, 152–161.

R. Faudree and M. Simonovits (1983), On a set of degenerate extremal graph problems, *Combinatorica* 3, 83–94.

G. Fejes Tóth (1972), Covering the plane by convex discs, *Acta Mathematica Academiae Scientiarum Hungarica* 23, 263–270.

G. Fejes Tóth (1976), Multiple packing and covering of the plane with circles, *Acta Mathematica Academiae Scientiarum Hungarica* 27, 135–140.

G. Fejes Tóth (1977a), On a Dowker-type theorem of Eggleston, *Acta Mathematica Academiae Scientiarum Hungarica* 29, 131–148.

G. Fejes Tóth (1977b), On the intersection of a convex disc and a polygon, *Acta Mathematica Academiae Scientiarum Hungarica* 29, 149–153.

G. Fejes Tóth (1980), On the section of a packing of equal balls, *Studia Scientiarum Mathematicarum Hungarica* 15, 487–489.

G. Fejes Tóth (1983), New results in the theory of packing and covering, in: *Convexity and Its Applications* (P. Gruber and J. Wills, eds.), Birkhäuser Verlag, Basel, 318–359.

G. Fejes Tóth and L. Fejes Tóth (1973a), Remark on a paper of C.H. Dowker, *Periodica Mathematica Hungarica* 3, 271–274.

G. Fejes Tóth and L. Fejes Tóth (1973b), On totally separable domains, *Acta Mathematica Academiae Scientiarum Hungarica* 24, 229–232.

G. Fejes Tóth and W. Kuperberg (1993a), Recent results in the theory of packing and covering, in: *New Trends in Discrete and Computational Geometry* (J. Pach, ed.), Springer-Verlag, New York, 251–279.

G. Fejes Tóth and W. Kuperberg (1993b), Blichfeldt's density bound revisited, *Mathematische Annalen.* 295, 721–727.

G. Fejes Tóth and W. Kuperberg (1993c), Thin non-lattice covering with an affine image of a strictly convex body, to appear in *Mathematika.*

G. Fejes Tóth and W. Kuperberg (1993d), Packing and covering with convex sets, in: *Handbook of Convex Geometry*, vol. B (P. Gruber and J. Wills, eds.), North-Holland, Amsterdam, 799–860.

L. Fejes Tóth (1950), Some packing and covering theorems, *Acta Scientiarum Mathematicarum (Szeged)* 12/A, 62–67.

L. Fejes Tóth (1959a), Annäherung von Eibereichen durch Polygone, *Mathematisch-Physikalische Semesterberichte* 6, 253–261.

L. Fejes Tóth (1959b), Kugelunterdeckungen und Kugelüberdeckungen in Räumen konstanter Krümmung, *Archiv der Mathematik* 10, 307–313.

L. Fejes Tóth (1964), *Regular Figures*, Pergamon Press, New York.

L. Fejes Tóth (1967), On the arrangement of houses in a housing estate, *Studia Scientiarum Mathematicarum Hungarica* 2, 37–42.

L. Fejes Tóth (1969), Über die Nachbarschaft eines Kreises in einer Kreispackung, *Studia Scientiarum Mathematicarum Hungarica* 4, 93–97.

L. Fejes Tóth (1953, 1972), *Lagerungen in der Ebene, auf der Kugel und im Raum*, 1st and 2nd eds., Springer-Verlag, New York.

L. FEJES TÓTH (1975), On Hadwiger numbers and Newton numbers of a convex body, *Studia Scientiarum Mathematicarum Hungarica* 10, 111–115.

L. FEJES TÓTH (1976), Close packing and loose covering with balls, *Publ. Mathematicae Debrecen* 23, 323–326.

L. FEJES TÓTH (1977), Research problem, *Periodica Mathematica Hungarica* 8, 103–104.

L. FEJES TÓTH (1978–1982), Egy kört eltakaró körökről, *Matematikai Lapok* 30, 317–320.

L. FEJES TÓTH (1983), On the densest packing of convex discs, *Mathematika* 30, 1–3.

L. FEJES TÓTH (1984a), Density bounds for packing and covering with convex discs, *Expositiones Mathematicae* 2, 131–153.

L. FEJES TÓTH (1984b), Compact packing of circles, *Studia Scientiarum Mathematicarum Hungarica* 19, 103–107.

L. FEJES TÓTH (1985), Densest packing of translates of a domain, *Acta Mathematica Hungarica* 45, 437–440.

L. FEJES TÓTH (1986), Densest packing of translates of the union of two circles, *Discrete and Computational Geometry* 1, 307–314.

L. FEJES TÓTH AND A. FLORIAN (1982), Packing and covering with convex discs, *Mathematika* 29, 181–193.

L. FEW (1964), Multiple packings of spheres, *J. London Mathematical Society* 39, 51–54.

P. FISHBURN AND J. REEDS (1992), Unit distances between vertices of a convex polygon, *Computational Geometry: Theory and Applications* 2, 81–91.

M. FORMANN, T. HAGERUP, J. HARALAMBIDES, M. KAUFMANN, T. LEIGHTON, A. SIMVONIS, E. WELZL, AND G. WOEGINGER (1993), Drawing planar graphs in the plane with high resolution, *SIAM J. Computing* 22, 1035–1052.

P. FRANKL AND J. PACH (1983), On the number of sets in a null-t-design, *European J. Combinatorics* 4, 21–23.

P. FRANKL AND J. PACH (1984), On disjointly representable sets, *Combinatorica* 4, 39–45.

P. FRANKL AND R. WILSON (1981), Intersection theorems with geometric consequences, *Combinatorica* 1, 259–286.

P. FRANKL, Z. FÜREDI, AND J. PACH (1987), Bounding one-way differences, *Graphs and Combinatorics* 3, 341–347.

H. DE FRAYSSEIX, P. DE MENDEZ, AND P. ROSENSTIEHL (1994), On triangle contact graphs, *Combinatorics, Probability and Computing* 3, 233–246.

H. DE FRAYSSEIX, J. PACH, AND R. POLLACK (1990), Drawing a planar graph on a grid, *Combinatorica* 10, 41–51.

G. FREDERICKSON (1987), Fast algorithms for shortest paths in planar graphs, with applications, *SIAM J. Computing* 16, 1004–1022.

G. FREIMAN (1966), *Foundations of a Structural Theory of Set Addition* (in Russian), Kazanskiĭ Gosudarstvennïĭ Pedagogicheskiĭ Institut, Kazan.

G. FREIMAN (1973), *Foundations of a Structural Theory of Set Addition*, Translations of Mathematical Monographs 37, American Mathematical Society, Providence, RI.

G. FREIMAN (1987), What is the structure of K if $K + K$ is small, in: *Number Theory, Lecture Notes in Mathematics*, vol. 1240 (S. Benkoski, ed.), Springer-Verlag, New York, 109–134.

Z. FÜREDI (1990), The maximum number of unit distances in a convex n-gon, *J. Combinatorial Theory, Series A* 55, 316–320.

Z. FÜREDI (1991a), Turán type problems, in: *Survey in Combinatorics, London Mathematical Society Lecture Notes Series*, vol. 166 (A. Keedwell, ed.), Cambridge University Press, Cambridge, 253–300.

Z. FÜREDI (1991b), On a Turán type problem of Erdős, *Combinatorica* 11, 75–79.

Z. FÜREDI AND P. HAJNAL (1992), Davenport–Schinzel theory of matrices, *Discrete Mathematics* 103, 233–251.

T. GALLAI (1944), Solution of problem 4065, *American Mathematical Monthly* 51, 169–171.

M. GAREY AND D. JOHNSON (1979), *Computers and Intractability: A Guide to the Theory of NP-Completeness*, Freeman, New York.

C. GAUSS (1836), Untersuchungen über die Eigenschaften der positiven ternären quadratischen Formen von Ludwig August Seeber, *Göttingische gelehrte Anzeigen* 1831, Juli 9, see *Werke II*, Königliche Gesellschaft der Wissenschaften, Göttingen, 188–196.

H. GAZIT AND G. MILLER (1990), Planar separators and the Euclidean norm, in: *Algorithms, Lecture Notes in Computer Science*, vol. 450 (T. Asano et al., eds.), Springer-Verlag, New York, 338–347.

J. GILBERT (1980), *Graph Separator Theorems and Sparse Gaussian Elimination*, Ph.D. Thesis, Stanford University, Stanford, CA.

J. GILBERT AND R. TARJAN (1987), The analysis of a nested dissection algorithm, *Numerische Mathematik* 50, 377–404.

J. GILBERT, J. HUTCHINSON, AND R. TARJAN (1984), A separator theorem for graphs of bounded genus, *J. Algorithms* 5, 391–407.

W. GODDARD, M. KATCHALSKI, AND D. KLEITMAN (1993), Forcing disjoint segments in the plane, *European J. Combinatorics*, to appear.

R. GRAHAM (1981), *Rudiments of Ramsey Theory*, CBMS Regional Conference Series in *Mathematics*, No. 45, American Mathematical Society, Providence, RI.

R. GRAHAM, B. ROTHSCHILD, AND J. SPENCER (1990), *Ramsey Theory*, 2nd ed., Wiley, New York.

P. GRITZMANN (1985), Lattice covering of space with symmetric convex bodies, *Mathematika* 32, 311–315.

P. GRITZMANN, B. MOHAR, J. PACH, AND R. POLLACK (1991), Embedding a planar triangulation with vertices at specified points (solution to problem E3341), *American Mathematical Monthly* 98, 165–166.

H. GROEMER (1963), Existenzsätze für Lagerungen im Euklidischen Raum, *Mathematische Zeitschrift* 81, 260–278.

H. GROEMER (1968), Existenzsätze für Lagerungen in metrischen Räumen, *Monatshefte für Mathematik* 72, 325–334.

H. GROEMER (1986), Multiple packings and coverings, *Studia Scientiarum Mathematicarum Hungarica* 21, 189–200.

M. GRÖTSCHEL, L. LOVÁSZ, AND A. SCHRIJVER (1987), *Geometric Algorithms and Combinatorial Optimization*, Springer-Verlag, New York.

P. GRUBER (1979), Geometry of numbers, in: *Contributions to Geometry* (J. Tölke and J. Wills, eds.), Birkhäuser Verlag, Basel, 186–225.

P. GRUBER AND C. LEKKERKERKER (1987), *Geometry of Numbers*, 2nd ed., North-Holland, Amsterdam.

B. GRÜNBAUM (1956), A proof of Vázsonyi's conjecture, *Bull. Research Council Israel, Section A* 6, 77–78.

B. GRÜNBAUM (1961), On a conjecture of Hadwiger, *Pacific J. Mathematics* 11, 215–219.

B. GRÜNBAUM (1963), Borsuk's problem and related questions, in: *Convexity, Proc. Symposia in Pure Applied Mathematics*, vol. 7 (V. Klee, ed.), American Mathematical Society, Providence, RI, 271–284.

B. GRÜNBAUM (1967), *Convex Polytopes*, Wiley, New York.

B. GRÜNBAUM (1971), *Arrangements and Spreads*, CBMS Regional Conference Series in *Mathematics*, No. 10, American Mathematical Society, Providence, RI.

A. GYÁRFÁS AND J. LEHEL (1969), A Helly-type problem in trees, in: *Combinatorial Theory and Its Applications, Colloquia Mathematica Societatis János Bolyai*, vol. 4 (P. Erdős, ed.), North-Holland, Amsterdam, 571–584.

A. GYÁRFÁS AND J. LEHEL (1983), Hypergraph families with bounded edge cover or transversal numbers, *Combinatorica* 3, 351–358.

A. GYÁRFÁS AND J. LEHEL (1985), Covering and coloring problems for relatives of intervals, *Discrete Mathematics* 55, 167–180.

E. GYŐRI, J. PACH, AND M. SIMONOVITS (1991), On the maximal number of certain subgraphs in K_r-free graphs, *Graphs and Combinatorics* 7, 31–37.

H. HADWIGER (1945), Überdeckung einer Menge durch Mengen kleineren Durchmessers, *Commentarii Mathematica Helvetici* 18 (1945/46), 73–75.

H. HADWIGER (1946), Mitteilung betreffend meine Note: Überdeckung einer Menge durch Mengen kleineren Durchmessers, *Commentarii Mathematica Helvetici* 19 (1946/47), 71–73.

H. HADWIGER (1969), Überdeckung des Raumes durch translationsgleiche Punktmengen und Nachbarzahlen, *Monatshefte für Mathematik* 73, 213–217.

H. HADWIGER, H. DEBRUNNER, AND V. KLEE (1964), *Combinatorial Geometry in the Plane*, Holt, Rinehart and Winston, New York.

P. HALL (1935), On representatives of subsets, *J. London Mathematical Society* 10, 26–30.

H. HANSEN (1978), Et isoperimetrisk problem, *Nordisk Mathematisk Tidskrift*, 25-26, 181–184.

H. HARBORTH (1974), Solution to problem 664A, *Elemente der Mathematik* 29, 14–15.

G. HARDY AND E. WRIGHT (1960), *The Theory of Numbers*, 4th ed., Oxford University Press, London.

D. HAUSSLER (1991), Sphere packing numbers for subsets of the Boolean n-cube with bounded Vapnik–Chervonenkis dimension, Tech. Rept. UCSC-CRL-91-41, University of California at Santa Cruz. Also: *J. Combinatorial Theory, Series A* 69 (1995), 217–232.

D. HAUSSLER AND E. WELZL (1987), ϵ–nets and simplex range queries, *Discrete and Computational Geometry* 2, 127–151.

D. HEATH-BROWN (1987), Integer sets containing no arithmetic progressions, *J. London Mathematical Society* 35, 385–394.

E. HELLY (1923), Über Mengen konvexer Körper mit gemeinschaftlichen Punkten, *Jahresbericht Deutschen Mathematiker-Vereinigung* 32, 175–176.

A. HEPPES (1956), Beweis einer Vermutung von A. Vázsonyi, *Acta Mathematica Academiae Scientiarum Hungarica* 7, 463–466.

A. HEPPES (1963), Filling a convex domain by discs, *Magyar Tudományos Akadémia Matematikai Kutató Intézet Közleményei* 8, 363–370.

A. HEPPES (1964), Filling a domain by discs, *Periodica Mathematica Hungarica* 8, 363–371.

A. HEPPES (1990), On the packing density of translates of a domain, *Studia Scientiarum Mathematicarum Hungarica* 25, 117–120.

A. HEPPES AND P. RÉVÉSZ (1956), Zum Borsukschen Zerteilungsproblem, *Acta Mathematica Academiae Scientiarum Hungarica* 7, 159–162.

E. HLAWKA (1944), Zur Geometrie der Zahlen, *Mathematische Z.* 49, 285–312.

H. HOPF AND E. PANNWITZ (1934), Aufgabe Nr. 167, *Jahresbericht der Deutschen Mathematiker-Vereinigung* 43, 114.

R. HOPPE (1874), Bemerkung der Redaktion, *Archiv Mathematical Physik (Grunert)* 56, 307–312.

W.-Y. HSIANG (1993), On the sphere packing problem and the proof of Kepler's conjecture, *International J. Mathematics* 4, 739–831.

A. HURWITZ (1891), Über die Approximationen von zwei komplexen inhomogenen Linearformen, *Monatshefte für Mathematik* 56, 61–74.

W. IMRICH (1984), Explicit construction of graphs without small cycles, *Combinatorica* 4, 53–59.

B. JACKSON AND G. RINGEL (1984), Colorings of circles, *American Mathematical Monthly* 91, 42–49.

S. JÓZSA AND E. SZEMERÉDI (1975), The number of unit distances in the plane, in: *Infinite and Finite Sets, Colloquia Mathematica Societatis János Bolyai,* vol. 10 (A. Hajnal et al., eds.), North-Holland, Amsterdam, 939–950.

G. KABATJANSKIĬ AND V. LEVENŠTEIN (1978), Bounds for packing on a sphere and in space, *Problems in Information Transmission,* 14, 1–17.

J. KAHN AND G. KALAI (1993), A counterexample to Borsuk's conjecture, *Bull. American Mathematical Society* 29, 60–62.

J. KARAMATA (1932), Sur une inégalité relative aux fonctions, *Publications Mathématiques Univ. Belgrade* 1, 145–148.

G. KÁROLYI, J. PACH, AND G. TÓTH (1995), Ramsey-type results for geometric graphs, Unpublished manuscript.

G. KATONA (1969), Graphs, vectors and inequalities in probability theory (in Hungarian), *Matematikai Lapok* 20, 123–127.

G. KATONA (1978), Continuous versions of some extremal hypergraph problems, in: *Combinatorics, Colloquia Mathematica Societatis János Bolyai,* vol. 18 (A. Hajnal and V. T. Sós, eds.), North-Holland, Amsterdam, 653–678.

G. KATONA, T. NEMETZ, AND M. SIMONOVITS (1964), On a problem of Turán in the theory of graphs (in Hungarian), *Matematikai Lapok* 15, 228–238.

G. KATONA, R. MAYER, AND W. WOYCZYŃSKI (1993), Unpublished manuscript.

L. KELLY AND W. MOSER (1958), On the number of ordinary lines determined by n points, *Canadian J. Mathematics* 10, 210–219.

R. KERSHNER (1939), The number of circles covering a set, *American J. Mathematics* 61, 665–671.

G. KERTÉSZ (1987), Lecture given at the Conference on Discrete Geometry, Oberwolfach.

L. KHACHIYAN (1993), Complexity of polytope volume computation, in: *New Trends in Discrete and Computational Geometry* (J. Pach, ed.), Springer-Verlag, New York, 91–101.

V. KLEE AND S. WAGON (1991), *Old and New Unsolved Problems in Plane Geometry and Number Theory,* Dolciani Mathematical Expositions, vol. 11, Mathematical Association of America, Washington, DC.

P. KOEBE (1936), Kontaktprobleme der konformen Abbildung, *Berichte über die Verhandlungen der Sächsischen Akademie der Wissenschaften, Leipzig, Mathematische–Physische Klasse* 88, 141–164.

J. KOMLÓS, J. PACH, AND G. WOEGINGER (1992), Almost tight bounds for epsilon-nets, *Discrete and Computational Geometry* 7, 163–173.

D. KÖNIG (1936), *Theorie der endlichen und unendlichen Graphen,* Leipzig. (English translation: *Theory of Finite and Infinite Graphs,* Birkhäuser Verlag, Boston, 1990.)

A. KOSTOCHKA (1982), A class of constructions for Turán's $(3, 4)$-problem, *Combinatorica* 2, 187–192.

D. KOTTWITZ (1991), The densest packing of equal circles on a sphere, *Acta Crystallographica* A 47, 158–165.

T. KŐVÁRI, V. SÓS, AND P. TURÁN (1954), On a problem of K. Zarankiewicz, *Colloquium Mathematicum* 3, 50–57.

L. KUIPERS AND H. NIEDERREITER (1974), *Uniform Distribution of Sequences,* Wiley-Interscience, New York.

G. KUPERBERG AND W. KUPERBERG (1990), Double lattice packings of convex bodies in the plane, *Discrete and Computational Geometry* 5, 389–397.

W. KUPERBERG (1982), Packing convex bodies in the plane with density greater than 3/4, *Geometriae Dedicata* 13, 149–155.

W. KUPERBERG (1987), An inequality linking packing and covering densities of plane convex bodies, *Geometriae Dedicata* 23, 59–66.

W. KUPERBERG (1989), Covering the plane with congruent copies of a convex body, *Bull. London Mathematical Society* 21, 82–86.

Y. KUPITZ (1979), *Extremal Problems in Combinatorial Geometry, Aarhus University Lecture Notes Series*, No. 53, Aarhus University, Denmark.

Y. KUPITZ (1984), On pairs of disjoint segments in convex position in the plane, *Ann. Discrete Mathematics* 20, 203–208.

G. LANG (1955), The dual of a well-known theorem, *Mathematical Gazette* 39, 314.

D. LARMAN, J. MATOUŠEK, J. PACH, AND J. TÖRŐCSIK (1994), A Ramsey-type result for planar convex sets, *Bull. London Mathematical Society* 26, 132–136.

M. LAS VERGNAS AND L. LOVÁSZ (1972), see D. Woodall, Property B and the four colour problem, in: *Combinatorics, Proc. Conference on Combinatorial Mathematics* (D. Welsh and D. Woodall, eds.), Institute of Mathematics and Its Applications, Southend-on-Sea, Essex, England, 322–340.

D. LÁZÁR (1947), Sur l'approximation des courbes convexes par des polygones, *Acta Scientiarum Mathematicarum (Szeged)* 11, 129–132.

J. LEECH (1967), Notes on sphere packing, *Canadian J. Mathematics* 19, 251–267.

J. LEECH AND N. SLOANE (1971), Sphere packing and error-correcting codes, *Canadian J. Mathematics* 23, 718–745.

H. LEFMANN AND T. THIELE (1995), Point sets with distinct distances, to appear in *Combinatorica* 5 15(2).

T. LEIGHTON (1983), *Complexity Issues in VLSI, Foundations of Computing Series*, MIT Press, Cambridge, MA.

C. LEISERSON (1983), *Area Efficient VLSI Computation, Foundations of Computing Series*, MIT Press, Cambridge, MA.

H. LENZ (1955), Zur Zerlegung von Punktmengen in solche kleineren Durchmessers, *Archiv der Mathematik* 6, 413–416.

V. LEVENŠTEIN (1975), Maximal packing density of equal spheres in n-dimensional Euclidean space, *Mathematical Notes of the Academy of Sciences of the USSR* 18, 765–771.

V. LEVENŠTEIN (1979), On bounds for packings in n-dimensional Euclidean space, *Doklady Akademii Nauk SSSR* 245, 1299–1303. [English translation appears in *Soviet Mathematics, Doklady* 20 (1979), 417–421.]

J. LINDSEY II (1986), Sphere packing in $\mathbb{R}^3$, *Mathematika* 33, 137–147.

J. VAN LINT (1982), *Introduction to Coding Theory*, Springer-Verlag, New York.

R. LIPTON AND R. TARJAN (1979), A separator theorem for planar graphs, *SIAM J. Applied Mathematics* 36, 177–189.

R. LIPTON AND R. TARJAN (1980), Applications of a planar separator theorem, *SIAM J. Computing* 9, 615–627.

R. LIPTON, D. ROSE, AND R. TARJAN (1979), Generalized nested dissection, *SIAM J. Numerical Analysis* 16, 346–358.

L. LOVÁSZ (1971), On the number of halving lines, *Annales Universitatis Scientiarum Budapest, Eötvös, Sectio Mathematica* 14, 107–108.

L. LOVÁSZ (1972), On the sieve-formula (in Hungarian), *Matematikai Lapok* 23, 53–69.

L. LOVÁSZ (1975), On the ratio of optimal integral and fractional cover, *Discrete Mathematics* 13, 383–390.

L. LOVÁSZ (1979), *Combinatorial Problems and Exercises*, North-Holland, Amsterdam, and Akadémiai Kiadó, Budapest.

L. LOVÁSZ AND M. SIMONOVITS (1990), The mixing rate of Markov chains, an isoperimetric inequality, and computing the volume, *Proc. 31st Annual IEEE Symposium on Foundations of Computer Science*, 346–354. See also: Random walks in a convex body and an improved volume algorithm, *Random Structures and Algorithms* 4 (1993), 359–412.

L. LOVÁSZ, J. PACH, AND M. SZEGEDY (1995), On Conway's thrackle conjecture, *Proc. 11th Annual Symposium on Computational Geometry*, 147–151.

A. LUBOTZKY, R. PHILIPS, AND P. SARNAK (1988), Ramanujan graphs, *Combinatorica* 8, 261–277.

E. LUTWAK (1979), Inequalities related to a problem of Moser, *American Mathematical Monthly* 86, 476-477.

A. MACBEATH (1951), An extremal property of the hypersphere, *Proc. Cambridge Philosophical Society* 47, 245–247.

K. MAHLER (1946a), The theorem of Minkowski–Hlawka, *Duke Mathematical Journal* 13, 611–621.

K. MAHLER (1946b), On lattice points in n-dimensional star bodies I. Existence theorems, *Proc. Royal Society London A* 187, 151–187.

K. MAHLER (1947), On the minimum determinant and the circumscribed hexagons of a convex domain, *Proc. K. Ned. Akad. Wet. Amsterdam* 50, 692–703.

E. MAKAI, JR. (1987), Five-neighbor packing of convex plates, in: *Intuitive Geometry, Colloquia Mathematica Societatis János Bolyai*, vol. 48 (K. Böröczky and G. Fejes Tóth, eds.), North-Holland, Amsterdam, 373–381.

S. MALITZ AND A. PAPAKOSTAS (1992), On the angular resolution of planar graphs, *Proc. 24th Annual ACM Symposium on Theory of Computing*, 527–538. Also: *SIAM J. Discrete Mathematics* 7 (1994), 172–183.

A. MARDEN AND B. RODIN (1990), On Thurston's formulation and proof of Andreev's theory, in: *Lecture Notes in Mathematics*, vol. 1435 (St. Ruscheweyh et al., eds.), Springer-Verlag, New York, 103–115.

G. MARGULIS (1982), Explicit constructions of graphs without short cycles and low density codes, *Combinatorica* 2, 71–78.

J. MATOUŠEK (1990), Construction of ϵ-nets, *Discrete and Computational Geometry* 5, 427–448.

J. MATOUŠEK (1992), Efficient partition trees, *Discrete and Computational Geometry* 8, 315–384.

J. MATOUŠEK (1994), On discrepancy bounds via dual shatter function, *Mathematika*, to appear.

J. MATOUŠEK (1995), Tight upper bounds for the discrepancy of half-spaces, *Discrete and Computational Geometry*, 13, 593–601.

J. MATOUŠEK AND J. SPENCER (1994), Discrepancy of arithmetic progressions, Tech. Rept., KAM Series, Dept. Applied Mathematics, Charles University, Prague.

J. MATOUŠEK, E. WELZL, AND L. WERNISCH (1993), Discrepancy and approximations for bounded VC-dimension, *Combinatorica* 13, 455–466.

K. MEHLHORN (1985), *Multi-dimensional Searching and Computational Geometry*, Springer-Verlag, Berlin.

E. MELCHIOR (1940), Über Vielseite der projektiven Ebene, *Deutsche Mathematik* 5, 461–475.

G. MILLER (1986), Finding small simple cycle separators for 2-connected planar graphs, *J. Computer and Systems Sciences* 32, 265–279.

G. MILLER AND W. THURSTON (1990), Separators in two and three dimensions, *Proc. 22nd Annual ACM Symposium on Theory of Computing*, 300–309.

G. MILLER, S. TENG, AND S. VAVASIS (1991), A unified geometric approach to graph separators, *Proc. 32nd Annual IEEE Symposium on Foundations of Computer Science*, 538–547.

H. MINKOWSKI (1896), *Geometrie der Zahlen*, Leipzig and Berlin.

H. MINKOWSKI (1904), Dichteste gitterförmige Lagerung kongruenter Körper, *Nachrichten der Gesellschaft der Wissenschaften zu Göttingen*, 311–355.

H. MINKOWSKI (1905), Diskontinuitätsbereich für arithmetische Äquivalenz, *J. reine und angewandte Mathematik* 129, 220–274.

J. MOLNÁR (1955), Konvex tartományok beírt és körülírt poligonjairól, *Matematikai Lapok* 6, 210–218.

J. MOON AND L. MOSER (1962), On a problem of Turán, *Magyar Tudományos Akadémia Matematikai Kutató Intézet Közleményei* 7, 283–286.

J. MOON AND L. MOSER (1965), On cliques in graphs, *Israel J. Mathematics* 3, 23–28.

M. MÖRS (1981), A new result on the problem of Zarenkiewicz, *J. Combinatorial Theory, Series A* 31, 126–130.

L. MOSER (1952), On different distances determined by *n* points, *American Mathematical Monthly* 59, 85–91.

W. MOSER AND J. PACH (1986), *Research Problems in Discrete Geometry*, Preprint, McGill University, to be published by Academic Press. See also *DIMACS Tech. Report* 93-32.

W. MOSER AND J. PACH (1993), Recent developments in combinatorial geometry, in: *New Trends in Discrete and Computational Geometry* (J. Pach, ed.), Springer-Verlag, New York, 281–302.

T. MOTZKIN (1951), The lines and planes connecting the points of a finite set, *Trans. American Mathematical Society* 70, 451–464.

T. MOTZKIN AND E. STRAUSS (1965), Maxima for graphs and a new proof of a theorem of Turán, *Canadian J. Mathematics* 17, 533–540.

D. MOUNT (1991), The densest double-lattice packing of a convex polygon, in: *Discrete and Computational Geometry: Papers from DIMACS Special Year, DIMACS Series*, vol. 6 (J. Goodman et al., eds.), American Mathematical Society, Providence, RI, 245–262.

D. MOUNT AND R. SILVERMAN (1990), Packing and covering the plane with translates of a convex polygon, *J. Algorithms* 11, 564–580.

D. MUDER (1988), Putting the best face on a Voronoi polyhedron, *Proc. London Mathematical Society* (3) 56, 329–348.

D. MUDER (1993), A new bound on the local density of sphere packings, *Discrete and Computational Geometry* 10, 351–375.

A. MÜLLER (1953), Auf einem Kreis liegende Punktmengen ganzzahliger Entfernungen, *Elemente der Mathematik* 8, 37–38.

K. MULMULEY (1994), *Computational Geometry: An Introduction Through Randomized Algorithms*, Prentice Hall, Englewood Cliffs, NJ.

D. NAIMAN AND H. WYNN (1994), On the growth function of Euclidean balls, to appear in *Computational Geometry: Theory and Applications*.

C. NASH-WILLIAMS (1959), Random walk and electric current in networks, *Proc. Cambridge Philosophical Society* 55, 181–199.

C. NEADERHOUSER AND G. PURDY (1982), On finite sets in $\mathbb{E}^k$ in which the diameter is frequently achieved, *Periodica Mathematica Hungarica* 13, 253–257.

A. NILLI (1993), On Borsuk's problem, to appear in *AMS Series Contemporary Mathematics* (Proc. Combinatorics Conference, Jerusalem, 1993; H. Barcelo and G. Kalai, eds.).

I. NIVEN AND H. ZUCKERMAN (1966), *An Introduction to the Theory of Numbers*, Wiley, New York.

A. ODLYZKO AND N. SLOANE (1979), New bounds on the number of unit spheres that can touch a unit sphere in *n* dimensions, *J. Combinatorial Theory, Series A* 26, 210–214.

P. O'DONNELL AND M. PERLES (1990), Every geometric graph with *n* vertices and $3.6n + 3.4$ edges contains 3 pairwise disjoint edges, manuscript, Rutgers University, New Brunswick, NJ.

O. ORE (1961), Arc covering of graphs, *Annali di Mathematica Pura ed Applicata IV* 55, 315–321.

J. O'ROURKE (1994), *Computational Geometry in C*, Cambridge University Press, Cambridge.

F. ÖSTERREICHER AND J. LINHART (1981), Packungen kongruenter Stäbchen mit konstanter Nachbarnzahl, *Elemente der Mathematik* 37, 5–16.

J. PACH (1980), Decomposition of multiple packing and covering, *Proc. 2nd Kolloquium über Diskrete Geometrie*, Salzburg, 169–178.

J. PACH (1991), Notes on geometric graph theory, in: *Discrete and Computational Geometry: Papers from the DIMACS Special Year, DIMACS Series*, vol. 6 (J. Goodman et al., eds.), American Mathematical Society, Providence, RI, 273–285.

J. PACH (1995), A remark on transversal numbers, in: *The Mathematics of Paul Erdős* (R. Graham and J. Nešetřil, eds.), to appear.

J. PACH AND M. SHARIR (1990), Repeated angles in the plane and related problems, *J. Combinatorial Theory, Series A* 59, 12–22.

J. PACH AND J. TÖRŐCSIK (1993), Some geometric applications of Dilworth's theorem, *Proc. 9th Annual Symposium on Computational Geometry*, 264–269. Also: *Discrete and Computational Geometry* 12 (1994), 1–7.

J. PACH AND G. WOEGINGER (1990), Some new bounds for epsilon-nets, *Proc. 6th Annual Symposium on Computational Geometry*, 10–15.

J. PACH, F. SHAHROKHI, AND M. SZEGEDY (1994), Applications of the crossing number, *Proc. 10th Annual Symposium on Computational Geometry*, 198–202.

J. PACH, W. STEIGER, AND E. SZEMERÉDI (1992), An upper bound on the number of planar k-sets, *Discrete and Computational Geometry* 7, 109–123.

C. PAPADIMITRIOU AND K. STEIGLITZ (1982), *Combinatorial Optimization: Algorithms and Complexity*, Prentice Hall, Englewood Cliffs, NJ.

R. POLLACK (1985), Increasing the minimum distance of a set of points, *J. Combinatorial Theory, Series A* 40, 450.

L. PÓSA (1962), A theorem concerning Hamiltonian lines, *Magyar Tudományos Akadémia Matematikai Kutató Intézet Közleményei* 7A, 225–226.

F. PREPARATA AND M. SHAMOS (1985), *Computational Geometry: An Introduction*, Springer-Verlag, New York.

G. PURDY (1974), Some extremal problems in geometry II, *Discrete Mathematics* 7, 305–315.

G. PURDY (1988), Repeated angles in E^4, *Discrete and Computational Geometry* 3, 73–75.

F. RAMSEY (1930), On a problem of formal logic, *Proc. London Mathematical Society, Series 2*, 30, 264–286.

V. REDDY AND S. SKIENA (1995), Frequencies of large distances in integer lattices, *Proc. Seventh International Conference on Graph Theory, Combinatorics, Algorithms, and Applications, Kalamazoo, MI* (Y. Alavi and A. Schwenk, eds.), Wiley, New York, 1995, to appear.

I. REIMAN (1958), Über ein Problem von K. Zarankiewicz, *Acta Mathematica Academiae Scientiarum Hungarica* 9, 269–278.

K. REINHARDT (1934), Über die dichteste gitterförmige Lagerung kongruente Bereiche in der Ebene und eine besondere Art konvexer Kurven, *Abh. Mathematischen Seminar, Hamburg, Hansischer Universität, Hamburg*, 10, 216–230.

D. REUTTER (1972), Problem 664A, *Elemente der Mathematik* 27, 19.

A. RIESLING (1971), Borsuk's problem in three-dimensional spaces of constant curvature, *Ukrainskiĭ Geometričeskiĭ Sbornik* 11, 78–83.

B. RODIN (1987), Schwarz's lemma for circle packings, *Inventiones Mathematicae* 89, 271–289.

B. RODIN (1989), Schwarz's lemma for circle packings II, *J. Differential Geometry* 30, 539–554.

B. RODIN AND D. SULLIVAN (1987), The convergence of circle packings to Riemann mapping, *J. Differential Geometry* 26, 349–360.

V. RÖDL (1985), Personal communication.

C. ROGERS (1947), Existence theorems in the geometry of numbers, *Ann. Mathematics* 48, 994–1002.

C. ROGERS (1951), The closest packing of convex two-dimensional domains, *Acta Mathematica* 86, 309–321.

C. ROGERS (1957), A note on coverings, *Mathematika* 4, 1–6.

C. ROGERS (1958), The packing of equal spheres, *Proc. London Mathematical Society*, Series 3, 8, 609–620.

C. ROGERS (1959), Lattice coverings of space, *Mathematika* 6, 33–39.

C. ROGERS (1964), *Packing and Covering*, Cambridge University Press, New York.

C. ROGERS (1971), Symmetrical sets of constant width and their partitions, *Mathematika* 18, 105–111.

C. ROGERS AND G. SHEPHARD (1957), The difference body of a convex body, *Archiv der Mathematik* 8, 220–233.

C. ROGERS AND G. SHEPHARD (1958), Convex bodies associated to a given convex body, *J. London Mathematical Society* 33, 270–281.

J. RUSH (1989), A lower bound on packing density, *Inventiones Mathematicae* 98, 499–509.

J. RUSH (1992), Thin lattice coverings, *J. London Mathematical Society* (2) 45, 193–200.

J. RUSH AND N. SLOANE (1987), An improvement to the Minkowski–Hlawka bound for packing superballs, *Mathematika* 34, 8–18.

I. RUZSA (1978), On the cardinality of $A+A$ and $A-A$, in: *Combinatorics, Colloquia Mathematica Societatis János Bolyai*, vol. 18 (A. Hajnal and V. T. Sós, eds.), North-Holland, Amsterdam, 933–938.

I. RUZSA (1992), Arithmetical progressions and the number of sums, *Periodica Mathematica Hungarica* 25, 105–111.

H. SACHS (1993), Coin graphs, polyhedra, and conformal mapping, *Discrete Mathematics* 134, 133–138.

L. SANTALÓ (1952), *Introduction to Integral Geometry*, Actualités Scientifiques et Industrielles 1198, Hermann, Paris.

L. SANTALÓ (1976), *Integral Geometry and Geometric Probability, Encylopaedia of Mathematics and Its Applications*, vol. 1, Addison-Wesley, Reading, MA.

E. SAS (1993), Über eine Extremumeigenschaft der Ellipsen, *Compositio Mathematica* 6, 468–470.

N. SAUER (1971), A generalization of a theorem of Turán, *J. Combinatorial Theory, Series B* 10, 109–112

N. SAUER (1972), On the density of families of sets, *J. Combinatorial Theory, Series A* 13, 145–147.

L. SCHLÄFLI (1901), Theorie der vielfachen Kontinuität, *Gesammelte Mathematische Abhandlungen I*, Birkhäuser Verlag, Basel (1950), 209.

W. SCHMIDT (1963), On the Minkowski–Hlawka theorem, *Illinois J. Math.* 7, 18–23.

P. SCHMITT (1991), Disks with special properties of densest packings, *Discrete and Computational Geometry* 6, 181–190.

R. SCHNEIDER (1967), Eine allgemeine Extremaleigenschaft der Kugel, *Monatshefte für Mathematik* 71, 231–237.

R. SCHNEIDER (1971), Zwei Extremalaufgaben für konvexe Bereiche, *Acta Mathematica Academiae Scientiarum Hungaricae* 22, 379–383.

R. SCHNEIDER (1993), *Convex Bodies: The Brunn–Minkowski Theory*, Encyclopedia of Mathematics and Its Applications, vol. 44, Cambridge University Press, Cambridge.

O. SCHRAMM (1988), Illuminating sets of constant width, *Mathematika* 35, 180–199.

O. SCHRAMM (1991), Rigidity of (infinite) circle packings, *J. American Mathematical Society* 4, 127–149.

M. SHARIR AND P. K. AGARWAL (1995), *Davenport-Schinzel Sequences and Their Geometric Applications*, Cambridge University Press, New York.

S. SHELAH (1972), A combinatorial problem; stability and order for models and theories in infinitary languages, *Pacific J. Mathematics* 41, 247–261.

V. SIDEL'NIKOV (1973), On the densest packing of balls on the surface of the *n*-dimensional Euclidean sphere, and the number of vectors of a binary code with prescribed code distances, *Doklady Akademii Nauk SSSR* 213, 1029–1032. [English translation appears in *Soviet Mathematics, Doklady* 14 (1973), 1851–1855.]

V. SIDEL'NIKOV (1974), New estimates for the packing of spheres in *n*-dimensional Euclidean space, *Matematicheskiĭ Sbornik* 95 (137), 148–158. [Also in *USSR Matematicheskiĭ Sbornik* 24 (1974), 147–157.]

M. SIMONOVITS (1966), A method for solving extremal problems in graph theory, stability problems, *Proc. Colloquium on Graph Theory*, Akadémiai Kiadó, Budapest, 279–319.

M. SIMONOVITS (1983), Extremal graph theory, in: *Selected Topics in Graph Theory* (L. Beineke and R. Wilson, eds.), Vol. 2, Academic Press, London, 161–200.

M. SIMONOVITS AND V. SÓS (1991), Szemerédi's partition and quasirandomness, *Random Structures and Algorithms* 2, 1–10.

R. SINGLETON (1966), On minimal graphs of maximum even girth, *J. Combinatorial Theory* 1, 306–332.

V. SÓS (1970), On extremal problems in graph theory, in: *Combinatorial Structures and Their Applications* (R. Guy et al., eds.), Gordon and Breach Science Publishers, New York, 407–410.

V. SÓS (1989), (Ed.) *Irregularities of Partitions*, Springer-Verlag, Berlin.

J. SPENCER (1972), Turán's theorem for *k*-graphs, *Discrete Mathematics* 2, 183–186.

J. SPENCER (1985), Six standard deviations suffice, *Trans. American Mathematical Society* 289, 679–706.

J. SPENCER (1987), *Ten Lectures on the Probabilistic Method*, Society for Industrial and Applied Mathematics, Philadelphia.

J. SPENCER (1990), Uncrowded graphs, in: *Mathematics of Ramsey Theory* (J. Nešetřil and V. Rödl, eds.), Springer-Verlag, New York, 253–262.

J. SPENCER, E. SZEMERÉDI, AND W. TROTTER, JR. (1984), Unit distances in the Euclidean plane, in: *Graph Theory and Combinatorics* (B. Bollobás, ed.), Academic Press, New York, 293–303.

E. SPERNER (1928), Ein Satz über Untermengen einer endlichen Menge, *Mathematische Zeitschrift* 27, 544–548.

K. STEPHENSON (1990), Circle packings in the approximation of conformal mappings, *Bull. American Mathematical Society* 23, 407–415.

S. STRASZEWICZ (1957), Sur un problème géométrique de Erdős, *Bull. Académie Polonaise des Sciences*, Cl. III 5, 39–40.

J. SUTHERLAND (1935), Solution to problem 167, *Jahresbericht det Deutschen Mathematiker-Vereinigung* 45, 33–34.

J. SYLVESTER (1893), Mathematical question 11851, *Educational Times* 59, 98–99.

G. SZEKERES (1941), On an extremum problem in the plane, *American J. Mathematics* 63, 208–210.

E. SZEMERÉDI (1978), Regular partitions of graphs, in: *Problèmes Combinatoires et Théorie de Graphes* (J. Bermond et al., eds.), Coll. Internationaux C.N.R.S., 260, C.N.R.S., Paris, 399–401.

E. SZEMERÉDI (1990), Integer sets containing no arithmetic progressions, *Acta Mathematica Academiae Scientiarum Hungarica* 56, 155–158.

E. SZEMERÉDI AND W. TROTTER, JR. (1983a), A combinatorial distinction between the Euclidean and projective planes, *European J. Combinatorics* 4, 385–394.

E. SZEMERÉDI AND W. TROTTER, JR. (1983b), Extremal problems in discrete geometry, *Combinatorica* 3, 381–392.

P. TAMMELA (1970), Ocenka kritičeskogo opredelitelja dvumernoĭvypukloĭ simmetričnoĭ oblasti (An estimate of the critical determinant of a two–dimensional convex symmetric domain), *Izvestija Vysših Učebnyh Zavedenii Matematika* 103, 103–107.

R. TAMMES (1930), On the origin of number and arrangement of the places of exit on the surface of pollen grains, *Recueil des Travaux Botaniques Neerlandais* 27, 1–84.

S. TENG (1991), *Points, Spheres, and Separators: A Unified Geometric Approach to Graph Partitioning*, Ph.D. Thesis, Carnegie Mellon University, Pittsburgh, PA.

T. THIELE (1993), Point sets with distinct slopes or lengths, unpublished manuscript.

A. THUE (1892), Om nogle geometrisk taltheoretiske Theoremer, *Forandlingerneved de Skandinaviske Naturforskeres* 14, 352–353.

A. THUE (1910), Über die dichteste Zusammenstellung von kongruenten Kreisen in der Ebene, *Christinia Vid. Selsk. Skr.* 1, 1–9.

W. THURSTON (1985a), *The Geometry and Topology of Three-Manifolds*, Princeton Lecture Notes, Princeton University Press, Princeton, NJ, Chapter 13.

W. THURSTON (1985b), The finite Riemann mapping theorem, Lecture delivered at the *International Symposium in Celebration of the Proof of the Bieberbach Conjecture*, Purdue University, West Lafayette, IN.

D. TODOROV (1983), On Turán numbers, in: *Mathematics and Mathematical Education* (W. Mills, ed.), Bulgarian Academy of Sciences, Sofia, 123–128.

G. TÓTH (1995), The shortest distance among points in general position, manuscript.

W. TROTTER, JR. (1992), *Combinatorics and Partially Ordered Sets*, Johns Hopkins University Press, Baltimore.

P. TURÁN (1941), Egy gráfelméleti szélsőértékfeladatról, *Matematikai és Fizikai Lapok* 48, 436–452.

P. TURÁN (1954), On the theory of graphs, *Colloquium Mathematicum* 3, 19–30.

P. TURÁN (1970a), Applications of graph theory to geometry and potential theory, in: *Combinatorial Structures and Their Applications* (R. Guy et al., eds.), Gordon and Breach Science Publishers, New York, 423–434.

P. TURÁN (1970b), Remarks on the packing constants of the unit sphere (in Hungarian), *Matematikai Lapok* 21, 39–44.

W. TUTTE (1963), How to draw a planar graph, *Proc. London Mathematical Society* 13, 743–768.

W. TUTTE (1970), Toward a theory of crossing number, *J. Combinatorial Theory* 8, 45–53.

J. ULLMAN (1984), *Computational Aspects of VLSI*, Computer Science Press, Rockville, MD.

V. VAPNIK AND A. CHERVONENKIS (1971), On the uniform convergence of relative frequencies of events to their probabilities, *Theory of Probability and Its Applications* 16, 264–280.

W. LEVEQUE (1956), *Topics in Number Theory*, Addison-Wesley, Reading, MA.

K. VESZTERGOMBI (1985), On the distribution of distances in finite sets in the plane, *Discrete Mathematics* 57, 129–146.

K. VESZTERGOMBI (1987), On large distances in planar sets, *Discrete Mathematics* 67, 191–198.

E. WELZL (1988), Partition trees for triangle counting and other range searching problems, *Proc. 4th Annual Symposium on Computational Geometry*, 23–33.

E. WELZL (1992), On spanning trees with low crossing numbers, *Lecture Notes in Computer Science*, vol. 594, Springer-Verlag, New York, 233–249.

R. WENGER (1991), Extremal graphs with no C_4's, C_6's, or C'_{10}'s, *J. Combinatorial Theory, Series B* 52, 113–116.

L. WERNISCH (1994), Note on stabbing numbers and sphere packing numbers, manuscript.

D. WILLARD (1982), Polygon retrieval, *SIAM J. Computing* 11, 149–165.

D. WOODALL (1971), Thrackles and deadlock, in: *Combinatorial Mathematics and Its Applications* (D. Welsh, ed.), Academic Press, London, 335–347.

D. WOODALL (1972), Two results on infinite transversals, in: *Combinatorics, Proc. Conference on Combinatorial Mathematics* (D. Welsh and D. Woodall, eds.), Institute of Mathematics and Its Applications, Southend-on-Sea, Essex, England, 341–350.

I. YAGLOM AND V. BOLTYANSKY (1951), *Convex Figures*, Moscow (in Russian).

K. ZARANKIEWICZ (1951), Problem 101, *Colloquium Mathematicum* 2, 301.

Z. ZHANG (1993), A note on arrays of dots with distinct slopes, *Combinatorica* 13, 127–128.

A. ZYKOV (1943), On some properties of linear complexes, *Matematicheskii Sbornik N.S.* 24 (66), 163–188. [Also in *American Mathematical Society Translations* 79 (1952).]

Index of Symbols

Author Index

Subject Index

WILEY-INTERSCIENCE
SERIES IN DISCRETE MATHEMATICS AND OPTIMIZATION